Extreme Man-Made and Natural Hazards in Dynamics of Structures

NATO Security through Science Series

This Series presents the results of scientific meetings supported under the NATO Programme for Security through Science (STS).

Meetings supported by the NATO STS Programme are in security-related priority areas of Defence Against Terrorism or Countering Other Threats to Security. The types of meeting supported are generally "Advanced Study Institutes" and "Advanced Research Workshops". The NATO STS Series collects together the results of these meetings. The meetings are co-organized by scientists from NATO countries and scientists from NATO's "Partner" or "Mediterranean Dialogue" countries. The observations and recommendations made at the meetings, as well as the contents of the volumes in the Series, reflect those of participants and contributors only; they should not necessarily be regarded as reflecting NATO views or policy.

Advanced Study Institutes (ASI) are high-level tutorial courses to convey the latest developments in a subject to an advanced-level audience

Advanced Research Workshops (ARW) are expert meetings where an intense but informal exchange of views at the frontiers of a subject aims at identifying directions for future action

Following a transformation of the programme in 2004 the Series has been re-named and re-organised. Recent volumes on topics not related to security, which result from meetings supported under the programme earlier, may be found in the NATO Science Series.

The Series is published by IOS Press, Amsterdam, and Springer, Dordrecht, in conjunction with the NATO Public Diplomacy Division.

Sub-Series

A. Chemistry and Biology	Springer
B. Physics and Biophysics	Springer
C. Environmental Security	Springer
D. Information and Communication Security	IOS Press
E. Human and Societal Dynamics	IOS Press

http://www.nato.int/science
http://www.springer.com
http://www.iospress.nl

Series C: Environmental Security

Extreme Man-Made and Natural Hazards in Dynamics of Structures

edited by

Adnan Ibrahimbegovic
Ecole Normale Supérieure de Cachan,
France

and

Ivica Kozar
University of Rijeka,
Croatia

Published in cooperation with NATO Public Diplomacy Division

Proceedings of the NATO Advanced Research Workshop on
Extreme Man-Made and Natural Hazards in Dynamics of Structures
Opatija, Croatia
28 May - 1 June 2006

A C.I.P. Catalogue record for this book is available from the Library of Congress.

ISBN-10 1-4020-5655-9 (PB)
ISBN-13 978-1-4020-5655-0 (PB)
ISBN-10 1-4020-5654-0 (HB)
ISBN-13 978-1-4020-5654-3 (HB)
ISBN-10 1-4020-5656-7 (e-book)
ISBN-13 978-1-4020-5656-7 (e-book)

Published by Springer,
P.O. Box 17, 3300 AA Dordrecht, The Netherlands.

www.springer.com

Printed on acid-free paper

TABLE OF CONTENTS

PREFACE

There is currently ever pressing need to provide a critical assessment of the current knowledge and indicate new challenges which are brought by the present time in fighting the man-made and natural hazards in transient analysis of structures. The latter concerns both the permanently fixed structures, such as those built to protect the people and/or sensitive storage material (e.g. military installations) or the special structures found in transportation systems (e.g. bridges, tunnels), and the moving structures (such as trains, plains, ships or cars). The present threat of the terrorist attacks or accidental explosions, the climate change which brings strong stormy winds or yet the destructive earthquake motion that occurs in previously inactive regions or brings about tsunamis, are a few examples of the kind of applications we seek to address in this work.

The common ground for all the problems of this kind from the viewpoint of structural integrity, which also justifies putting them on the same basis and addressing them within the same context, is their sudden appearance, their transient nature and the need to evaluate the consequence for a high level of uncertainty in quantifying the cause. The problems of such diversity cannot be placed within a single traditional scientific discipline, but they call for the expertise in probability theory for quantifying the cause, interaction problems for better understanding the physical nature of the problems, as well as modeling and computational techniques for improving the representation of inelastic behavior mechanisms and providing the optimal design.

The present time of high uncertainty is very likely to increase (rather than decrease) the frequency or severe intensity of the high-risk situations that a very few engineering structures have been built to sustain. It is therefore important to understand any potential reserve, which might exist in engineering structures for taking on a higher level of risk. The complementary goal, also of great importance, pertains to providing the best way of reducing the negative impact of high-risk situations that cannot be avoided, by resorting to a more sound design procedure.

Never before have we had the same level of development of scientific and technological achievements, which can be brought to bear on the present problem of high complexity. First, the constant progress in computational tools ought to be exploited to construct the more refined structure models than those used previously, which can provide a more detailed information and explore all potential reserves in a more old-fashioned design. Second, one can nowadays understand much better the particular physical nature of the loading conditions, by simulating the actual physical process that is at its origin, and in that manner providing a more reliable estimate of the parameters governing the processes of this kind. The quantitative information can be bracketed between the probabilistic bounds, which can nowadays be constructed for more and more complex processes, thanks to significant advances in modern computational probability research.

The present work is the outcome of a lively exchange of ideas among the world leading scientists dealing with different facets of this class of complex problems. Among them, specialists in probability, in structural engineering, in interaction problems and in development of computer models, as well as related experimental works, have all contributed to the successful accomplishment of the hazard reduction goal set for our meeting. The lectures presented in this book are regrouped on any single topic to provide the most detailed presentations, seeking to reach eventually complementary points of view. A fair number of pages is allocated to each chapter in order to provide the complete presentation of any given facet of the problem and a sufficiently detailed exposition to any important idea to be grasped. Among several illustrative applications which are discussed herein we find: quantifying the plane-crash or explosion induced impact loading, quantifying the effects of a strong earthquake motion, quantifying the impact and long-duration effects of strong stormy winds, providing the most efficient tools to construct the probabilistic bounds and computational tools for probabilistic analysis, constructing refined models for nonlinear dynamic analysis and optimal design and presenting modern computational tools for that purpose.

All the papers collected in this book are first presented as the keynote lectures at NATO-ARW No. 981641, which was held in the city of Opatija in Croatia, from May 28 to June 1, 2006. We would like to thank all the participants for their important contributions to the successful outcome of this meeting, and in particular to the keynote lecturers, Professors K.J. Bathe, N. Bicanic, F. Dias, P. Fajfar, M. Geradin, A. Ibrahimbegovic, P. Leger, H.G. Matthies,, D.R.J Owen, M. Papadrakakis, D. Peric and E.L. Wilson, for ensuring a more lasting impact of this meeting in terms of the present book.

Last but not least, we would also like to thank NATO Science Committee for selecting our meeting, NATO-ARW No. 981641, for the financial support by NATO.

NATO-country co-director:
Professor Adnan Ibrahimbegovic
ENS-Cachan, Paris, France

Partner country co-director:
Professor Ivica Kozar
FGZ-Rijeka, Croatia

Part I: NUMERICAL MODELLING AND DYNAMICS OF COMPLEX STRUCTURES

COMPUTATIONAL ISSUES IN THE SIMULATION OF BLAST AND IMPACT PROBLEMS: AN INDUSTRIAL PERSPECTIVE

D. R. J. Owen (d.r.j.owen@swansea.ac.uk) and Y. T. Feng
Civil & Computational Engineering Centre, School of Engineering University of Wales Swansea, Swansea, SA2 8PP, U.K.

M. G. Cottrell and J. Yu
Rockfield Technology Ltd., Technium, Swansea, SA1 8PH, U.K.

Abstract. Computational strategies involved in the numerical simulation of large scale industrial problems, characterised by a transition from a continuous to a discontinuous state, are reviewed in the present work. Particular attention is focused on the numerical modelling of the progressive damage that precedes crack initiation for both brittle/ductile materials and the introduction of discrete fractures within an initially continuum formulation. For problems restricted to relatively small deformations, the use of continuum based methods may be suitable, but for situations involving large geometric changes with possible post-fracture particle flow, there are compelling advantages in employing combined finite/discrete element solution strategies. The need for rigorous consideration of both theoretical and algorithmic issues is emphasised, particularly in relation to the computational treatment of finite strain elasto-plastic (viscoplastic) deformation and element technology capable of dealing with plastic incompressibility. Other important aspects such as adaptive mesh refinement strategies are discussed and procedures are presented for undertaking the transition from continuously distributed fracture states to discrete crack systems, followed by post-failure modelling. The practical application of the above methodology is illustrated in the numerical solution of industrially relevant problems.

Key words: Combine finite/discrete element approach, Multi-fracture, Element technology, Mesh adaptation.

1. Introduction

The last decade has experienced a major advance in the development of computational based tools for the simulation of a wide range of industrial problems. Progress has been made on several fronts and key developments in the treatment of both continuum and discrete problems have achieved a degree of maturity. The use of commercial software in the design and optimisation of many complex industrial processes is now undertaken on a routine basis, particularly in highly competitive sectors of manufacturing industry. Such simulations, which can be

A. Ibrahimbegovic and I. Kozar (eds.),Extreme Man-Made and Natural Hazards in Dynamics of Structures, 3–35.

usually obtained through relatively inexpensive numerical analysis, are of great assistance to the designer, allowing an optimisation of process parameters and manufactured product properties as well as dramatically reducing conception-to-production times.

However, for problems in which material failure takes place due to progressive damaging resulting in the formation of either single or multiple fractures, the current position of computational modelling is not so advanced. Examples of such situations include:

- Problems dominated by the development of predominantly a single fracture. Although eventual failure is manifested by such a discrete event, the conditions leading to the onset of crack propagation are controlled by damage based micro-cracking mechanisms. Examples include high speed machining operations and the penetration of ductile targets.
- The multi-fracturing of quasi-brittle materials such as ceramics, rock and concrete. Industrial applications include the response of ceramics under high velocity impact, such as occurs in many defence problems related to armour design, the response of reinforced concrete structures to missile and explosive loading and rock blasting operations within the minerals recovery industry.

The numerical treatment of the above classes of problems necessitates consideration of a blend of continuous and discrete computational processes to provide adequate solution. For the first class of problems, crucial issues include constitutive modelling and the use of low order finite element technology for near-incompressibility and adaptive mesh refinement. The use of low order elements in the presence of high nearly-isochoric plastic strains is addressed with the introduction of a methodology whereby the volumetric constraint is enforced over patches of simplex elements. This allows the effective use of linear tetrahedra without the volumetric locking typically associated with conventional low order displacement-based elements. It is important to emphasise that the industrial processes we wish to model are often characterised by intricate contact conditions as well as complex evolving geometries. The use of simplex elements in the treatment of such problems, including the associated mesh generation issues, is highly desirable due to their known robustness. Another important aspect of the overall computational framework is the requirement of an efficient re-meshing procedure. In the present context, most problems of practical industrial interest involve distortions of such an extent that sufficiently accurate solutions cannot be obtained unless a re-meshing procedure is applied at several stages of the process.

An extra degree of complexity is included in the second class of problems, in which failure in the form of multi-fracturing is the major phenomenon controlling the process and its accurate modelling is of the utmost importance to the success of numerical simulations. These problems are initially represented by a small number of continuous regions prior to the deformation process. During the loading phase, the bodies are progressively damaged and modelling of the subsequent fragmentation may result in possibly three to four orders of magnitude more bodies by the end of the simulation. The overall system response is governed rstly by appropriate constitutive mechanisms which control the material separation process followed by description of the inter-element interaction forces which govern the subsequent motion of particles.

For modelling multi-fracturing phenomena in particular, the existing strategies range from continuum based nite element approaches (de Borst, 2001; Ruiz et al., 2001; Dolbow et al., 2001), such as cohesive zone models and discontinuous Galerkin formulations, to discontinuum driven formulations, such as discrete discontinuous analysis (DDA) techniques (Shi, 1988) and distinct/discrete element approaches (Cundall and Strack, 1979). Originally, each element was assumed to be *rigid* in the classic discrete element method (Cundall and Strack, 1979). The incorporation of deformation kinematics into the discrete element formulation has lead naturally to combined finite /discrete element approaches (Owen et al., 2001; Owen et al., 2004). For situations involving multi-fracturing phenomena, such coupled strategies offer a natural solution route to modelling the continuum/discontinuum transformation involved. It not only provides a better description of the physical processes, but also often renders the constitutive material description more tractable. It may be argued that by modelling the continuous to discrete transformation explicitly, a physically more realistic representation is obtained, which provides advantages in allowing a constitutive description of the material that requires a minimum number of parameters which may be identified from standard experimental tests. Consequently, there are compelling advantages in employing combined finite/discrete element solution strategies to model discrete/discontinuous systems.

In several applications of industrial relevance, the presence of an additional phase, either gaseous, liquid or both, often controls the behaviour of the system. Examples of such interaction include (i) Fluid flow within fractured rock masses, and (ii) Generation of gas pressure fields which drives the fracturing process in rock blasting. Although a generic modelling strategy, and accompanying data structures, can be developed for a broad range of such problems, nevertheless, some applications require an individual approach to solution.

Besides their discrete/discontinuous nature, the problems concerned are often characterised by the following additional features: they are highly *dynamic* with rapidly *changing domain configurations*; *sufficient resolution* is required; and *multi-physics phenomena* are involved. The domination of contact/impact behaviour also gives rise to a very strong *nonlinear* system response. These factors dictate that there is almost no alternative to employing time integration schemes of an *explicit* nature to numerically simulate such problems. The necessity of frequent introduction of new physical cracks and/or adaptive re-meshing at both local and global levels adds another dimension of complexity (Cottrell, 2002).

2. Continuum Modelling Issues

Crucial to the appropriate characterisation of material response in the finite strain range is the rigourous formulation of constitutive models. In industrial applications of practical interest, finite bulk deformations are almost invariably accompanied by large inelastic strains and the formulation of consistent finite strain elasto-plasticity models is necessary. Since the models are to be incorporated within a finite/discrete environment suited to large scale simulations, the formulation of stable and accurate numerical integration algorithms for path-dependent constitutive equations have to be carefully addressed. The discussion of such issues is beyond the scope of this paper and the interested reader is referred to (Perić and Owen, 1998; Perić and Owen, 2004). In problems where inelastic deformations are volume preserving, appropriate finite element technology has to be adopted to produce accurate results. Other advances in the treatment of industrial problems include significant developments in model building and visualisation tools to permit interpretation of the results of complex large non-linear problems and both sequential and parallel solution strategies for very large- scale equation systems.

2.1. ELEMENT TECHNOLOGY

As the incompressibility limit is approached, conventional displacement-based finite elements with low order shape functions are known to perform poorly, showing the typical volumetric locking behaviour which may completely invalidate the finite element solution. In spite of such a major drawback, low order elements are often preferred for industrial applications due to their inherent simplicity. In order to allow the use of low order elements near the incompressible limit, many formulations have been proposed which, for explicit transient dynamic analysis, are usually based on the use of reduced integration in conjunction with hourglass control to avoid spurious zero energy modes (Schweizerhof et al., 1992; Belytschko et al., 1984). However, the lowest order elements produced by such formulations are the bi-linear (4-noded) quadrilateral in 2-D and the tri-linear (8-noded) hexahedron in 3-D. In many situations, the use of even lower order elements – the simplex (3-noded) triangle and (4-noded) tetrahedron in 3-D – is far more desirable. This is particularly true in the context of industrial processes, where complex geometry and contact conditions are typically present and many problems are characterised by extremely high strains. Even when the initial geometry is simple, the presence of very large strains may lead to poorly shaped elements unless some form of adaptive remeshing is used. The lack of robustness of currently available hexahedral mesh generators may impose severe

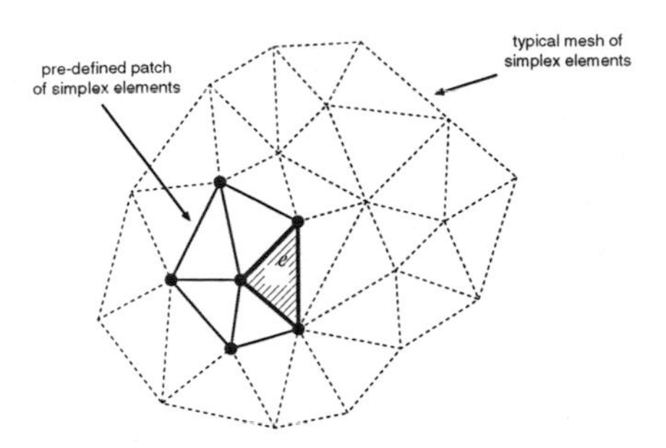

Figure 1. Definition of a patch of simplex elements.

limitations on the use of such elements in this context. This issue is crucial to the robustness of the overall finite element framework. The ideal alternative would be the use of simplex elements with added special techniques to handle finite nearly isochoric deformations. Some approaches to adapt simplex elements to this class of problems have been proposed (see, for instance the elements proposed by (Zienkiewicz et al., 1998; Bonet and Burton, 1998; Taylor, 2000) and (Bonet et al., 2001)), but the issue still remains largely open.

A possible solution to this problem is the so-called *F-bar-Patch* methodology recently proposed in (De Souza Neto et al., 2003; De Souza Neto et al., 2005). The idea underlying the technique is simple. It is based on relaxation of the excessive volumetric constraint by the enforcement of (near-)incompressibility over a *patch* of elements, rather than in a point-wise manner as in conventional elements.

Consider a mesh of simplex finite elements split into a number of non-overlapping patches of elements and $\mathcal{P}$ let be one such patch (Fig. 1). The technique defines a modified deformation gradient, $\bar{\mathbf{F}}^k$, for any element $k \in \mathcal{P}$ as:

$$\bar{\mathbf{F}}^k \equiv \left(\frac{\bar{J}}{J^k} \right)^{\frac{1}{3}} \mathbf{F}^k , \qquad (1)$$

where

$$J^k \equiv \det \mathbf{F}^k, \qquad \bar{J} \equiv \frac{v^{\text{patch}}}{V^{\text{patch}}}, \qquad (2)$$

with v^{patch} and V^{patch} denoting, respectively, the total volume of the patch $\mathcal{P}$ of elements in its current (deformed) and initial (undeformed) configurations:

$$v^{\text{patch}} = \sum_{i \in \mathcal{P}} v^i , \qquad V^{\text{patch}} = \sum_{i \in \mathcal{P}} V^i . \qquad (3)$$

In the above, v^i and V^i stand, respectively, for the deformed and undeformed volumes of a generic element i.

The required constraint relaxation is obtained by simply replacing the standard deformation gradient, $\mathbf{F}^k$, obtained from the linear displacement interpolation functions of element k with an assumed modified deformation gradient, $\bar{\mathbf{F}}^k$, in the computation of the Cauchy stress tensor. For an elasto-plastic material model, for example, the Cauchy stress tensor is typically the outcome of an implicit (incremental) constitutive function $\hat{\boldsymbol{\sigma}}$ defined by means of some numerical algorithm for integration of the elasto-plastic evolution equations. For a conventional displacement-based element k, considering the typical time interval $[t_n, t_{n+1}]$, the stress at t_{n+1} is computed as:

$$\boldsymbol{\sigma}_{n+1}^k = \hat{\boldsymbol{\sigma}}(\mathbf{F}_{n+1}^k, \boldsymbol{\alpha}_n)\,, \tag{4}$$

where $\boldsymbol{\alpha}_n$ is the set of internal variables relevant to the model considered evaluated at t_n. For the F-bar-Patch element the stress is computed instead as:

$$\boldsymbol{\sigma}_{n+1}^k = \hat{\boldsymbol{\sigma}}(\bar{\mathbf{F}}_{n+1}^k, \boldsymbol{\alpha}_n)\,. \tag{5}$$

The stress computed by the above expression is then used in the assembly of the element internal force vector:

$$\mathbf{f}_{n+1}^k = \int_{\Omega_k} \mathbf{B}^T \boldsymbol{\sigma}_{n+1}^k \,\mathrm{d}v\,, \tag{6}$$

where $\mathbf{B}$ is the standard discrete symmetric gradient operator.

3. Continuum To Discrete Transition

Key issues that need to be addressed for the successful modelling of continuous/discontinuous transformation include (i) the development of constitutive models which govern the material failure; (ii) the ability of numerical approaches to introduce discontinuities such as shear bands and cracks generated during the material failure and fracture process; and (iii) the effective simulation of contact between the region boundaries and crack surfaces during the failure process and (iv) particle behaviour motion of fragments in post-failure phases.

Internal damage is characterised by the presence and evolution of voids and micro-cracks which may, eventually, lead to material failure. In order to predict the localised continuum-discrete transition associated with the formation of a macro crack, appropriate local and non-local failure criteria have to be adopted. The ability of a finite element methodology to accommodate the continuum to discrete transition is of paramount importance in the modelling of post failure interaction. A blend of localisation and discrete element concepts is employed whereby discrete fracture is initiated post localisation of damage (Yu., 1999). Subsequently, physical fractures or cracks are inserted into the finite element mesh such that the initial continuum is gradually degraded into discrete bodies. The primary motivation for this approach is to correctly model post-failure interaction of fractures and the motion of particles created during the failure process.

3.1. FAILURE STRATEGIES

Several fracture criteria available from the literature have been employed to predict material failure that may result from the gradual internal deterioration associated with high straining. This prediction remains, to a great extent, relegated to post-simulation analyses. The adoption of a methodology, whereby the coupling between material behaviour and deterioration is considered at the constitutive level during the process simulation offers a more scientifically-based alternative to empirical methods with a potential improvement in predictive capability. Prediction of ductile fracture onset in damaged materials usually adopts the damage variable itself as indicator by assuming that failure takes place when $D = D_{cr}$. Recent comparative analyses suggest that damage-based measures are more reliable in predicting the correct site of fracture initiation (Andrade Pires et al., 2004) and the use of fracture criteria based on *total damage work*, generally defined as:

$$I_{\omega_D} = \int_0^t (-Y)\dot{D}\,dt = \int (-Y)\,dD, \tag{7}$$

in which $\dot{D}$ is the damage rate and Y denotes the damage energy release rate, offer a promising alternative, due to the high gradient exhibited by the indicator near the critical failure zone.

Strain rate effects: Material characteristics can be significantly influenced when loading rates become high, resulting in high strain rates within the material. For material fracture, the key observations from high strain rate crack propagation experiments are (Grady and Hollenbach, 1977; Ravi-Chandar and Knauss, 1984a; Ravi-Chandar and Knauss, 1984b): (a) The crack initiation time is time dependent, (b) The crack velocity is dependent on the loading rate, (c) The stress intensity at the crack tip is independent of the loading rate at lower strain rates but becomes rate dependent at higher loading rates and (d) The time to fracture is dependent on the loading rate but converges asymptotically to a minimum time to fracture. Rate dependence effects can be incorporated within the current strain softening based failure model by making both the failure stress and softening slope a function of the local strain rate (Yu., 1999). Under dynamic conditions the area under the softening slope is no longer equal to the fracture energy G_f due to the effects of inertia on the micro-mechanical response (Yu., 1999; Grady and Hollenbach, 1977).

Fracture in quasi-brittle materials is generally an anisotropic phenomenon, with the coalescence and growth of micro-cracks occurring in the directions that attempt to maximise the subsequent energy release rate and minimise the strain energy density. The optimum propagation paths maintain an orientation normal to the maximum extension strains. The localisation of micro-cracking into effective crack bands results in softening normal to the crack direction. On a continuum basis, fracture is considered in the form of a rate dependent rotating smeared-crack model utilising tensile strain-softening to represent material degradation. The smeared crack model provides a mechanism for directional softening within a continuum framework by envisaging a cracked solid as an equivalent anisotropic continuum with degraded properties in directions normal to crack band orientation. After initial yield the rotating crack formulation introduces anisotropic damage by degrading the elastic modulus in the direction of the current principal stress invariant. The model enforces coincident rotation of the principal axes of orthotropy and the principal strain axes. It provides an effective mechanism for eliminating stress locking and excess shear stress and has been shown to yield a more reliable lower bound response compared to fixed crack models (Klerck, 2000; Klerck et al., 2004).

For the description of material degradation and subsequent discrete fracturing of quasi-brittle materials under multi-axial stress states the Mohr-Coulomb failure criterion is adapted and coupled with the fully anisotropic tensile smeared crack model (Klerck et al., 2004). The Mohr-Coulomb criterion is able to recover the salient features of the quasi-brittle response within engineering accuracy, including dilation and has the advantage of being based on parameters that are easily determined experimentally. Fracturing due to dilation is accommodated by introducing an explicit coupling between the inelastic strain accrued by the Mohr-Coulomb yield surface and the anisotropic degradation of the mutually orthogonal tensile yield surfaces. The proposed model represents a phenomenological approach in which micromechanical processes are only considered in terms of the average global response. Isotropy of strength in compression is justified by assuming uniform material heterogeneity, while accrual of inelastic strain and associated degradation of the tensile strength is necessarily anisotropic and dependent on the loading direction. The above approach is adopted primarily for rock and concrete. For other materials, predominantly ceramics and steels subjected to ordnance levels of high strains and strain rates, specialist constitutive models are employed, such as the Johnson-Cook and Johnson- Holmquist descriptions. The interested reader is directed to (Cottrell, 2002) for further details.

Mesh objectivity: Although energy dissipation in the crack band model is rendered objective by normalising the softening curve with respect to the specific fracture energy, the spatial localisation is necessarily arbitrary. Localisation occurs in individual elements, resulting in the width of localisation and the crack band spacing depending on the mesh discretisation. Furthermore, the mesh orientation gives rise to a directional bias of propagating crack bands due to the fact that strain discontinuities exist at the element boundaries. Consequently, a non-local averaging of the damage measure is adopted in each orthotropic direction to ensure discretisation objectivity by introducing a length scale to govern the width of the localisation zone. More detailed description of various fracture models can be found in (Yu., 1999; Klerck, 2000). When the unloading process within a localisation zone is complete, a discrete (and physical) crack is inserted as described below.

3.2. DISCRETE CRACK INSERTION

The transition of a body from a continuum to discrete description is developed from dispersed micro cracks coalescing into macroscopic fractures. The appearance of a discrete fracture within the material results in the global realisation of inelastic strains and the associated unloading of the surrounding material. The process of inserting a discrete fracture into a continuum based finite element mesh follows three key steps: (i) the creation of a non local failure map, that is based upon the weighted nodal averages of the damage within the finite element system is required; (ii) the failure map is used to determine the likelihood of fracture within the domain; and (iii) a numerical code to perform the topological update whereby a fracture is inserted in the domain, and any additional nodes are inserted and necessary elemental connectivities are updated.

The so-called *failure factor* is typically defined as the ratio of the inelastic fracturing strain ϵ^f and the critical fracturing strain ϵ_c^f or the ratio of damage and the critical damage. The elemental or local failure factor F_k that is associated with the Gauss point of an element k is given by

$$F_k = \left(\frac{\epsilon^f}{\epsilon_c^f}\right)_k \quad \text{or} \quad F_k = \left(\frac{D}{D_{\mathrm{cr}}}\right)_k \tag{8}$$

where F_k is associated with the elemental local fracture direction θ_k which is defined as being normal to the direction to the local failure softening direction. Discrete fracture is realised through the failure factor reaching unity. In addition, the nodal basis of a finite element system leads to a simpler and more efficient approach for the insertion and creation of discrete fractures. The associated failure factor $\bar{F}_p$ and the corresponding direction of failure $\bar{\theta}_p$ for the nodal point p are given by,

$$\bar{F}_p = \frac{\sum_{k=1}^{ngauss_{adj}} F_k w_k}{\sum_{k=1}^{ngauss_{adj}} w_k}, \bar{\theta}_p = \frac{\sum_{k=1}^{ngauss_{adj}} \theta_k w_k}{\sum_{k=1}^{ngauss_{adj}} w_k}, \tag{9}$$

where the summation is calculated over the number of element Gauss integration points that are immediately adjacent and w_k is the elemental weighting factor, which is normally taken as the elemental volume. When the associated failure factor $\bar{F}_p$ and the direction of failure for the considered node p have been determined, a discrete fracture of the given orientation will be inserted into the finite element mesh, passing through the associated nodal point. Having identified that a nodal point is to undergo the fracture insertion process, it is necessary to identify the actual discrete crack orientation within the finite element mesh. Generally, there are one of two choices to be made at this stage. Firstly, the fracture plane can be aligned in the exact orientation of the weighted average nodal failure direction, thereby following a process known as intra-element fracturing where a series of new nodal points and elements are systematically created, as shown in Fig. 2(b).

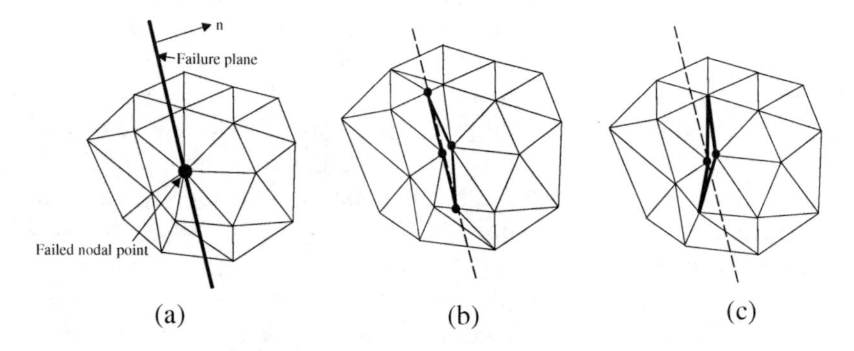

Figure 2. (a) The weighted average nodal failure direction; (b) The intra-element fracture description; (c) The inter-element fracture description.

In the second approach, which is usually preferred from the computational standpoint particularly in 3D problems, the discrete fracture orientation is aligned with the best orientated element boundary attached to the node considered , thereby following a process known as inter-element fracturing where a series of new nodal points are systematically created but no new elements are generated. This is schematically shown in Fig. 2(c). The process of inter-element fracturing,provided a fine mesh is utilised, will usually yield satisfactory results for most numerical simulations. The reasoning behind the preference lies in the ability to effectively maintain control over the critical time step used in the explicit time integration algorithm. Also, in the case of intra-element fracture, local mesh refinement is necessary in the vicinity of the newly introduced crack in order to provide an adequate element topology.

In addition, an element erosion procedure is applied to deal with the situation where the material represented by the discrete element formed no longer contributes to the physical response for the problem, such as the case where the material melts or vapourises at high temperature or is transformed into very small particles.

3.3. CONTACT AND POST-FAILURE MODELLING

The presence of contact phenomena during fracturing and post failure in particular requires an effective and robust contact simulation procedure. In concept, the surfaces of the original continuous regions and the subsequently created crack surfaces are represented by discretised contact surfaces/elements on the basis of the embedded finite element mesh. These contact elements are 2-noded segments in 2D cases and 3-noded triangular facets in 3D cases. Following the general contact detection procedure outlined below, the contact between continuous regions and/or the interaction between the large number of subsequent fragments in the post failure stage may be effectively handled. The most challenging aspects of the contact treatment include modelling of sharp corner to corner contact and maintaining total energy conservation.

4. Discrete Element Modelling (DEM)

Following continuum to discrete transformation appropriate computational strategies must be introduced to model the subsequent inter-particle flow. Space prohibits a full description of this area of numerical modelling and the interested reader is directed to (Owen et al., 2004). In summary, the principal issues involved relate to:

- Contact detection procedures to efficiently monitor contact between neighbouring particles. This is a computationally intensive procedure for industrial scale problems involving large numbers of particles. Consequently, the location of objects is described in terms of bounding boxes in order to simplify this global search and hence based on this approximate geometrical description only potential contact is determined. Procedures adopted for searching include tree-based methods (Feng and Owen, 2002) with complexity O(NlogN) which limits their application as problem sizes become increasingly large and cell-based procedures (Han, 2004) which display O(N) complexity.
- Contact resolution. Based on the potential contactor list determined in the global contact search, the actual geometry of objects in potential contact must be used to determine whether the objects are in contact or not.
- Interaction laws and contact modelling. After the local interaction resolution, the interaction forces between each actual interacting pair are determined according to a constitutive relationship or interaction law. Most commonly, interaction laws are based on the framework of a penalty method, which only approximately satisfies the contact displacement constraints for finite values of the penalty coefficients, allowing a small amount of overlap to occur. In general, the interaction laws describe the relationship between the overlap

and the corresponding repulsive force between a contact pair and may be developed on the basis of the physical phenomena involved. The surfaces of the original continuous regions and the subsequently created crack surfaces are represented by discretised contact surfaces on the basis of the embedded finite element mesh. These contact elements are 2-noded segments in 2D cases and 3-noded triangular facets in 3D cases. The most challenging aspects of the contact treatment include modelling of sharp corner to corner contact and maintaining total energy conservation (Feng and Owen, 2004; Feng et al., 2005).

5. Adaptivity

For many industrial problems the nature of the strains that characterise the deformation implies that a mesh of high quality in the initial configuration of the solid, with optimal element aspect ratios throughout, degenerates to such an extent during the simulation that the numerical procedure may fail to produce any results of practical interest. In addition to acceptable element aspect ratio, mesh quality implies sufficient refinement in areas where the relevant fields exhibit steeper gradients. These areas are the regions of the solid where most of the deformation localises, i.e. where the relevant dissipative mechanisms are most active. Such areas are not known in advance and, in addition, their location within the domain may change as the simulation proceeds. In this context, the use of an effective mesh refinement procedure, capable of ensuring that high quality meshes are used over the entire simulation process is essential to allow the practical use of a finite element framework within an industrial environment.

However, it should be emphasised that for the solution of large scale industrial problems involving gross deformations the primary aim of adopting mesh refinement strategies, at present, is to preserve the integrity of the mesh and not to control the error in the solution. The underlying theory of adaptive mesh refinement strategies for nonlinear problems is still an open issue, with problems related to error estimation, variable mapping of discontinuous fields, dispersion of mapped quantities, etc. currently being the subject of extensive research. Consequently, the use of adaptive strategies for this class of problem should at present be treated with caution and employed only for selected applications in which the nonlinear material behaviour is well understood.

In the present context, remeshing is undertaken when the quality of the mesh, as measured by some distortion index, becomes unacceptable. At this stage standard adaptive mesh refinement procedures (Zienkiewicz and Zhu, 1987; Ortiz and Quigley IV, 1991; Perić et al., 1994; Owen et al., 1995) are invoked, whereby the local error throughout the mesh is estimated by use of an appropriate measure – the damage dissipation in the present case – which is then employed to predict a new element size distribution. Selected variables are then transferred between the old and new mesh through established transfer operators (Perić et al., 1994; Morançay et al., 1997).

It is worth remarking that for use of adaptive mesh refinement strategies in the solution of large scale industrial problems, there are several crucial issues seldom addressed in academic circles. Firstly, it is important that the entire problem definition (geometry, material regions, loading and boundary conditions) be undertaken within a solid modelling environment and not prescribed on the basis of the original finite element mesh. During subsequent remeshing the quantities necessary for mesh generation must be recovered from the solid modeller entities in order to preserve the true geometric characteristics of the model and other data specification. This implies a need for a transparent link between the solid modeller, mesh generation module and analysis code throughout the entire solution. This preservation of geometry is crucial in many applications. For example, when solids are loaded through specific shaped impactors or die systems, it is important that in contact areas the mesh preserves the geometrical features of these loading surfaces. In this case, element size predictions given by the error estimator have to be overridden by the need to have elements of a dimension capable of capturing detailed geometric features. Also, in the meshing of free surfaces it is essential that the volume of the solid is preserved between remeshing sequences, since in forming problems, for example, a cumulative volume change of less than 1% can result in unacceptable die closure force prediction. This generally requires definition of the free surface of the old mesh in terms of NURBS, or other interpolation methods, to provide an accurate geometric description upon which to base the nodal values of the new mesh.

6. Coupled Fluid/Strucutre Modelling: Rock Blasting

In this application coupling takes place through interaction between the gas pressure due to explosive detonation and the progressively fracturing rock. The most appropriate route to solution is provided by superposing a background Eulerian grid over the Lagrangian mesh used for fracture modelling. Within this regular Eulerian grid the gas pressure modelling is based on the mass conservation and momentum equations for gas flow employing directional porosities derived from the rock fracture modelling. The coupling takes place through an interdependence between the evolving gas pressure distribution driving the fracturing process which, in turn, provides the porosity distribution which controls the gas pressure. Computationally, solution can be effectively provided through use of a staggered solution scheme based upon time integration of the two fields with partitioned time stepping.

6.1. THEORY OF GAS FLOW THROUGH ROCK CRACKS

The 2D mass conservation equation for gas flow through rock cracks can be expressed as

$$\frac{\partial}{\partial t}(\rho\alpha) + \frac{\partial}{\partial x}(\rho v_x \alpha_x) + \frac{\partial}{\partial y}(\rho v_y \alpha_y) = 0 \tag{10}$$

where v_x and v_y are the component velocities for the gas, ρ is the gas density, α is the porosity of the rock mass and α_x and α_y are the directional porosity in the x and y directions, respectively. The directional porosities may be evaluated in terms of the average crack heights per unit distance in the x and y directions, denoted $\bar{h_x}$ and $\bar{h_y}$, as

$$\alpha_x = \bar{h_y} = h_y/dy; \quad \alpha_y = \bar{h_x} = h_x/dx \tag{11}$$

The gas velocities are obtained through the momentum equation:

$$v_x = -\frac{\alpha_x^2}{12\mu}\frac{\partial p}{\partial x}; \quad v_y = -\frac{\alpha_y^2}{12\mu}\frac{\partial p}{\partial y} \tag{12}$$

where μ is the gas viscosity for flow through the rock cracks and p is the gas pressure. Since the gas pressure arises from the detonation process, the pressure is obtained using the equation of state for the explosive, which may be defined generally as:

$$p = p(e, t, \rho) \tag{13}$$

Eq. (10) may be solved by adopting the finite difference method. By dividing the whole domain into a structured grid in the x and y directions Eq. (10) can be discretised as

$$\frac{\partial}{\partial t}(m_{i,j}) + \frac{F_{x_{i+1/2,j}} - F_{x_{i-1/2,j}}}{\Delta x} + \frac{F_{y_{i,j+1/2}} - F_{y_{i,j-1/2}}}{\Delta y} = 0 \tag{14}$$

where $m_{i,j}$ is the gas mass, Δ_x and Δ_y are respectively the grid spacings in the x and y directions and F_x and F_y are the flux in the x and y directions; i.e.

$$F_x = \rho v_x \Delta v; \quad F_y = \rho v_y \Delta v_y \tag{15}$$

and subscripts i and j refer to the grid point number in the x and y directions respectively. Evaluation of the component flux defined by equation (15) necessitates evaluation of the velocities, porosities and gas densities at the grid lines. The porosities are evaluated from the topology of the mechanical field. The velocities are defined as:

$$v_x|_{i+1/2,j} = -\frac{\alpha_x^2|_{i+1/2,j}}{12\mu}\frac{p_{i+1,j} - p_{i,j}}{\Delta x}; \; v_y|_{i,j+1/2} = -\frac{\alpha_y^2|_{i,j+1/2}}{12\mu}\frac{p_{i,j+1} - p_{i,j}}{\Delta y} \tag{16}$$

For stability consideration gas densities at the grid lines are evaluated using the 'upstream' values rather than a local average, i.e.

$$\rho_{i+1/2,j} = \begin{Bmatrix} \rho_{i+1,j} & \text{if } v_x < 0 \\ \rho_{i,j} & \text{if } v_x > 0 \end{Bmatrix}; \quad \rho_{i,j+1/2} = \begin{Bmatrix} \rho_{i,j+1} & \text{if } v_y < 0 \\ \rho_{i,j} & \text{if } v_y > 0 \end{Bmatrix} \tag{17}$$

The time integration of (14) is carried out by the forward Euler method.

6.2. GAS PRESSURE CALCULATIONS

If a borehole is fired but the detonation process has not finished, the borehole pressure is calculated based on an empirical formulation. The pressure subsequent to detonation is defined as:

$$p_0 = c s_1 \rho_0^{g_0} e^{s_2} \tag{18}$$

where

$$s_1 = g_0 + \rho_0(g_1 + g_2\rho_0) - 1 \tag{19}$$

$$s_2 = \rho_0(g_1 + 0.5\, g_2\rho_0) \tag{20}$$

and c, g_o, g_1, g_2, are material parameters. The definition of p_0 in equation (18) is a function of gas density and not related to time.

7. Numerical Validations and Illustrations

The performance of the algorithms discussed in this paper is validated and demonstrated in this section. Their practical application is illustrated in the numerical solution of a representative set of industrially relevant problems.

7.1. 2D VALIDATION: BOREHOLE BREAKOUT MODELLING

As a test of the ability of the continuum/discrete model for multi-fracturing solids to capture the contrasting failure mechanisms exhibited by strong and weak rocks under different confining pressure conditions, Fig. 3 illustrates the problem of borehole breakout modelled as a 2D problem. The initial continuum mesh is shown and the horizontal confining pressure is maintained at a constant value whilst the vertical pressure is incrementally increased at the rate indicated. The properties that describe the behaviour of the material are indicated, where it is seen that the parameters required are readily identified from standard triaxial and tension tests. In addition to the usual elastic properties, the essential parameters are the tensile strength, compressive strength, fracture energy release rate and the standard Mohr-Coulomb data. An additional parameter is the frictional sliding coefficient on newly created discrete cracks. The material parameters presented

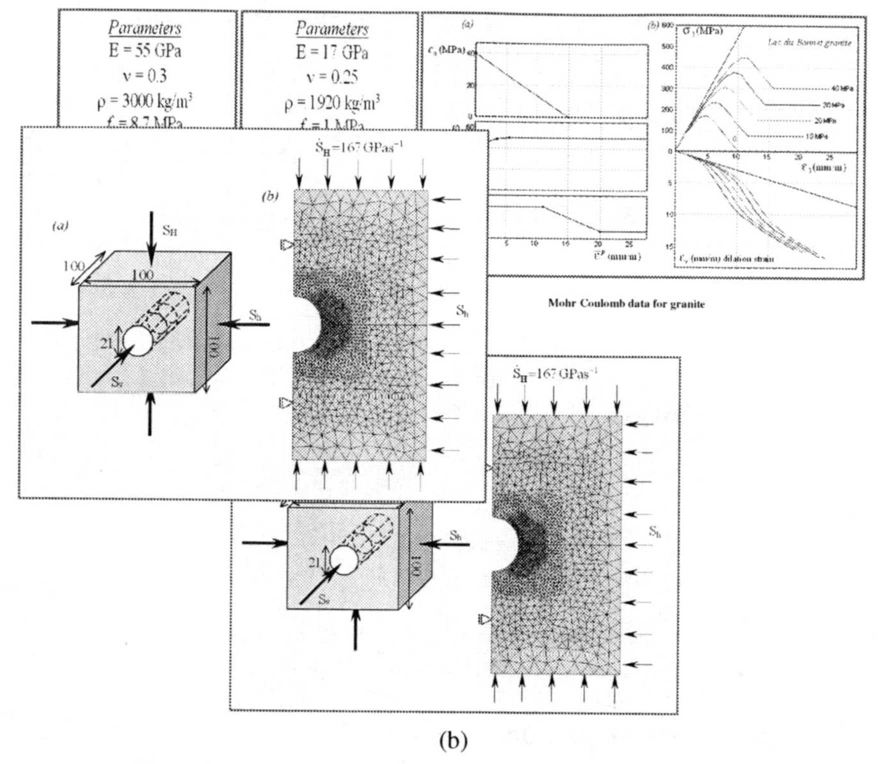

(b)

Figure 3. Borehole test: (a) material properties; (b) numerical model.

correspond to Lac du Bonnet granite (Lee and Haimson, 1993) and the weak sedimentary rock Cardova cream limestone (Lee and Haimson, 1993).

Figure 4 illustrate the fracture patterns developed for the two materials where it is seen that fundamentally different mechanisms are involved. For the granite specimen -Fig. 4(a)- failure takes place by the development of sub-vertical fractures at the regions indicated and very good agreement is evident between the experimental observations of Lee and Haimson (Lee and Haimson, 1993) and the numerical predictions. In Fig. 4(b) similar comparison is made for the Cardova cream limestone, for which failure takes place by the development of fracture shear bands. In the computational model these manifest themselves as distinct crack bands formed by en-echelon systems of tensile fractures. Again, there is excellent correspondence between the experimental failure mode presented by Haimson and Song (Haimson and Song, 1993) and the numerical simulation. It is important to note that the computational model has been able to reproduce these two fundamentally different failure mechanisms by only changing the relevant material parameters.

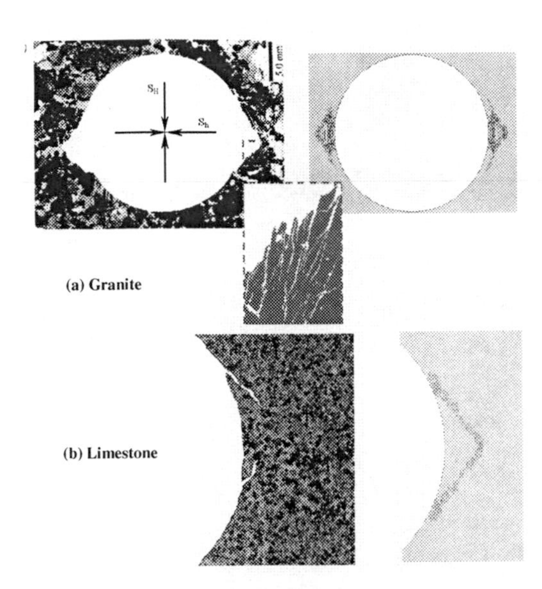

Figure 4. Borehole test: Comparison of experimental and numerically predicted failure patterns.

7.2. 3D VALIDATION: LOW SPEED BEAM IMPACT

Figure 5. 3D Beam Impact Validation: View of impact test rig.

The methodology of the combined discrete/finite element approach is further validated by comparing the numerical simulation with a carefully designed 3D reinforced concrete beam test under low speed and high mass impact conditions.

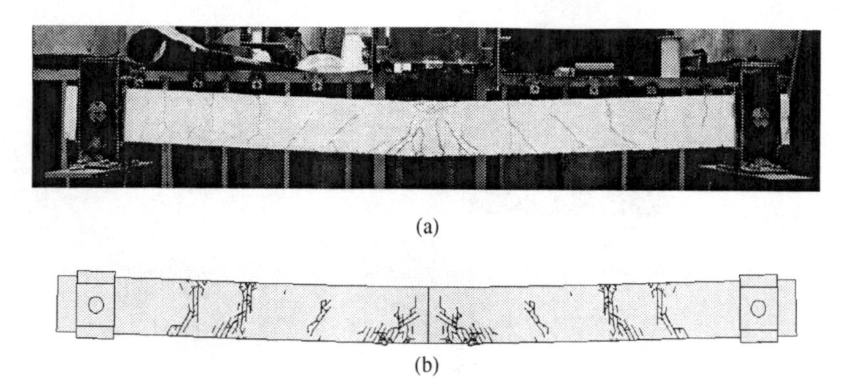

(a)

(b)

Figure 6. 3D Beam Impact Validation - final fracture pattern: (a) experimental results; (b) numerical simulation.

The test facility comprises a drop weight system that has been used to allow weights of up to 200 kg to fall through heights of up to 4 m. Specimens tested to date have been beams with a maximum span of 3.0 m and slabs of 2.5 m square. A general view of the test facility is shown in Fig. 5.

In the beam test considered, the span of the beams was 2.7m, the dropping weight used was 98 kg and the impact velocity was 7.3 m/s. The aspects of particular interest are the global behaviour of the beam, the order of crack formation and spallation, the local fracture behaviour in the vicinity of the impact, and the impact force history. The final fracture pattern of the beam is shown in Fig. 6(a).

In addition to the numerical models described above for the concrete regions of the beam, the reinforcement bars are modelled as 3D beams with elasto-plastic material properties and a perfect bond condition is assumed between the bars and the concrete. The simulated final fracture pattern, depicted in Fig. 6(b), appears to agree very well with the experiment. The impact force history also exhibits a high degree of agreement between the experimental and computational results (May et al., 2005). A distinct advantage of the numerical simulation is its ability to reveal the detailed fracture formation pattern and the behaviour of the reinforcing bars during the whole course of the impact, which can not be readily observed experimentally.

7.3. INTERFACE DEFEAT FOR MISSILE IMPACT

The design of lightweight vehicle armour has lead to the use of coupled ceramic and metal composites. This approach can potentially be used in highly weight efficient armour systems, subjected to a large range of impact velocities. The purpose of this example is to numerically demonstrate that techniques based on adaptive remeshing and discrete element fracture can be coupled to give useful insights to the impact of metallic deformable projectiles into ceramics. The experimental

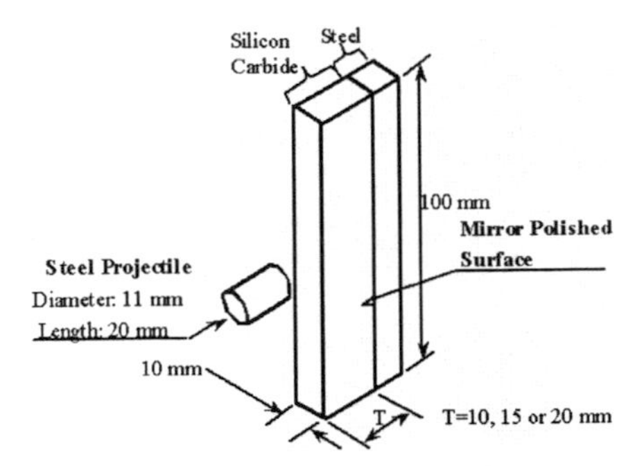

Figure 7. Two dimensional impact: problem geometry.

testing procedure can be found in Ref. (Riou et al., 1998). A schematic of the impact configuration for an unconfined and for a RHA steel backed silicon carbide 'beam' is given in Fig.7. The impacting projectile has a length of 20 $[mm]$ and is 11 $[mm]$ in diameter and impacts an unconfined silicon carbide 'beam' with a velocity of 203 $[m/s]$. The 'beam' has a length of 100 $[mm]$, is 10 $[mm]$ wide, and the thickness T is either 10, 15 or 20 $[mm]$. The ceramic element of the system is

Table 1. Material parameters for silicon carbide.

Description	Symbol	Value
Density	ρ	3,150 $[kg/m^3]$
Elastic bulk modulus	K	192,000 $[MN/m^2]$
Elastic shear modulus	G	175,000 $[MN/m^2]$
Specific fracture energy	G_f	24.7 $[J/m]$
Maximum tensile strength	f_t	43$[MPa]$
Coulomb friction coefficient	m	0.25

numerically modelled using the rotating crack tensile softening plasticity model followed by discrete crack insertion (Cottrell, 2002; Cope et al., 1980; Yu., 1999; Klerck, 2000). The corresponding material data are given in Table 1. In addition, the RHA impactor is numerically modelled using the Johnson-Cook viscoplastic constitutive model (Johnson and Cook, 1983).

The governing equation for the yield stress associated with the Johnson-Cook viscoplastic constitutive model is given by

$$\bar{\sigma} = [A + B\,(\bar{\epsilon}^{\mathrm{p}})^{n}]\,[1 + C\ln(\dot{\epsilon}^{\mathrm{p}})]\left[1 - \left(\frac{T - T_{\mathrm{ref}}}{T_{\mathrm{melt}} - T_{\mathrm{ref}}}\right)^{m}\right] \tag{21}$$

where $\bar{\epsilon}^{\rm p}$ is the accrued inelastic strain, $\dot{\epsilon}^{\rm p}$ is the inelastic strain rate, T is the material temperature which is accounted for through use of adiabatic heating and $A, B, C, m, n, T_{\rm ref}$ and $T_{\rm melt}$ are user defined material parameters whose values are given in Table. 2.

Table 2. Material parameters for RHA steel (Westerling et al., 2001)

Description	Symbol	Value
Density	ρ	7,830 $[kg/m^3]$
Elastic bulk modulus	K	159,000 $[MN/m^2]$
Elastic shear modulus	G	81,800 $[MN/m^2]$
Strain hardening exponent	n	0.26
Strain rate coefficient	C	0.014
Thermal softening exponent	m	1.13
Initial reference temperature	$T_{\rm ref}$	300 $[K]$
Specific heat capacity	C_v	477 $[J/Kg/K]$
Melting temperature	$T_{\rm melt}$	1,793$[K]$
Energy conversion coefficient	f	90 %

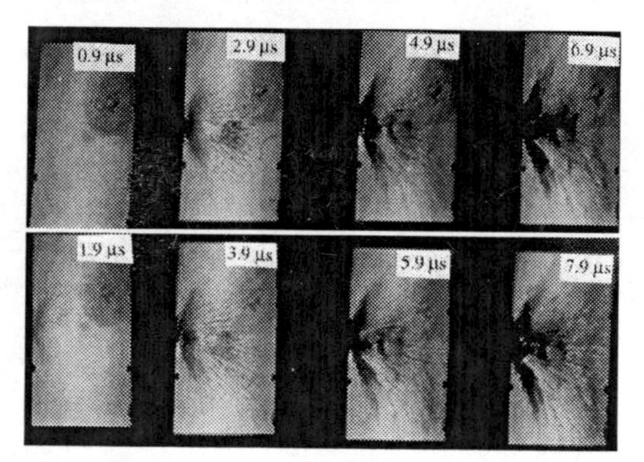

Figure 8. Two dimensional impact: experimental discrete fracture evolution (Riou et al., 1998) for a velocity of 203 $[m/s]$.

Figure 8 shows the experimental findings of Riou *et al.* (Riou et al., 1998) for the unconfined target configuration at an impact velocity of 203 $[m/s]$ and Fig. 9 shows the corresponding numerical predictions.

Upon comparison of Fig.8 and Fig.9, it is clear that the proposed technology is capable of capturing the salient features present within the experimental tests.

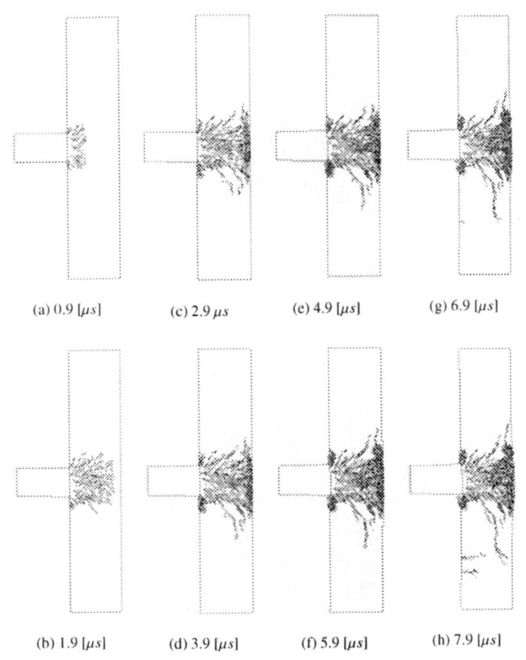

Figure 9. Discrete fracture evolution obtained from the numerical simulation impacted at 203 $[m/s]$.

The numerical simulations correctly identify keys features, such as conical type fractures dominating the fragmentation process. In addition, a degree of spalling type fracture is also accurately predicted to occur on the distal face of the ceramic as well as at the impact interface. Besides the fracture profiles presented in Fig. 9, it is of the utmost importance that the finite element mesh maintains a good aspect ratio and the adaptive scheme produces sufficiently refined meshes where the underlying physical mechanisms require. The driving motivation behind the proposed technology is the ability to attain a numerical solution with a high level of accuracy allied to an economical computational cost. Figure 10 illustrates that the evolving finite element mesh can be effectively controlled, providing a high resolution in regions of high strain rate activity and fracture and coarsened elsewhere. The means of controlling the finite element mesh size using a criterion based upon the total incremental strain rate term appears to behave well for the fragmenting silicon carbide. In addition when this is combined with the inelastic strain criterion (Yu., 1999) it provides the analyst with a flexible strategy capable of modelling both metallic and ceramic systems.

This second example provides further assessment of the (partial and complete) defeat of a high velocity KE penetrator against a ceramic target system. In this case the penetrator velocity is much greater resulting in extensive plas-

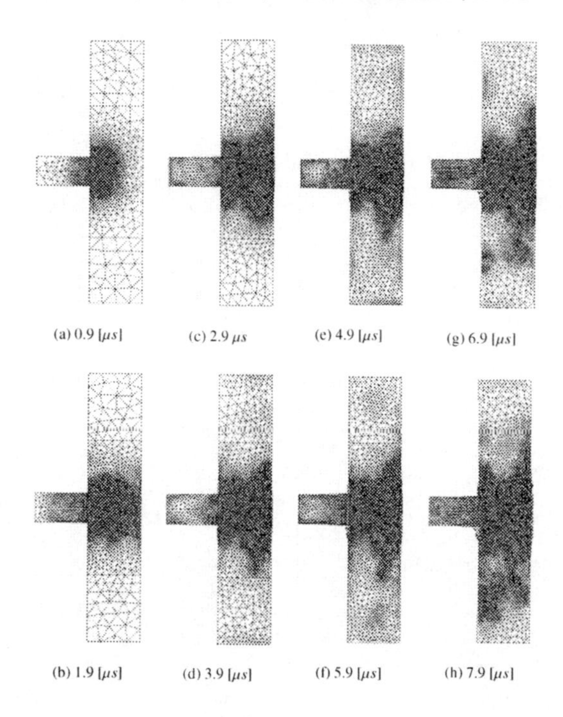

Figure 10. Finite element mesh evolution for an impact velocity of 203 $[m/s]$.

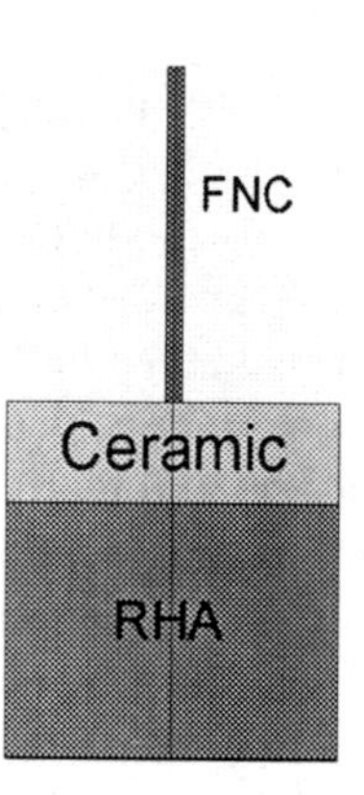

Figure 11. Schematic of the target configuration.

tic deformation of the penetrator. For this high velocity application, constitutive modelling of the ceramic is based on the pressure dependent Johnson-Holmquist JH1 model (Holmquist and Johnson, 2002) suitable for brittle materials subjected to high strain rates, strains and pressures. The evolution of the failure surface is controlled by an accrued damage function which is a function of the accumulated inelastic strain increment and the inelastic fracture strain for a constant pressure.

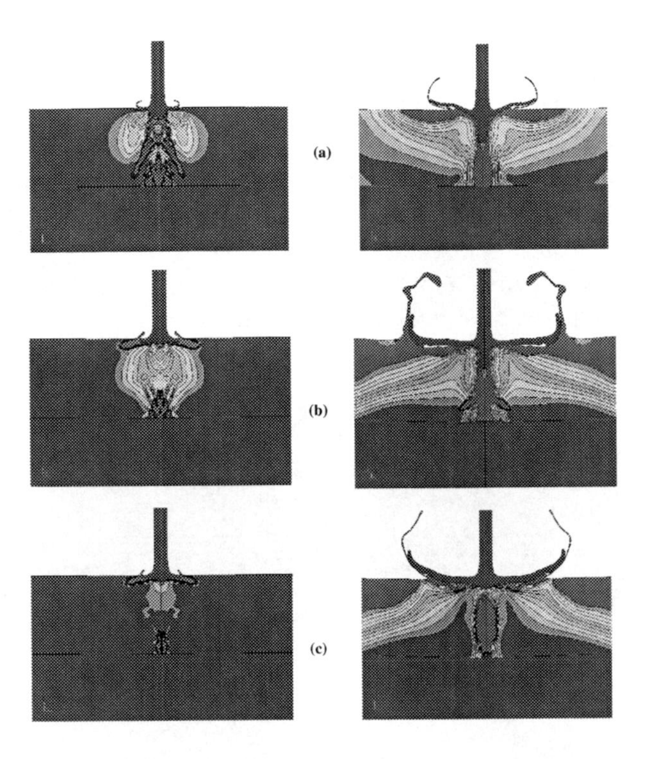

Figure 12. The evolution of JH1 material damage at 5 and 30 μs for 1410 m/s impact with finite element mesh size 0.20 mm: (a) D_1 =0.012 GPa^{-1}; (b) D_1 = 0.020 GPa^{-1}; and (c) D_1 = 0.12 GPa^{-1};.

The standard damage accumulation function is expressed by the equation,

$$D = \sum \frac{\Delta \bar{\epsilon}^p}{\epsilon_p^f} \tag{22}$$

where $\Delta \bar{\epsilon}^p$ is the inelastic strain increment developed during a single time step. The calculation of the inelastic fracture strain measure ϵ_p^f follows the formulation detailed by Lundberg (Lundberg, 2004), in which the inelastic fracture strain is defined by the relation

$$\epsilon_p^f = D_1 (P + T)^{D_2} \qquad \epsilon_{min}^f < \epsilon_p^f < \epsilon_{max}^f \tag{23}$$

where the pressure P and the tensile strength T are not normalised by the Hugoniot pressure term P_{HEL}. The selection of the damage parameters D_1 and D_2 is of critical importance in the definition of the damage accrual path.

The target configuration is illustrated in Fig. 11, where a FNC tungsten rod penetrator, with a 5 [mm] diameter and a 20:1 (L/D) ratio, is impacted at 1410 [m/s] at normal incidence into a SiC-B ceramic target material comprising 100× 100×30 [mm] tiles supported on a RHA steel backing plate. Depending on the parameter D_1 in equation (23) controlling the damage accrual, significantly different penetration characteristics (or dwell phenomena) are exhibited.

Figure 12(a) shows the case for a low damage resistance $D_1 = 0.012\ \text{GPa}^{-1}$, where it is seen that significant penetration of the ceramic component takes place. For the intermediate case ($D_1 = 0.012\ \text{GPa}^{-1}$) shown in Fig. 12(b), the penetration is reduced and for $D_1 = 0.12\ \text{GPa}^{-1}$ (Fig. 12(c)) it is seen that the penetrator is completely defeated. It is evident for this application that the use of adaptive remeshing is essential. Typically each problem requires approximately 500 remeshings in order to arrive at the final configuration.

7.4. ROCK BLASTING - HPE TEST

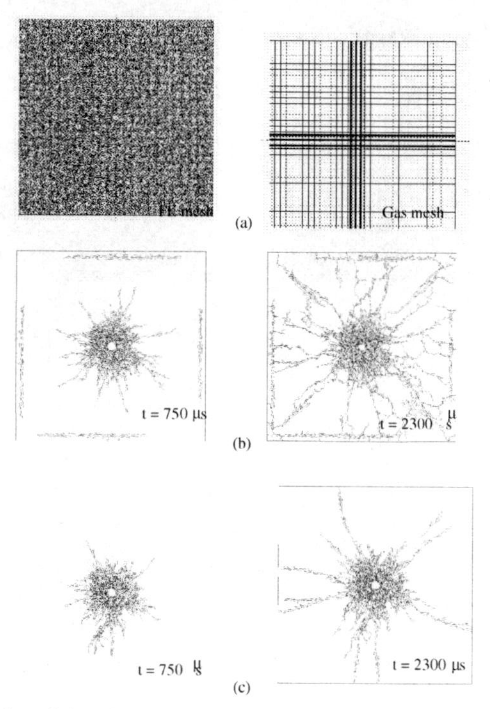

Figure 13. Rock blasting: finite element and gas meshes (a); and fracture development at different times - (b) free boundaries and (c) non-reflecting boundaries.

Figure 13 illustrates a benchmark test case of an idealised plane strain explosive detonation within a rock mass. A borehole of 165 mm diameter is located at the centre of a 3m × 3m square rock mass. The material properties of the rock are listed in Fig. 14(a) and the variation of the tensile strength with strain rate is summarised in Fig. 14(b). The bore hole data (which is fully stemmed) and explosive properties are given in Fig. 14(c). The initial continuum finite element

(a)

Material Properties	
Property	Value
Young's Modulus	7.0E+10 N/m^2
Poisson's Ratio	0.25
Density	2900 Kg/m^3
Tensile Strength	20.0E+6 N/m^2
Fracture Energy	120 N/m
t_f^{min}	1.0E-7

(b)

Tensile Strength Data	
Strain Rate	Tensile Strength
1	20.0E+6N/m^2
1,000	80.0E+6N/m^2
2,000	106.0E+6N/m^2
6000	108.0E+6N/m^2
1000000	350.0E+6N/m^2

(c)

Bore Hole & Explosive Properties	
Property	Value
Explosive Diameter	165 mm
VOD	3033 m/s
Density	500 Kg/m^3
CJ Density	668 Kg/m^3
Von Neumann pressure PVN	2.2E+9 N/m^2
CJ Pressure	1.0E+9 N/m^2
CJ time	50.0E-06 s
CJ velocity	1000 m/s
g0	1.33
g1	0.49
g2	0.21
Exp. length	1.0 m
Stem length	0.0 m
Height	1 m
delay	0.0 s
Mesh size	30 mm

Figure 14. Rock blasting: Material properties of rock mass.

mesh and the background Eulerian grid employed for modelling the gas pressure evolution are illustrated in Fig. 13(a). In the central region the grid size is 0.02 m and in the outer zones a 0.2 m grid spacing is used, since it is anticipated that significant gas flow will be limited to the central region only.

Two cases are considered: (i) where the external boundaries are considered to be free and (ii) where the block is considered to be part of a continuous region and consequently non-reflecting (or energy absorbing) boundary conditions are applied to the four boundary edges. For Case (i) (shown in Fig. 13(b)), it is seen that when the compressive stress wave arising from the explosive detonation reaches the free boundaries, it is reflected as a tensile wave resulting in tensile fractures occurring at these locations. On the other hand for Case (ii) (shown in Fig. 13(c)), the stress wave passes through the boundaries causing no spurious crack propagation. In each case, there is extensive fracturing in the vicinity of the bore hole with discrete (or semi-discrete) fractures propagating away from this region.

7.5. THREE DIMENSIONAL HIGH VELOCITY OBLIQUE IMPACT

In applications involving ballistic type experiments that are frequently performed within the defence industry (Cottrell, 2002), the associated finite element discretisation fails to provide an accurate description of the material response whether this is due to large finite strains that characterise ductile materials, or degradation into multiple bodies, typically observed for brittle materials. Therefore the objective of this example is to demonstrate how a mesh adaptive approach, combined with an appropriate damage indicator to predict material failure, can provide useful insights to the response of metallic systems when subjected to ultra ordnance velocity loadings. This example consists of a reverse impact in which an oblique RHA steel target system interacts with a stationary, pitched tungsten long rod penetrator. A sketch of the impact configuration is given in Fig. 15.

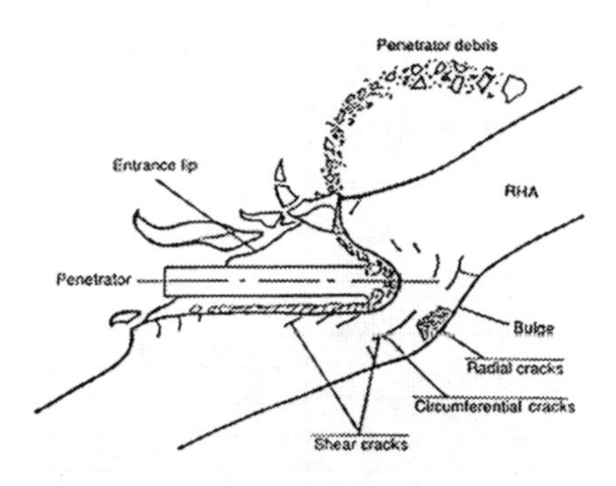

Figure 15. Sketch of the impact configuration.

The oblique impact of long rod penetrators upon target systems at ultra ordnance velocities produces extensive local contact deformations as well as the propagation of elastic and inelastic waves throughout the impactor. A number of generic aspects that characterise interaction of the long rod with a hard target are depicted in Figure 15. The schematic highlights the fragmentation and erosion of the target entrance. In addition, penetrator comminution, bulging and cracking occur at the distal surface prior to target perforation. The resultant cross-sectional cutaway is shown in Fig.16 and it is possible to identify the formation of a 'lip' on the lower side of the distal surface of the target plate. This photograph is in agreement with the expected phenomenological events reported by Goldsmith (Goldsmith, 1999) and Zukas (Zukas, 1990).

The application of predictive modelling techniques for assessing the initial stages of interaction of high velocity KE penetrators upon complex metallic target systems is a challenging field of research. It has been repeatedly observed that the overall response of the interacting target and penetrator is highly influenced by their initial interaction, and that the accurate representation of the material flow in the region of the impact face is of paramount importance. The initial finite element mesh is presented in Fig. 17, where the length of the rod is 90 $[mm]$ and the diameter is 3 $[mm]$ with a negative pitch of 3.40 degrees (tail down). The target has a diameter of 40 $[mm]$ and thickness of 5 $[mm]$. The initial angle of the target is 50 degrees and the initial impact velocity is 1,743 $[m/s]$.

Both rod and target are numerically modelled using the Johnson-Cook viscoplastic constitutive model (Johnson and Cook, 1983) defined by Eq. (21). In addition, the temperature increase of the material is assumed to be the result of adiabatic heat generation. The volumetric response of the material under high strain rates demands the proper modelling of the pressure evolution. In this example, this effect was taken into account by introducing a Mie-Grüneisen equation of state. The data specific for the viscoplastic response of the FNC tungsten penetrator and RHA target plate are given in Table 3.

Table 3. Viscoplastic material parameters.

Description	Symbol	FNC	RHA
Density	ρ	17,600 $[kg/m^3]$	7,830 $[kg/m^3]$
Elastic bulk modulus	K	289,640 $[MPa]$	164,000 $[MPa]$
Elastic shear modulus	G	133,680 $[MPa]$	82,000 $[MPa]$
Static yield limit	A	1,365 $[MPa]$	1,040 $[MPa]$
Strain hardening	B	900 $[MPa]$	650 $[MPa]$
Hardening exponent	n	0.65	0.40
Strain rate coefficient	C	0.025	0.013
Thermal exponent	m	0.85	0.82
Initial temperature	T_{ref}	300 $[K]$	300 $[K]$
Specific heat capacity	C_v	140 $[J/Kg/K]$	450 $[J/Kg/K]$
Melting temperature	T_{melt}	1,750 $[K]$	1,800 $[K]$
Energy conversion	f	90 %	90 %

Besides the Johnson-Cook viscoplastic constitutive model (Johnson and Cook, 1983), a damage indicator that allows the prediction of failure regions, was also included in the analysis. The authors (Johnson and Cook, 1983), proposed the

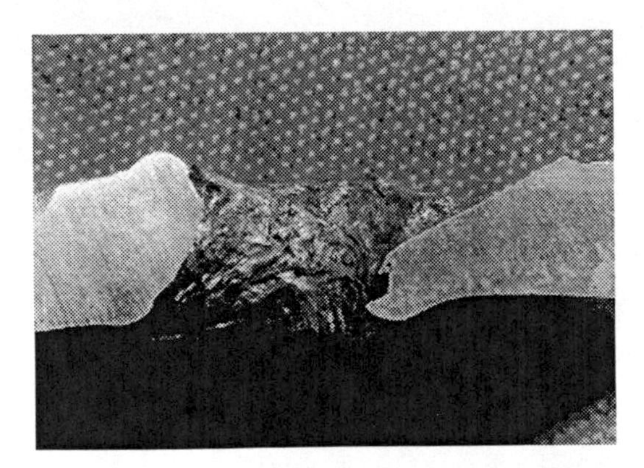

Figure 16. Cross-sectional cutaway of target remnants.

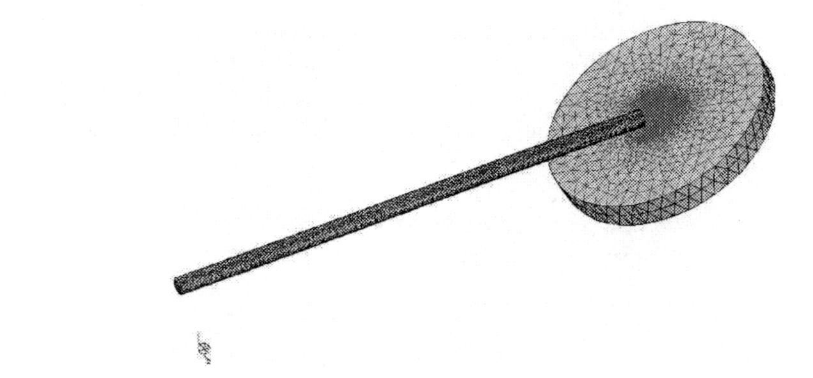

Figure 17. Initial finite element discretisation.

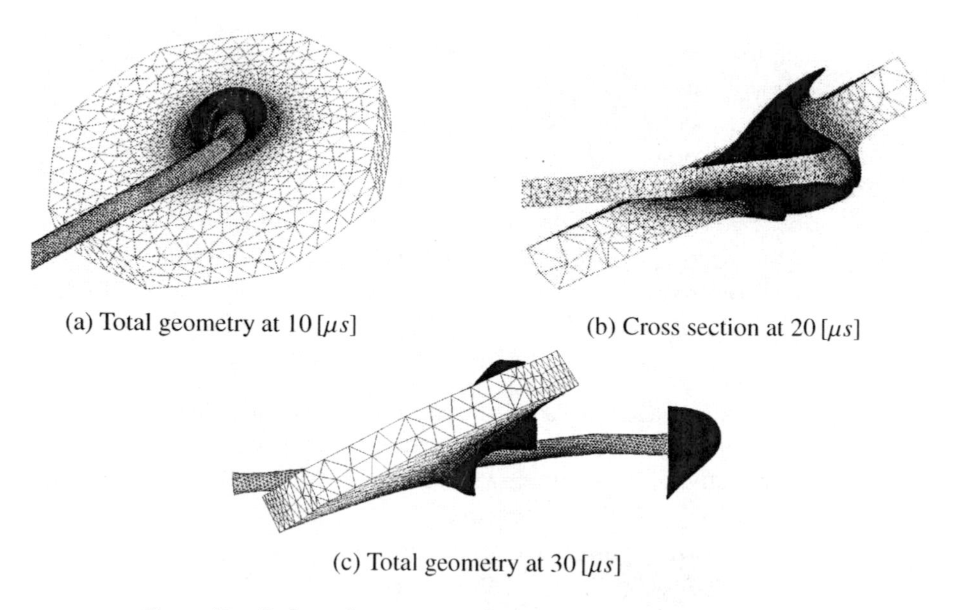

(a) Total geometry at 10 [μs]

(b) Cross section at 20 [μs]

(c) Total geometry at 30 [μs]

Figure 18. Deformed geometry configurations at various time intervals.

following expression to evaluate the inelastic strain to fracture,

$$\epsilon^f = [D_1 + D_2 \cdot \exp(D_3 \check{\sigma})] [1 + D_4 \ln(\dot{\epsilon^p})] \left[1 + D_5 \left(\frac{T - T_{\text{ref}}}{T_{\text{melt}} - T_{\text{ref}}}\right)\right] \tag{24}$$

where the normalised equivalent stress, $\check{\sigma}$ is given by

$$\check{\sigma} = \frac{p}{\bar{\sigma}} \tag{25}$$

in which, p is the hydrostatic pressure and $\bar{\sigma}$ stands for the yield flow stress. The decoupled post-processed damage measure is computed accordingly to

$$D = \sum \frac{\Delta \bar{\epsilon}^p}{\epsilon^f} \qquad 0 \le D \le 1 \tag{26}$$

The model data specific to the damage indicator for the FNC tungsten penetrator and RHA target plate are given in Table 4.

Table 4. Damage material parameters for the Johnson-Cook viscoplastic damage model.

Description	Symbol	FNC penetrator	RHA target
Damage parameter	D_1	0.025	-0.80
Damage parameter	D_2	0.60	2.10
Damage parameter	D_3	-4.00	-0.50
Damage parameter	D_4	-0.033	0.002
Damage parameter	D_5	8.00	0.61
Minimum fracture strain	ϵ^p_{min}	0.022	0.035

Interaction is accounted for between the penetrator and plate in the finite element system by using contact algorithms reported in Ref. (Cottrell, 2002; Yu., 1999). The remeshing algorithms are triggered by the element distortion, and the mesh prediction is based upon the current inelastic strain distribution.

The use of adaptive remeshing techniques allows the simulation of problems where evolving boundaries emerge from unconstrained material flow. The prediction of material rupture is estimated using the Johnson-Cook viscoplastic damage model. The evolution of both geometries and damage variable field obtained in the finite element analysis is illustrated in the contour plots shown in Fig. 18.

The example presented outlines the assembly of different numerical techniques, where the robustness and reliability of three dimensional adaptive remeshing is emphasised.

8. Concluding Remarks

This paper presents an integrated discrete/finite element simulation strategy for modelling practical problems where fracture is the major phenomenon controlling the process. A simple technique to permit the use of linear triangular and tetrahedral elements under finite strain near-incompressibility conditions has been presented. The method is based on the enforcement of the volumetric constraint over patches of simplex elements. This has significant implication in the solution of

3D problems with complex geometries and the introduction of discrete fractures. For quasi-brittle materials, fracture is considered in the form of a rate dependent rotating smeared-crack model utilising tensile strain-softening to represent material degradation. For tensile/compressive states the model is complemented by a modified Mohr-Coulumb model (Klerck, 2000) or other failure surfaces such as the JH1 or JH2 models (Cottrell, 2002). A topology updating scheme is employed to introduce new physical cracks. For discrete element modelling, an effective contact detection procedure for monitoring interaction between large numbers of discrete objects is employed. Penalty based interaction laws are used to represent the contact between discrete elements.

Although considerable progress has been made in the computational modelling of multi-fracturing solids, several outstanding issues still require attention. One particular area of development is the coupling of multi-fracturing solids or particulates, with fluid regimes for use in applications such as rock blasting (gas dynamic equations), stability of rock masses (seepage flow) or fluidized beds (turbulent flows). Each application represents a further specific topic of future research.

Due to the intensive computations involved in hybrid continuous/discrete simulations of industrial applications, parallelisation becomes an obvious option for significantly increasing existing computational capabilities. Significant advances in the development of parallel computer hardware, particularly the emergence of commodity PC clusters, make such a parallel computing option feasible and attractive. The effective implementation of parallel procedures for conventional finite element problems, generally based on static domain decomposition procedures, are well established and solution strategies have been developed for both shared memory and distributed memory platforms. Unfortunately, for problems in which progressive fracturing takes place the situation is complicated by a continuing change in problem topology, resulting from the creation of new surfaces and objects. A similar difficulty exists when mesh adaption processes are implemented, with or without fracture. While parallel implementation on shared memory machines may be tractable, considerable difficulties exist for distributed memory platforms arising from such dynamically changing geometric configurations and boundary conditions. Essentially, processor workload, inter-processor communication and local data storage requirements undergo continuous evolution during the simulation. For this class of problems dynamic domain decomposition parallel strategies must be adopted in which the domain configurations are incrementally evolved throughout the analysis to both minimise inter-processor data communication and maintain a balance of processor workload. Although some progress has been made in the development of appropriate algorithms, considerable further work is required to provide general robust and efficient solution procedures that account for all the technical complexities encountered in industrial scale problems.

References

Andrade Pires, F. M., De Souza Neto, E. A., and Owen, D. R. J. (2004) On the Finite Element Prediction of Damage Growth and Fracture Initiation In Finitely Deforming Ductile Materials.

Belytschko, T., Ong, J., Liu, W., and Kennedy, J. (1984) Hourglass Control in Linear and Non Linear Problems, *Comp. Meth. App. Mech. Engng.* **43**, 251–276.

Bonet, J. and Burton, A. (1998) A Simple Average Nodal Pressure Tetrahedral Element for Incompressible and Nearly Incompressible Dynamic Explicit Applications, *Comm. Num. Meth. Engng.* **14**, 437–449.

Bonet, J., Marriot, J., and Hassan, O. (2001) An Average Nodal Deformation Gradient Linear Tetrahedral Element For Large Strain Explicit Dynamic Applications, *Comm. Num. Meth. Engng.* **17**, 551–561.

Cope, R., Rao, P., Clark, L., and Norris, P. (1980) Modelling of reinforced concrete behaviour for finite element analysis of bridge slabs, In *Numerical Methods For Nonlinear Problems*, pp. 457–470, Pineridge Press, Swansea.

Cottrell, M. G. (2002) The Development of Rational Computational Strategies for the Numerical Modelling of High Velocity Impact, Ph.D. thesis, University of Wales Swansea, Swansea.

Cundall, P. A. and Strack, O. D. L. (1979) A discrete numerical model for granular assemblies, *Géotechnique* **29**, 47–65.

de Borst, R. (2001) Some recent issues in computational failure mechanics, *Int. J. Numer. Meth. Engng.* **52**, 63–96.

De Souza Neto, E., Andrade Pires, F., and Owen, D. R. J. (2003) A New F-bar Based Method For Linear Triangles and Tetrahedra in The Finite Element Analysis of Nearly Incompressible Solids, In E. Onate and D. R. J. Owen (eds.), *Computational Plasticity: Fundamentals and Applications*, Vol. **7**, Barcelona, Spain, CIMNE.

De Souza Neto, E., Andrade Pires, F., and Owen, D. R. J. (2005) F-bar-based linear triangles and tetrahedra for finite strain analysis of nearly incompressible solids. Part I: Formulation and benchmarking, *Int. J. Numer. Meth. Engng.* **62**, 353–383.

Dolbow, J., Moes, N., and Belytschko, T. (2001) An extended finite element method for modelling crack growth with frictional contact, *Comp. Meth. Appl. Mech. Engng.* **190**, 6825–6846.

Feng, Y. T., Han, K., and Owen, D. R. J. (2005) An energy based polyhedron-to-polyhedron contact model, In 3rd *M.I.T. Conference of Computational Fluid and Solid Mechanics*, MIT, USA.

Feng, Y. T. and Owen, D. R. J. (2002) An augmented spatial digital tree algorithm for contact detection in computational mechanics, *Int. J. Numer. Meth. Engng.* **55**, 556–574.

Feng, Y. T. and Owen, D. R. J. (2004) 'A 2D polygon/polygon contact model: algorithmic aspects, *Engineering Computations* **21**, 265–277.

Goldsmith, W. (1999) Review: Non-ideal projectile impact on targets, *Int. J. Imp. Engng.* **22**, 95–395.

Grady, D. E. and Hollenbach, R. E. (1977) Rate-controlling Mechanisms in Brittle Failure of Rock, Technical Report SAND 76-0559, University of Higher Education, Sandia Laboratories.

Haimson, B. and Song, I. (1993) Laboratory study of borehole breakouts in cordova cream: a case of shear failure mechanism, *Int. J. Mech. Min. Sci. & Geomech. Abstr.* **30**, 1047–1056.

Han, K. (2004) Algorithmic and Modelling Aspects of Discrete Element Methods With Applications to Shot Peening and Peen Forming Processes, Ph.D. thesis, University of Wales Swansea, Swansea.

Holmquist, T. J. and Johnson, G. R. (2002) Response of silicon carbide to high velocity impact, *Int. J. Applied Physics* **91**, 5858–5866.

Johnson, G. and Cook, W. (1983) A constitutive model and data for metals subjected to large strains, high strain rates and high temperatures, In *7th International Symposium On Ballistics*, The Hague, Netherlands.

Klerck, P. A. (2000) The Finite Element Modelling of Discrete Fracture in Quasi-Brittle Materials, Ph.D. thesis, University of Wales Swansea, Swansea.

Klerck, P. A., Sellers, E. J., and Owen, D. R. J. (2004) Discrete fracture in quasi-brittle materials under compressive and tensile stress states, *Comp. Meth. Appl. Mech. Engng.* **193**, 3035–3056.

Lee, M. and Haimson, B. (1993) Laboratory study of borehole breakouts in Lac du Bonnet granite: a case of extensile failure mechanism, *Int. J. Mech. Min. Sci. & Geomech. Abstr.* **30**, 1039–1045.

Lundberg, P. (2004) Interface defeat and penetration: two modes of interaction between metallic projectiles and ceramic targets, *Comprehensive summaries of Uppsala dissertations for the faculty of science and technology*.

May, I. M., Chen, Y., Owen, D. R. J., Feng, Y. T., and Bere, A. (2005) Behaviour of reinforced concrete beams and slabs under drop-weight impact loads, In 5th *Asia-Pacific Conf. on SHOCK and IMPACT LOADS ON STRUCTURES*, Perth, Western Australia.

Morançay, L., Homsi, H., and Roelandt, J. (1997) Application of Remeshing Techniques to the Simulation of Metal Cutting by Punching, In E. Onate, D. R. J. Owen, and E. Hinton (eds.), *Computational Plasticity: Theory and Applications*, CIMNE, Barcelona.

Ortiz, M. and Quigley IV, J. (1991) Adaptive Mesh Refinement in Strain Localization Problems, *Comp. Meth. Appl. Mech. Engng.* **90**, 781–804.

Owen, D. R. J., Feng, Y. T., Neto, E. D., Wang, F., Cottrell, M. G., and Yu, J. (2004) The modelling of multi-fracturing solids and particulate media., *Int. J. Numer. Meth. Engng.* **60**, 317–340.

Owen, D. R. J., Feng, Y. T., Yu, J., and Perić, D. (2001) Finite/discrete element analysis of multi-fracture and multi-contact phenomena, *Lecture Notes in Computer Science* **1981**, 484–505.

Owen, D. R. J., Perić, D., Crook, A. J. L., De Souza Neto, E., Yu, J., and Dutko, M. (1995) Advanced Computational Strategies for 3-D Large Scale Metal Forming Simulations, In S.-F. Shen and P. Dawson (eds.), *Fifth International Conference on Numerical Methods in Industrial Forming Processes*, Ithaca, New York, Balkema, Rotterdam.

Perić, D. and Owen, D. R. J. (1998) Finite element applications to the nonlinear mechanics of solids, *Reports on Progress in Physics* **161**, 1495–1574.

Perić, D. and Owen, D. R. J. (2004) Computational modelling of forming processes, In E. Stein, R. de Borst, and T. J. R. Hughes (eds.), *Encyclopeadia of Computational Mechanics*, John Wiley & Sons Ltd.

Perić, D., Yu, J., and Owen, D. R. J. (1994) On Error Estimates and Adaptivity in Elastoplastic Solids: Application to the Numerical Simulation of Localization in Classical and Cosserat Continua, *Int. J. Num. Meth. Engng.* **37**, 1351–1379.

Ravi-Chandar, K. and Knauss, W. (1984a) An Experimental investigation into dynamic fracture – Part I. Crack Initiation and Arrest, *Int. J. Fract.* **25**, 247–262.

Ravi-Chandar, K. and Knauss, W. (1984b) An Experimental investigation into dynamic fracture – Part II.Microstructural Aspects, *Int. J. Fract.* **26**, 65–80.

Riou, P., Denoul, C., and Cottenot, C. (1998) Visualisation of the damage evolution in impacted silicon carbide ceramics, *Int. J. Imp. Engng.* **21**, 225–235.

Ruiz, G., Pandolfi, A., and Oritz, M. (2001) Three-dimensional cohesive modelling of dynamic mixed-mode fracture, *Int. J. Numer. Meth. Engng.* **52**, 97–111.

Schweizerhof, K., Nilsson, L., and Hallquist, J. (1992) Crashworthiness Analysis in the Automotive Industry, *Int. J. Comp. Appl. Techn.* **5**, 134–156.

Shi, G. (1988) Discontinuous deformation analysis: a new method for computing stress, strain and sliding of block systems, Ph.D. thesis, University of California, Berkeley.

Taylor, R. (2000) A Mixed-Enhanced Formulation for Tetrahedral Finite Elements, *Int. J. Num. Meth. Engng.* **47**, 205–227.

Westerling, L., Lundberg, P., and Lundberg, B. (2001) Tungsten long-rod penetration into confined cylinders of boron carbide at and above ordnance velocities, *Int. J. Imp. Engng.* **25**, 703–714.

Yu., J. (1999) A Contact Interaction Framework for Numerical Simulation of Multi-Body Problems and Aspects of Damage and Fracture for Brittle Materials, Ph.D. thesis, University of Wales Swansea, Swansea.

Zienkiewicz, O., Rojek, J., Taylor, R., and Pastor, M. (1998) Triangles and Tetrahedra in Explicit Dynamic Codes for Solids, *Int. J. Num. Meth. Engng.* **43**, 565–583.

Zienkiewicz, O. and Zhu, J. (1987) A Simple Error Estimator and Adaptive Procedure For Practical Engineering Analysis, *Int. J. Num. Meth. Engng.* **24**, 337–357.

Zukas, J. (1990) *High Velocity Impact Dynamics*, John Wiley and Sons.

NONLINEAR TRANSIENT ANALYSIS AND DESIGN OF COMPLEX ENGINEERING STRUCTURES FOR WORST CASE ACCIDENTS : EXPERIENCE FROM INDUSTRIAL AND MILITARY APPLICATIONS *

Adnan Ibrahimbegovic (ai@lmt.ens-cachan.fr)
D. Brancherie, J-B. Colliat, L. Davenne, N. Dominguez, G. Herve, P. Villon
Ecole Normale Supérieure de Cachan, LMT-Cachan,
61, avenue du président Wilson, 94235 Cachan, France

Abstract. In this work we address some of the present threats posed to engineering structures in placing them under extreme loading conditions. The common ground for the problems studied herein from the viewpoint of structural integrity is their transient nature characterized by different time scales and the need to evaluate the consequence for a high level of uncertainty in quantifying the cause. The pertinent issues are studied in detail for three different model problems: i) the worst-case scenario of system functioning failure accident in a nuclear power plant causing the loss of cooling liquid, ii) the terrorist attacks brought explosion and impact of large aeroplane on a massive structure, iii) devastating fire and sustained high temperatures effects on massive cellular structures. By using these case studies, we discuss the issues related to multi-scale modelling of inelastic damage mechanisms for massive structures, as well as the issues pertaining to the time integration schemes in presence of different scales in time variation of different sub-problems brought by a particular nature of loading (both for a very short and a very long loading duration) and finally the issues related to model reduction seeking to provide an efficient and yet sufficiently reliable basis for parametric studies employed within the framework of a design procedure. Several numerical simulations are presented in order to further illustrate the approaches proposed herein. Concluding remarks are stated regarding the current and future research in this domain.

Key words: complex structures, extreme conditions, transient response, impact, explosion, fire.

1. Introduction and Motivation

The present time challenge in fighting the man-made and natural hazards has brought about new issues in dealing with extreme transient conditions for engineering structures. The latter may concern on one side the permanently fixed structures, such as those built for energy source (e.g. nuclear power plants, dams) or the ones providing the storage of sensitive material (e.g. military installations) or the special structures found in transportation systems (e.g. bridges, tunnels),

* Work supported by French Ministry of Research, CEA, CTTB and EDF

A. Ibrahimbegovic and I. Kozar (eds.), Extreme Man-Made and Natural Hazards in Dynamics of Structures, 37–69.

and on the other side the moving structures (such as trains, plains, ships or cars). A number of questions pertain to the the problems of this kind from the viewpoint of structural integrity related to computer models of complex structures, a reliable representation of extreme loading conditions (which nowadays implies the need for solving a coupled problem) and uncertainty. In particular, we focus first and foremost on the first of those goals, which concerns the multi-scale modelling of inelastic behavior of complex structures. The main advantage of the multi-scale approach is in its by far the greatest capabilities to provide the basis for constructing a sufficiently predictive model.

The risk of present time is very likely to increase (rather than decrease) the frequency or severe or extreme conditions for which a very few engineering structures have been built to sustain. The worst case accident in a nuclear power plant associated with a loss of cooling liquid is one such case study. In this kind of situation the demand on refined prediction capabilities of the mechanics model will rise significantly, because we would need to know not only the usual information on 'extent' of cracking as defined by smeared models of plasticity or damage (Lemaitre and Chaboche, 1988), but we must also achieve a much more difficult goal of providing a sufficiently reliable information on crack spacing and opening, which is needed in order to compute the rate of leak. The model of this kind for predicting the detailed information in the fracture process of massive structure have been developed only recently, with applicability to metals (Ibrahimbegovic and Brancherie, 2003) and concrete (Brancherie and Ibrahimbegovic, 2005) or reinforced concrete (Dominguez et al., 2005). The main characteristic of the models of this kind, presented in Section 2, concerns not only the original theoretical formulation, but also the finite element implementation.

In trying to to validate our hypothesis on the governing damage mechanisms of a complex structure, we are prompted to carry out experimental studies. The latter is often sufficient to explain the inelastic behavior of a particular material, but much less capable of providing the interpretation of the inelastic behavior of a complex structure. The case in point is the testing procedure of fire resistance of masonry walls. In sharp contrast to simple material testing procedures, we are rarely (if ever) capable of providing a sufficient number of measurements in testing of the structures. For that reason we use the numerical modelling in order to supplement the experimental results and provide a more reliable interpretation. The main advantage of such an approach is its capability to integrate directly much more reliable experimental results obtained in the material testing, such as those on damage mechanisms and the required amount of fracture energy. The computational strategy for this kind of problems must also consider the multi-scale aspects pertaining to a large difference between the characteristic times between the evolution of mechanics damage (or internal variables) and heat transfer phenomena of convection and radiation. These issues are further discussed in Section 3.

The present terrorist danger has brought yet another extreme loading condition in terms of an impact of a massive structure by a large aeroplane, for which very few engineering structures have been designed to sustain. From the standpoint of nonlinear analysis, the main difficulty in such a problem pertains to a significant high frequency content typical of impact phenomena, as well as a quite likely presence of extensive structural damage. The first goal of the development we carried out pertains to the models for predicting inelastic behavior and damage for both aeroplane and massive structure. Yet another goal is to examine different impact scenarios in trying to quantify any potential reserve which might exist in the given design of an engineering structures for taking on higher level of risk. The complementary goal also of great importance pertains to providing the best way of reducing the negative impact of high-risk situation that cannot be avoided, by resorting to a more sound design procedure. One such procedure is illustrated in Section 4 in application to aeroplane impact problem.

Several representative numerical simulations are presented in Sections 2, 3 and 4 in order to further illustrate the performance of the nonlinear analysis and design procedures presented herein. Concluding remarks are stated in Section 5.

2. Multi-scale Modelling of Inelastic Behavior of Complex Structures

When using the multi-scale modelling of inelastic behavior our main goal is provide a more reliable and overall better explanation of inelastic damage mechanisms. This kind of strategy has already proved successful in modelling inelastic material behavior by using standard three point bending tests (Ibrahimbegovic and Markovic, 2003) and (Markovic and Ibrahimbegovic, 2004), or other experiments with non-homogeneous stress state. One such experiment, carried out by Nooru-Mohamed et TU Delft on a double notched concrete specimen under non-proportional loading, was used as one of the most demanding tests in a recent benchmark contest organized by French National Electrical Power Company (EDF). This benchmark competition has demonstrated that none of the presently available anisotropic models for concrete is capable of providing a very good result. In fact, the best results are obtained subsequently by using the discrete model representing an assembly of cohesive forces (Ibrahimbegovic and Delaplace, 2003) and (Delaplace and Ibrahimbegovic, 2005) (see Figure 1).

Although it is not conceivable to use such a refined discrete model throughout a complex structure, this kind of experience has still taught us something useful for constructing the predictive models for inelastic behavior of structures. First, the theoretical formulation with an ever increasing number of internal variables is not enough, and we also need to examine the issue of choosing the finite element interpolations. The latter can pertain to either incompatible mode enhancement of strain field (Ibrahimbegovic and Brancherie, 2003) or an assumed stress field

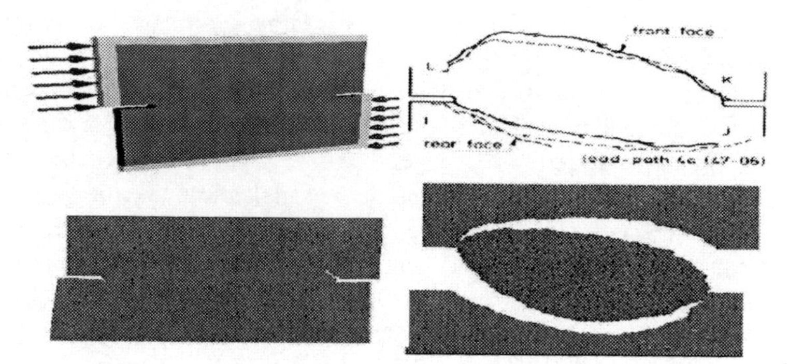

Figure 1. Nooru-Mohamed experiment with non-proportional loading on double-notched concrete specimen and numerical results computed by discrete model.

enhancement (Markovic and Ibrahimbegovic, 2005). Second, it is clearly advantageous to separate complexities and create a model as a building bloc where each mechanism is associated its own criterion of inelastic behavior. One such example of the anisotropic concrete damage model capable of representing both volumetric dissipation and surface dissipation is discussed next.

The proposed model is constructed for dealing with massive structures. Indeed, massive structures often posses at the crack tip a fairly large "process zone" where damage develops, induced by the creation of micro-cracks, until the macro-crack forms due to the coalescence of some micro-cracks. The dissipation produced in the process-zone is far from being negligible with respect to the total dissipation and thus it has to be taken into account by exploiting a continuum damage model. We present a model capable of taking into account both the contribution of diffuse dissipative mechanisms (e.g. micro-cracks) accompanied by strain hardening as well as localized damage modes with strain softening effect. This is achieved by introducing displacement discontinuities coupled with a continuum damage mechanics model. The combination of the two types of dissipative mechanisms is taken into account by building adequate continuum model to describe diffuse dissipation and an adequate localized or discrete model to deal with the dissipation taking place in localized zones. The combined result of two dissipative mechanisms can also be interpreted as an alternative approach to constructing an anisotropic damage model to a number of only partially successfully works using only the classical continuum mechanics framework, where the macro-crack creation is guided by the micro-cracking phase. One can thus obtain not only a more robust implementation, but also a more reliable estimate of the final orientation of the macro-crack which follows the corresponding stress redistribution in the micro-cracking phase. Finally, we present a multi-surface discrete damage model taking into account the contribution of the localized dissipative mechanism, which accounts both for normal interface and tangential interface damage modes. Two mechanisms of continuum and discrete damage are then forced to operate in a coupled manner in each element (Ibrahimbegovic and Brancherie, 2003). We first present the theoretical formulation for both of them, with the bulk damage model and the discontinuity damage model, as well as their modifications induced by the introduction of a displacement discontinuity. The key points of the finite element implementation are then presented with particular developments dedicated to the necessary modifications of the solution strategy (Ibrahimbegovic and Markovic, 2003) or (Markovic and Ibrahimbegovic, 2004) due to the introduction of a displacement discontinuity.

The key point of the theoretical formulation is in reinterpretation of the strain field through introduction of a surface of displacement discontinuity, which will represent all localized dissipative mechanisms due to the apparition and development of localization zones. We further develop the modification introduced by this displacement discontinuity and present more precisely how to build each model associated to each dissipative mechanism. We consider a domain of influence $\tilde{\Omega}$ which is split into two sub-domains by a surface of discontinuity Γ. The surface of discontinuity is characterized at each point by a unit normal vector, denoted as $\mathbf{n}$, and a unit tangential vector, denoted as $\mathbf{m}$. The discontinuous displacement field can then be written as:

$$\mathbf{u}(\mathbf{x}, t) = \bar{\mathbf{u}}(\mathbf{x}, t) + \bar{\bar{\mathbf{u}}}(t)(H_\Gamma(\mathbf{x}) - \varphi(\mathbf{x})) \; ; \; H_\Gamma(\mathbf{x}) := \begin{cases} 1 \; ; \; \mathbf{x} \in \Omega^+ \\ 0 \; ; \; \mathbf{x} \in \Omega^- \end{cases} \tag{1}$$

where H_Γ is the Heaviside function and φ is the correction function which reduces to zero the discontinuity contribution outside of the domain of influence. The strain field can also be decomposed into a regular part and a singular part, the latter accompanying the Dirac-delta function δ_Γ

$$\epsilon(\mathbf{x}, t) = \bar{\epsilon}(\mathbf{x}, t) + \nabla^s \varphi(\mathbf{x}) \bar{\bar{\mathbf{u}}}(t) + (\bar{\bar{\mathbf{u}}}(t) \otimes \mathbf{n})^s \delta_\Gamma(\mathbf{x}) \tag{2}$$

When considering a damage model, for which the strain-stress relation can be written by using the compliance tensor $\mathcal{D}$

$$\epsilon = \mathcal{D}\sigma \tag{3}$$

the compliance $\mathcal{D}$ must be decomposed into a regular and a singular part

$$\mathcal{D} = \bar{\mathcal{D}} + \bar{\bar{\mathcal{D}}} \delta_\Gamma \tag{4}$$

The last two expression ensure that the stress field will remain bounded and sufficiently smooth.

The same kind of additive decomposition is introduced for all internal variables, with any of them having both a regular and a singular part. Each part of the decomposition is associated with a damage dissipative mechanism, with the regular part for the continuum damage model and the singular one for the discrete damage model. We present subsequently the chosen formulation of the two damage models, the continuum one, associated to the bulk material, and the discrete one, associated to the localization zone. The continuum damage model, which takes into account the apparition and development of micro-cracks in the process zone with a quasi-homogeneous distribution, is considered as isotropic. The admissible stress domain is defined in terms of the damage function. The internal variables are the damage compliance denoted as $\bar{\boldsymbol{\mathcal{D}}}$ (with an initial value denoted as $\boldsymbol{\mathcal{D}}^e$ equal to the inverse of the elasticity tensor $\bar{\boldsymbol{\mathcal{D}}}^e = \boldsymbol{\mathcal{C}}^{-1}$) and the variable associated to hardening denoted as $\bar{\xi}$. The evolution equations for these internal variables are obtained by appealing to the principle of maximum damage dissipation. The main ingredients of the continuum damage model are summarized in Table 1.

Table 1. Main ingredients of continuum damage model

Helmholtz free energy	$\bar{\psi}(\boldsymbol{\sigma}, \bar{\boldsymbol{\mathcal{D}}}, \bar{\xi}) = \frac{1}{2}\boldsymbol{\sigma} \cdot \bar{\boldsymbol{\mathcal{D}}}\boldsymbol{\sigma} + \bar{\Xi}(\bar{\xi})$
Damage function	$\bar{\phi}(\boldsymbol{\sigma}, \bar{q}) := \underbrace{\sqrt{\boldsymbol{\sigma} \cdot \bar{\boldsymbol{\mathcal{D}}}^e \boldsymbol{\sigma}}}_{\Vert\sigma\Vert_D} - \frac{1}{\sqrt{E}}(\sigma_f - \bar{q}) \leq 0$
Constitutive equations	$\epsilon = \bar{\boldsymbol{\mathcal{D}}}\boldsymbol{\sigma}$; $\bar{q} = -\frac{d\bar{\Xi}(\bar{\xi})}{d\bar{\xi}}$
Dissipation	$\bar{D} = \frac{1}{2}\boldsymbol{\sigma} \cdot \dot{\bar{\boldsymbol{\mathcal{D}}}}\boldsymbol{\sigma} + \bar{q}\dot{\bar{\xi}}$
Evolution equations	$\dot{\bar{\boldsymbol{\mathcal{D}}}} = \dot{\bar{\gamma}}\frac{1}{\Vert\boldsymbol{\sigma}\Vert_D}\bar{\boldsymbol{\mathcal{D}}}^e$; $\dot{\bar{\xi}} = \dot{\bar{\gamma}}\frac{1}{\sqrt{E}}$

Discrete damage model is proposed to account for the dissipative behavior in the localization zones, corresponding to macro-cracks for damage materials. It is constructed in an equivalent manner to the continuum damage model, except for multi-surface format of the proposed damage criterion (Brancherie and Ibrahimbegovic, 2004), which is needed in order to control separately the normal and tangential components of the traction along the surface of discontinuity. This kind of damage surface coupling provides the proper influence of the normal interface pressure upon the tangential sliding resistance, as well as the reduction of cohesive force due to lateral sliding. The internal variables of the discrete damage model are the compliance of the interface, denoted as $\bar{\bar{\mathbf{Q}}} = \mathbf{n}\bar{\bar{\boldsymbol{\mathcal{D}}}}\mathbf{n}$, and the variable associated to softening, denoted as $\bar{\bar{\xi}}$. In Table 2 below we summarize the main ingredients of this model.

In the finite element implementation of the proposed damage model, the main difficulty pertains to accounting for displacement discontinuities, which can not be done by using the standard isoparametric finite elements. For that reason, a modified strain field including discontinuity is introduced by using the incompatible mode method (Ibrahimbegovic and Wilson, 1991a) or (Ibrahimbegovic

Table 2. Main ingredients of discrete damage model

Helmholtz free energy	$\bar{\bar{\psi}}(\bar{\mathbf{t}}_\Gamma, \bar{\bar{\mathbf{Q}}}, \bar{\bar{\xi}}) = \frac{1}{2}\mathbf{t}_\Gamma \cdot \bar{\bar{\mathbf{Q}}}\mathbf{t}_\Gamma + \bar{\bar{\Xi}}(\bar{\bar{\xi}})$
Damage functions	$\phi_n(\mathbf{t}_\Gamma, \bar{\bar{q}}) := \mathbf{t}_\Gamma \cdot \mathbf{n} - (\sigma_f - \bar{\bar{q}}) \leq 0$
	$\phi_m(\mathbf{t}_\Gamma, \bar{\bar{q}}) := \lvert\mathbf{t}_\Gamma \cdot \mathbf{m}\rvert - (\sigma_s - \frac{\sigma_s}{\sigma_f}\bar{\bar{q}}) \leq 0$
Constitutive equations	$\bar{\bar{\mathbf{u}}} = \bar{\bar{\mathbf{Q}}}\mathbf{t}_\Gamma \;;\; \bar{\bar{q}} = -\frac{d\bar{\bar{\Xi}}(\bar{\bar{\xi}})}{d\bar{\bar{\xi}}}$
Dissipation	$\bar{\bar{D}} = \frac{1}{2}\mathbf{t}_\Gamma \cdot \dot{\bar{\bar{\mathbf{Q}}}}\mathbf{t}_\Gamma + \bar{\bar{q}}\dot{\bar{\bar{\xi}}}$
Evolution equations	$\dot{\bar{\bar{\mathbf{Q}}}} = \frac{\dot{\bar{\bar{\gamma}}}_n}{\mathbf{t}_\Gamma \cdot \mathbf{n}}\mathbf{n} \otimes \mathbf{n} + \frac{\dot{\bar{\bar{\gamma}}}_m}{\lvert\mathbf{t}_\Gamma \cdot \mathbf{m}\rvert}\mathbf{m} \otimes \mathbf{m} \;;\; \dot{\bar{\bar{\xi}}} = \dot{\bar{\bar{\gamma}}}_n + \frac{\sigma_s}{\sigma_f}\dot{\bar{\bar{\gamma}}}_m$

and Kozar, 1995). A particular choice is made for the finite element interpolation according to (see Figure 2):

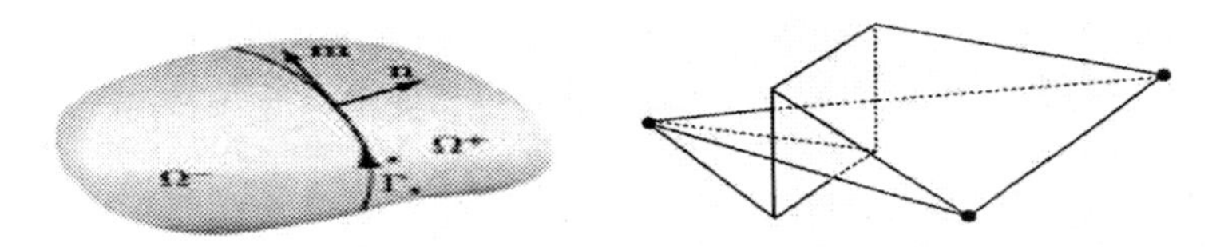

Figure 2. Discontinuity of displacement across Γ and its CST-based nite element approximation.

$$\mathbf{u}^h|_{\Omega^e} = \sum_{a=1}^{3} \mathbf{N}_a^e \mathbf{d}_a + \mathbf{M}^e \boldsymbol{\alpha}^e \tag{5}$$

where $\mathbf{N}_a^e$ is the standard shape function associated with the constant strain triangle and $\mathbf{M}^e$ is a discontinuous function presented in Figure 2. It is important to note that such a modified displacement interpolation has the same nodal value as the standard one, and hence $\mathbf{d}_a$ are still the nodal displacement vectors. With such an approximation, the finite element interpolation of the strain field can be written as

$$\boldsymbol{\epsilon}^h|_{\Omega^e} = \sum_{a=1}^{3} \mathbf{B}_a^e \mathbf{d}_a + \mathbf{G}^e \boldsymbol{\alpha}^e \;;\; \mathbf{B}_a^e - \mathbf{L}\mathbf{N}_a^e \;;\; \mathbf{G}^e = \mathbf{L}\mathbf{M}^e \tag{6}$$

where $\mathbf{L}$ is the matrix representation of the differential operator ∇. The finite element interpolation of the virtual strain field can be constructed in the same manner

$$\nabla^s \mathbf{w}^h|_{\Omega^e} = \sum_{a=1}^{3} \mathbf{B}_a^e \mathbf{w}_a + \hat{\mathbf{G}}^e \boldsymbol{\beta}^e \tag{7}$$

where $\mathbf{w}_a$ and $\boldsymbol{\beta}^e$ are, respectively, the virtual displacement nodal vector and element-wise virtual displacement jump. We indicate in (7) that the incompatible mode strain-displacement matrix $\hat{\mathbf{G}}^e$ is a modified form (Ibrahimbegovic and

Wilson, 1991a) of $\mathbf{G}^e$ in (6), which is needed in order to guarantee the satisfaction of the patch-test; In general, these two matrices are different from one another. The virtual strain field will thus be decomposed into a regular and a singular part, and incompatible mode strain-displacement matrix can be written as:

$$\hat{\mathbf{G}}^e = \tilde{\mathbf{G}}^e + \bar{\bar{\mathbf{G}}}^e \delta_{\Gamma^e} \tag{8}$$

With those interpolations on hand, the discrete problem can be written as:

$$\begin{aligned} \mathbf{0} &= \mathbf{r}(\mathbf{d}, \boldsymbol{\alpha}^e, [\boldsymbol{\mathcal{D}}]_{GNP}) := \mathop{\mathbb{A}}_{e=1}^{n_{el}}[\mathbf{f}^{int,e}] - \mathbf{f}^{ext} \ ; \ \mathbf{f}^{int,e} = \textstyle\int_{\Omega^e} \mathbf{B}^{e,T} \boldsymbol{\sigma} \, dV \\ \mathbf{0} &= \mathbf{h}^e(\mathbf{d}, \boldsymbol{\alpha}^e, [\boldsymbol{\mathcal{D}}]_{GNP}) := \textstyle\int_{\Omega^e} \tilde{\mathbf{G}}^{e,T} \boldsymbol{\sigma} \, dV - \int_{\Gamma^e} \bar{\bar{\mathbf{G}}}^{e,T} \mathbf{t}_{\Gamma^e} \, dV \ ; \ \forall e \in [1, n_{el}] \end{aligned} \tag{9}$$

For the frozen values of internal variables at Gauss numerical integration points $[\boldsymbol{\mathcal{D}}]_{GNP}$, the system of equations to be solved consists of two parts: the first one is the set of global equilibrium equations, which takes a standard form for the finite element method and the second one is a local equilibrium equation written in each localized element. The latter can also be interpreted as the weak form of the traction continuity condition along the surface of discontinuity (Brancherie and Ibrahimbegovic, 2005). The time evolution of the discrete problem is computed by an implicit time integration scheme, accounting for the evolution of internal variables. The consistent linearization of this system leads to the set of linear algebraic equations, which can be written for time step $n+1$ and iteration (i):

$$\begin{aligned} &\mathop{\mathbb{A}}_{e=1}^{n_{el}}(\mathbf{K}_{n+1}^{e,(i)} \Delta \mathbf{d}_{n+1}^{(i)} + \mathbf{F}_{n+1}^{e,(i)} \Delta \boldsymbol{\alpha}^{e,(i)}) = \mathbf{f}_{n+1}^{ext} - \mathop{\mathbb{A}}_{e=1}^{n_{el}}(\mathbf{f}_{n+1}^{int,e,(i)}) \\ &(\hat{\mathbf{F}}_{n+1}^{e,(i),T} + \hat{\mathbf{k}}_{n+1}^{(i)}) \Delta \mathbf{d}_{n+1}^{(i)} + (\mathbf{H}_{n+1}^{(i)} + \mathbf{k}_{n+1}^{(i)}) \Delta \boldsymbol{\alpha}_{n+1}^{e,(i)} = -\mathbf{h}^e \ ; \ \forall e \in [1, n_{el}] \end{aligned} \tag{10}$$

where:

$$\begin{aligned} \mathbf{K}_{n+1}^{e,(i)} &= \textstyle\int_{\Omega^e} \mathbf{B}^{e,T} \mathcal{C}_{n+1}^{ed,(i)} \mathbf{B}^e \, dV \ &; \ \mathbf{H}_{n+1}^{e,(i)} &= \textstyle\int_{\Omega^e} \tilde{\mathbf{G}}_{n+1}^{e,T} \mathcal{C}_{n+1}^{ed,(i)} \mathbf{G}^e \, dV \\ \mathbf{F}_{n+1}^{e,(i)} &= \textstyle\int_{\Omega^e} \mathbf{B}^{e,T} \mathcal{C}_{n+1}^{ed,(i)} \bar{\mathbf{G}}^e \, dV \ &; \ \hat{\mathbf{F}}_{n+1}^{e,(i)} &= \textstyle\int_{\Omega^e} \mathbf{B}^{e,T} \mathcal{C}_{n+1}^{ed,(i)} \tilde{\mathbf{G}}^e \, dV \\ \mathbf{k}_{n+1}^{(i)} &= \textstyle\int_{\Omega^e} \bar{\bar{\mathbf{G}}}^{e,T} \frac{\partial \mathbf{t}_{\Gamma^e}}{\partial \mathbf{d}} \, dA \ &; \ \hat{\mathbf{k}}_{n+1}^{(i)} &= \textstyle\int_{\Omega^e} \bar{\bar{\mathbf{G}}}^{e,T} \frac{\partial \mathbf{t}_{\Gamma^e}}{\partial \boldsymbol{\alpha}} \, dA \end{aligned} \tag{11}$$

By taking into account that the second equation holds independently in each localized element, for a given value of the displacement vector increment we can obtain the corresponding increment of the displacement jump on element-wise basis. We can then carry out the static condensation at the element level, reducing the system of equations to the standard form which is identical to the one obtained by isoparametric finite elements

$$\begin{aligned} &\mathop{\mathbb{A}}_{e=1}^{n_{el}}[\hat{\mathbf{K}}_{n+1}^{e,(i)}] \Delta \mathbf{d}_{n+1}^{(i)} = \mathbf{f}_{n+1}^{ext} - \mathop{\mathbb{A}}_{e=1}^{n_{el}}(\mathbf{f}_{n+1}^{int,e,(i)}) \\ &\hat{\mathbf{K}}_{n+1}^{e,(i)} = \mathbf{K}_{n+1}^{e,(i)} - \mathbf{F}_{n+1}^{e,(i)} [\mathbf{H}_{n+1}^{(i)} + \mathbf{k}_{n+1}^{(i)}]^{-1} (\hat{\mathbf{F}}_{n+1}^{e,(i),T} + \hat{\mathbf{k}}_{n+1}^{(i)}) \end{aligned} \tag{12}$$

We therefore obtain an enhanced performance of the finite element approximation without any change of the global structure of the code. The proposed damage model, which considers interplay of both continuum and discrete dissipation mechanisms, produces mesh-invariant results for any sufficiently refined mesh; see Figure 3.

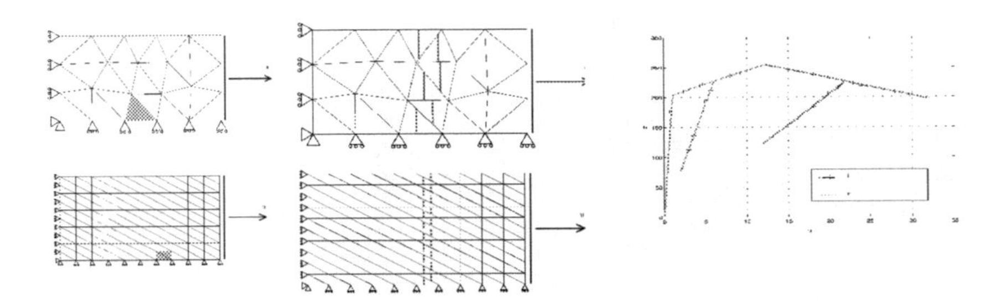

Figure 3. Force-displacement diagram for simple tension test on a specimen with imperfection: mesh-invariant results are obtained for two different approximation of macro-crack with a coarse or a ne mesh, because two dissipative mechanisms always remain coupled and are forced to store the total external energy input.

The procedure presented herein for an anisotropic damage model can be generalized to even a more general case. For example, we can obtain a coupled damage-plasticity model, capable of representing inelastic behavior of porous metals (Ibrahimbegovic et al., 2003) or concrete under compaction (Herve et al., 2005). The latter is used in application to a commercial aeroplane impact problem on the massive structure, which is further discussed in Section 4. Finally, the same kind of development can be carried out for yet the most general framework of reinforced concrete model, which is assembled from the anisotropic damage model for concrete, the elastoplastic model for steel and a coupled damage-plasticity model for bond-slip (Dominguez et al., 2005). The latter is implemented within the zero-thickness bond element with normal and tangential degrees of freedom, which is constructed in the same manner as the contact element described by (Ibrahimbegovic and Wilson, 1991b). One can therefore easily represent an increase in slip resistance due do the confinement pressure. The finite element approximation of such model can also be placed within the previously described framework, with the only difference that one must enforce the continuity of parameters representing bond slip throughout the length of the particular steel bar. The predictive capabilities with respect to crack spacing and opening for this kind of model are illustrated in Figure 4, where we demonstrated the important role played by bond-slip element in redistributing more evenly the stress transfer mechanism along the steel bar, as well as an excellent correlation between the experimental and numerical results, even in the case of dispersion of slip resistance (Dominguez et al., 2005).

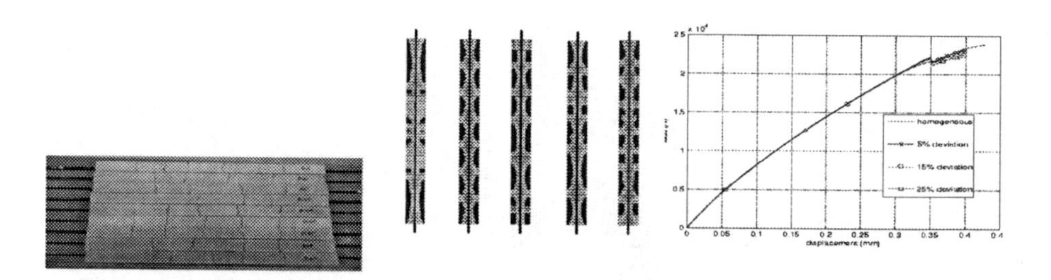

Figure 4. Pull-out test for reinforced concrete beam: experimental results, numerical results (with no bond-slip, with constant bond-slip and with variable bond-slip resistance using the standard variation of 0.05, 0.10 and 0.20 and resulting force-displacement diagram.

3. Multi-scale in Time and Load of Long Duration: Fire Resistance Computation of Massive Cellular Structures

Figure 5. Fire resistance experiment on masonry structure: the experimental set-up and computed results for damage with the full-size model of the complete wall.

Yet another worthy goal in constructing refined modelling approach to inelastic behavior of complex structures pertains to providing supplementary results for the experiments carried out on complex structures. Namely, in sharp contrast to the experiments carried out on a single structural component where very few experimental results are needed (e.g. limit of elasticity) and is therefore easy to obtain an excessive number of measurements, in the experiments on structures carried out under heterogeneous stress field and accounting for heterogeneities of mechanical and geometric properties, one can never have the same advantage of having an excessive number of measurements. It is much more likely not to have enough of the measurements, and to be obliged to appeal to numerical modelling in order to facilitate the identification procedure. One such problem is studied in this section, with the main motivation to develop predictive models capable of describing the inelastic behavior of cellular structures under sustained long term effect of high temperatures. The model of this kind should initially supplement and then eventually replace the standard testing procedure for evaluating the fire resistance of the cellular structures, built of clay or concrete hollow blocks. The model should be able to account for a number of complex phenomena of heat conduction and radiation, as well as the inelastic behavior of material with thermomechanical coupling. While the heat transfer and thermomechanical coupling appear to be firmly under control for solid bodies with inelastic behavior (Ibrahimbegovic and Chorfi, 2002), the present problem gives rise to a number of novel issues both in terms of describing pertinent heat transfer phenomena and of accounting for various aspects of thermomechanical coupling. We review briefly some of the main contributions, which have been presented in detail in our recent work on the subject (Ibrahimbegovic et al., 2005a).

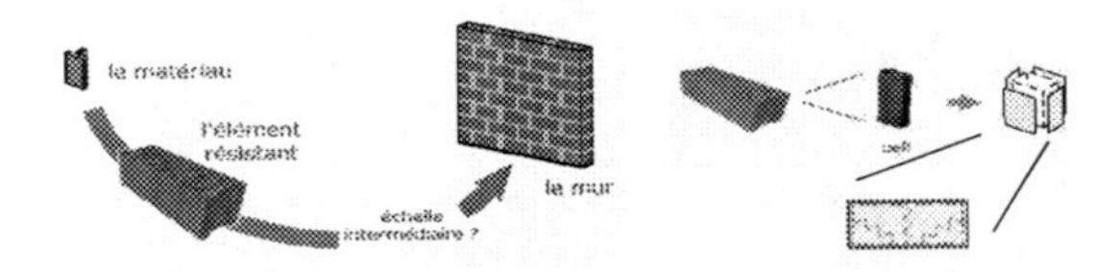

Figure 6. Multi-scale modelling of inelastic damage mechanisms in a masonry wall, with îmicro-scaleî represented with a piece of barrier between the cells and heat radiation element placed inside each cell.

In order to provide a more efficient representation of the inelastic damage mechanisms, the model of a cellular structure built of hollow units is constructed by using the non smooth shell finite element. The latter is a modified version of shell element of Ibrahimbegovic et al. (e.g. see (Ibrahimbegovic et al., 1990b), (Ibrahimbegovic and Frey, 1994) or (Ibrahimbegovic et al., 2005a)). Such an approach allows for a detailed modelling of a single hollow unit of a cellular structure, where the bending action in one shell element is coupled with the membrane action of its neighbor (see Figure 6). The coupling of this kind is provided by the added non-standard feature of shell element pertaining to rotations around the normal or so-called drilling degrees of freedom (Ibrahimbegovic et al., 1990b). The variational formulation of the mechanics part of the coupled problem can be written as

$$\begin{aligned} 0 &= \textstyle\int_{\overline{A}}\{[\varepsilon^*_{(\alpha\beta)}n_{(\alpha\beta)} + \chi^*_{(\alpha\beta)}m_{(\alpha\beta)} + \gamma^*_{(\alpha)}q_\alpha] + \varepsilon^*_{[\alpha\beta]}n_{[\alpha\beta]}\}\,dA - G_{ext} \\ 0 &= \textstyle\int_{\overline{A}}\{\varepsilon_{[\alpha\beta]} - \frac{1}{\gamma}n_{[\alpha\beta]}\}\,dA \end{aligned} \tag{13}$$

where $\varepsilon^*_{(\alpha\beta)}, \chi^*_{(\alpha\beta)}, \gamma^*_{(\alpha)}$ are, respectively, virtual strain measures for membrane, bending and shear deformations, which are defined in accordance with linear variation of virtual displacement field in through-the-thickness direction

$$u^*_\alpha(x_\alpha,\zeta) = u^*_\alpha(x_\alpha) - \zeta\widetilde{\theta}^*_\alpha(x_\alpha)\ ;\ u^*_3(x_\alpha,\zeta) = u^*_3(x_\alpha) \tag{14}$$

In (13) above we defined the stress resultants of membrane and shear forces and bending moments according to

$$n_{(\alpha\beta)} = \int_{-t/2}^{t/2}\sigma_{(\alpha\beta)}\,d\zeta\ ;\ q_\alpha = \int_{-t/2}^{t/2}\sigma_{\alpha 3}\,d\zeta\ ;\ m_{(\alpha\beta)} = \int_{-t/2}^{t/2}\zeta\sigma_{(\alpha\beta)}\,d\zeta \tag{15}$$

along with the skew-symmetric membrane force $n_{[12]}$, which is a crucial feature of this kind of shell elements (Ibrahimbegovic and Frey, 1994). The particular choice of the constitutive equations, as those addressed subsequently, will relate these stress resultants to the real strain measures defined in the same manner as the virtual ones. The regularized form of the variational formulation allows to eliminate the shew-symmetric membrane force and recover an alternative format which employs only kinematic variables, displacements and rotations, which gives a very convenient starting point for the finite element implementation as described next.

In order to provide the numerical model which can contribute in a sufficiently reliable manner to completing the experimental results established in the experiments on structures, we employ a multi-scale modelling approach, where the elementary results of either mode I or mode II fracture, obtained for a single component either numerically or experimentally, can be directly exploited in computations. From the standpoint of the modelling of the whole structure the shell element is placed at the " micro-scale", in the sense that very many of them are used to build the finite element model of a single unit. In trying to construct a predictive model for the whole structure, where at such "macro-scale" we consider a large number of units and therefore even larger number of shell elements, one must represent only fairly simple failure mechanisms at the level of a single shell element. In this work, the latter is selected in terms of Saint-Venant failure criterion (Colliat et al., 2005b), assuming that the inelastic behavior is associated with crossing the threshold of the principal elastic tensile strain, which can easily be calibrated with respect to brittle fracture of materials used for the structures of this kind. In the spirit of shell model the Saint-Venant failure criterion is recast first in terms of stress and then in terms of stress resultants in the format of multi-surface plasticity (see Figure 7). The resulting multi-surface yield criterion can be written in terms of stress resultants as

$$
\begin{aligned}
\phi_1 &:= \frac{K+\frac{4\mu}{3}}{2\mu}\left|\widehat{n}_{\alpha\beta}+\widehat{m}_{\alpha\beta}\right|_I - \frac{K-\frac{2\mu}{3}}{2\mu}\left|\widehat{n}_{\alpha\beta}+\widehat{m}_{\alpha\beta}\right|_{II} - (\sigma_y(\theta)-q(\theta)) \leq 0\\
\phi_2 &:= \frac{K+\frac{4\mu}{3}}{2\mu}\left|\widehat{n}_{\alpha\beta}+\widehat{m}_{\alpha\beta}\right|_{II} - \frac{K-\frac{2\mu}{3}}{2\mu}\left|\widehat{n}_{\alpha\beta}+\widehat{m}_{\alpha\beta}\right|_{I} - (\sigma_y(\theta)-q(\theta)) \leq 0\\
\phi_3 &:= \frac{K+\frac{4\mu}{3}}{2\mu}\left|\widehat{n}_{\alpha\beta}-\widehat{m}_{\alpha\beta}\right|_I - \frac{K-\frac{2\mu}{3}}{2\mu}\left|\widehat{n}_{\alpha\beta}-\widehat{m}_{\alpha\beta}\right|_{II} - (\sigma_y(\theta)-q(\theta)) \leq 0\\
\phi_4 &:= \frac{K+\frac{4\mu}{3}}{2\mu}\left|\widehat{n}_{\alpha\beta}-\widehat{m}_{\alpha\beta}\right|_{II} - \frac{K-\frac{2\mu}{3}}{2\mu}\left|\widehat{n}_{\alpha\beta}-\widehat{m}_{\alpha\beta}\right|_{I} - (\sigma_y(\theta)-q(\theta)) \leq 0
\end{aligned}
\tag{16}
$$

where the stress resultant and couples normalized values defined with

$$
\widehat{n}_{\alpha\beta} = \frac{n_{\alpha\beta}}{t}\ ;\ \widehat{m}_{\alpha\beta} = \frac{6m_{\alpha\beta}}{t^2} \tag{17}
$$

and $|\cdot|_{I/II}$ denoting the principal values of symmetric tensor, and K and μ denoting, respectively, the bulk and the shear moduli. With respect to the subsequent thermomechanical modification of the given model, the temperature dependence is assumed both for the limit of elasticity $\sigma_y(\theta)$ and for the variable which controls the evolution of the elastic domain $q(\theta)$. The latter is typically related to stress softening branch which eventually drives the stress to zero (Ibrahimbegovic and Brancherie, 2003). This kind of plasticity criterion is capable of representing the inelastic phenomena of cracking both in tension, with cracks orthogonal to load direction, and in compression, with cracks parallel to load direction (see Figure 7). A special provision is taken (Colliat et al., 2005b) to ensure a smooth transition between fracture energies in compression and in tension. We note in passing that the predictive model of this kind can also be constructed by directly incorporating experimental results obtained on the specimen corresponding to a particular shell element (e.g. traction, compression or three-point bending test performed on a plate-like piece of material).

The heat radiation element is developed (Colliat et al., 2005a) and placed inside each cell surrounded by four shell elements representing the barriers (see Figure 6). Sustained high temperatures impose that thermomechanical coupling must also be considered for this problem, and not only through usual dependence of the mechanical properties with respect to temperature. An additional concern in this work is the modification which must be made to heat conduction and radiation problems related to spreading of failure zone, where two (or more) adjacent cells connect upon the complete damage of the shell barriers in-between. We further give a brief description of our strategy.

In order to develop shell-like formulation, we split representation of the shell 3D domain into mid-surface $\overline{A}$ and thickness direction, and further assume (in

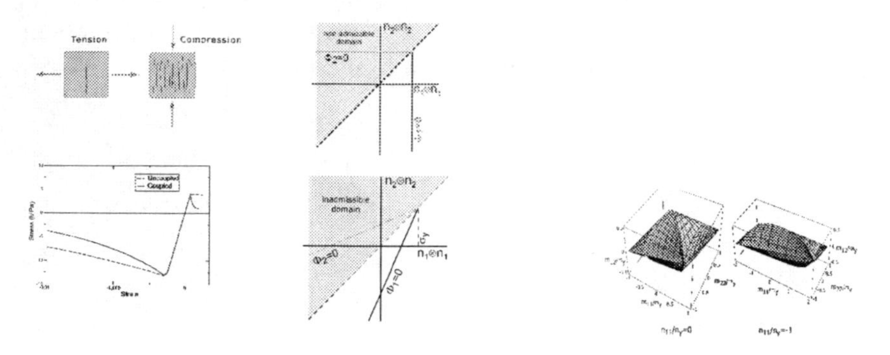

Figure 7. "Micro-scale" representation of inelastic damage mechanism by a multi-surface St. enant plasticity model in principal elastic strains, with different fracture energy values for cracking in traction vs. cracking in compression. Model is further recast in terms of principal stress values and nally in terms of stress resultants.

analogy with mechanics part) a linear variation of the weighting temperature field,

$$\theta^* \left(x_\alpha, \zeta\right) = \vartheta^* \left(x_\alpha\right) + \zeta\varphi^* \left(x_\alpha\right) \tag{18}$$

where ϑ^* is the mid-surface temperature and φ^* is through-the-thickness gradient. The weak form of the energy balance equation in a shell-like domain can then be written as,

$$\begin{aligned} 0 = G_\theta(\vartheta, \varphi) := \textstyle\int_{\overline{A}} \left(ct\, \vartheta^* \dot{\vartheta} + c\frac{t^3}{12}\, \varphi^* \dot{\varphi}\right) dA - \int_{\overline{A}} \left(\vartheta^*_{,\alpha} p_\alpha + \varphi^*_{,\alpha} r_\alpha + \varphi^* p_3\right) dA \\ + \textstyle\int_{\overline{A}} \Big((\vartheta^* + \frac{t}{2}\varphi^*) q_n^+ + (\vartheta^* - \frac{t}{2}\varphi^*) q_n^- \Big)\, dA \end{aligned} \tag{19}$$

In (19) above we denote the resultant heat fluxes by

$$p_\alpha(x_\alpha) = \int_{-\frac{t}{2}}^{+\frac{t}{2}} q_\alpha \, d\zeta \; ; \; r_\alpha(x_\alpha) = \int_{-\frac{t}{2}}^{+\frac{t}{2}} q_\alpha \, \zeta \, d\zeta \tag{20}$$

and by q_n^- and q_n^+ the heat fluxes on the upper and lower surface, respectively. The latter implies that even neglecting the heat sources of structural heating and dissipation, which is quite justified for such a problem of brittle fracture, the thermomechanical coupling still remains present through boundary conditions. The classical form of the radiative heat exchange is employed herein, enforcing that the resultant flux over each surface within each cell must be equal to zero.This can be written as

$$q_{n,i}^{\pm} \longleftarrow q_{n,i}^{\pm} - c_{SB} \sum_{j=1}^{n_{surf}} \epsilon_j A_j F_{ij} (\theta_j^{\pm})^4 \tag{21}$$

where the first term $q^{\pm}_{n,i}$ represents the outgoing flux, whereas the last term represents the contribution of incoming fluxes, which is dependent upon the Stefan-Boltzmann constant c_{SB}, the surface emitting ϵ_j, the area of the surface A_j, the relative shape factor of each pair of surfaces F_{ij} and the surface temperature to the power four. We note that each such equation is nonlinear in temperature terms and thus it requires several iterations at global level to converge. In addition, the modification of such an equation in (21) has to be done each time the fracture of the barriers will change the cell configuration. We refer to (Colliat et al., 2005a) for a more detailed discussion of this issue.

The finite element approximation of the coupled problem is carried out by using the isoparametric interpolations for the temperature field along with the incompatible mode based strain fields for the chosen 4 node shell element (Ibrahimbegovic and Frey, 1994); We note in passing that such a choice of finite element interpolations produces fully compatible representations for both mechanical and thermal strains, which helps to eliminate any potential locking phenomena. Hence, the finite element method will carry out the semi-discretisation procedure which allows us to present the problem of thermomechanical coupling of shells in terms of a set of nonlinear differential-algebraic equations, with the nonlinear algebraic equations expressing mechanics equilibrium equations and the differential equations describing the heat transfer. This can formally be written as

$$\begin{bmatrix} \mathbf{r}^u(\mathbf{d}^u, \mathbf{d}^\theta, \left[\varepsilon^p_{\alpha\beta}(\theta), \chi^p_{\alpha\beta}(\theta), \xi(\theta)\right]_{GNP}) \\ \mathbf{M}^{\theta\theta}\dot{\mathbf{d}}^\theta - \mathbf{r}^\theta(\mathbf{d}^\theta, \left[\varepsilon^p_{\alpha\beta}(\theta), \chi^p_{\alpha\beta}(\theta), \xi(\theta)\right]_{GNP}) \end{bmatrix} = \mathbf{0} \tag{22}$$

where $\mathbf{d}^u$ are nodal values of displacements and rotations and $\mathbf{d}^\theta$ are nodal valus of temperatures. We note that residual vectors $\mathbf{r}^u$ and $\mathbf{r}^{|theta}$ also depend on the values of internal variables (plastic strains and softening variable), which ought to be computed from their evolutions equations; The latter is a set of local equations, which pertains to each Gauss numerical integration point (GNP) of each element

$$\left[\dot{\varepsilon}^p_{\alpha\beta} = \sum_i \dot{\gamma}^i \frac{\partial \phi_i}{\partial n_{\alpha\beta}}\ ;\ \dot{\chi}^p_{\alpha\beta} = \sum_i \dot{\gamma}^i \frac{\partial \phi_i}{\partial m_{\alpha\beta}}\ ;\ \dot{\xi} = \sum_i \dot{\gamma}^i \frac{\partial \phi_i}{\partial q}\right]_{GNP} ;\ \begin{matrix} \forall GNP \in \Omega^e ; \\ \forall e \in [1, n_{el}] \end{matrix} \tag{23}$$

Since all the equations on mechanics and heat transfer side remain tightly coupled, the real time parameter is used throughout. We note in passing that such a time-parametrization will not affect the rate-independent constitutive response which is selected herein. The latter implies that the typical relaxation time for internal variables, which could be introduced into the selected model through a viscous regularization or viscoplasticity model (Ibrahimbegovic, 2006), will always remain much smaller than to characteristic time evolution of temperature and kinematic fields.

By integrating the evolution equations for internal variables along with the heat transfer equation with an implicit, one-step scheme (such as backward Euler scheme), the coupled problem can finally be recast as a set of nonlinear algebraic equations.

The latter is written as follows:

Given: $\mathbf{d}_n^u$, $\mathbf{d}_n^\theta$ and $\left[\varepsilon^p_{\alpha beta,n}, \chi^p_{\alpha beta,n}, \xi_n)\right]_{GNP}$; $\forall GNP$

Find: $\mathbf{d}_{n+1}^u$, $\mathbf{d}_{n+1}^\theta$ and $\left[\varepsilon^p_{\alpha beta,n+1}, \chi^p_{\alpha beta,n+1}, \xi_{n+1})\right]_{GNP}$

Such that:

$$\begin{array}{l} \mathbf{r}^u(\mathbf{d}^u_{n+1}, \mathbf{d}^\theta_{n+1}, \left[\varepsilon^p_{\alpha\beta,n+1}, \chi^p_{\alpha\beta,n+1}, \xi_{n+1}\right]_{GNP}) = \mathbf{0} \\ \mathbf{M}^{\theta\theta}\frac{1}{\Delta t}(\mathbf{d}^\theta_{n+1} - \mathbf{d}^\theta_n) + \mathbf{r}^\theta_{n+1}(\mathbf{d}^\theta_{n+1}, \left[\varepsilon^p_{\alpha\beta,n+1}, \chi^p_{\alpha\beta,n+1}, \xi_{n+1}\right]_{GNP}) = \mathbf{0} \end{array} \tag{24}$$

$$\begin{array}{l} \varepsilon^p_{\alpha\beta,n+1} - \varepsilon^p_{\alpha\beta,n} - \sum_i \lambda^i \frac{\partial \phi_i}{\partial n_{\alpha\beta,n+1}} = 0 \\ \chi^p_{\alpha\beta,n+1} - \chi^p_{\alpha\beta,n} - \sum_i \lambda^i \frac{\partial \phi_i}{\partial m_{\alpha\beta,n+1}} = 0 \\ \xi_{n+1} - \xi_n - \sum_i \lambda^i \frac{\partial \phi_i}{\partial q} = 0 \end{array} ; \begin{array}{l} \phi_i \leq 0 \, \lambda_i \geq 0 \, ; \forall i \\ \forall GNP ; \, \forall e \end{array} \tag{25}$$

Having neglected the plastic heating effect for the present case of brittle fracture, the themomechanical coupling in the discrete heat equation enters only through changing the boundary conditions for heat radiation contribution. Hence, having made the choice of the latter at any particular time step, the linearized form of the system of algebraic equations to be solved can be written as

$$\begin{bmatrix} \mathbf{K}^{uu} & \mathbf{K}^{u\theta} \\ \mathbf{0} & \frac{1}{\Delta t}\mathbf{M}^{\theta\theta} + \mathbf{K}^{\theta\theta} \end{bmatrix}^{(i)}_{n+1} \cdot \begin{bmatrix} \Delta \mathbf{d}^u \\ \Delta \mathbf{d}^\theta \end{bmatrix}^{(k)}_{n+1} = - \begin{bmatrix} \mathbf{r}^{u,(k)} \\ \mathbf{M}^{\theta\theta}\frac{1}{\Delta t}(\mathbf{d}^{\theta,(k)}_{n+1} - \mathbf{d}^\theta_n) + \mathbf{r}^{\theta,(k)}_{n+1} \end{bmatrix} \tag{26}$$

These equations are solved for the converged values of internal variables, which ensures the plastic admissibility of the computed stress state with $\phi_i \leq 0$, leading to an improved iterative values of displacements and temperatures

$$\begin{array}{l} \mathbf{d}^{u,(k+1)}_{n+1} \longleftarrow \mathbf{d}^{u,(k)}_{n+1} + \Delta \mathbf{d}^{u,(k)}_{n+1} \\ \mathbf{d}^{\theta,(k+1)}_{n+1} \longleftarrow \mathbf{d}^{\theta,(k)}_{n+1} + \Delta \mathbf{d}^{\theta,(k)}_{n+1} \end{array} \tag{27}$$

An operator split solution procedure for solving this coupled thermomechanical problem is developed (Ibrahimbegovic et al., 2005a) in order to reduced the computational cost with respect full-size non-symmetric system. The latter implies that the computations are performed in a sequential manner between heat transfer and mechanics, which also includes the solution to evolution equations for internal variables. More precisely, the operator split consists of the thermal phase at fixed configuration, followed by the mechanical equilibrium phase at fixed temperature. In a general case such an isothermal split leads to only conditionally stable solution scheme, and it should be replaced by an isentropic split where entropy rather than the temperature is kept constant in the mechanical step. However, for the present case of brittle materials, where we neglect the dissipation based and structural heating, two kinds of split problems remain the same.

In this work we take this development a step further by providing the operator split solution procedure where the time-integration schemes for mechanics and heat transfer do not necessarily need to use the same time step. In that way one can choose the optimal value of the time step for each sub-problem in order to adapt the obtained result accuracy to the time scale of the evolution process for either thermal or mechanics part. The choice which is made for the case of practical interest for this work of long term fire resistance, is to combine a large time step for thermal sub-problem with a number of smaller time steps for mechanical sub-problem. Moreover, in order to increase the robustness of the algorithm, we have tested the adaptive time-step procedure for the mechanical phase. The step size control is chosen on the basis of the evolution of the hardening/softening variable ξ, which is quite representative of the extent of the development of plastic process. In accordance with the suggestion of (Ibrahimbegovic et al., 2005a), we take the value of indicator R as the guide on the time-step refinement, which is defined as

$$R = \max_{GNP} \frac{(\Delta\xi_{n+1})_{GNP}}{\Delta\bar{\xi}_{n+1}} \tag{28}$$

where $\Delta\bar{\xi}_{n+1}$ is a target value. In each time step, we try to keep R close to 1, by using the following strategy:

i) IF $R > 1,25$ the solution is rejected and the new time-step size is set to: $\Delta t^{\star} = \frac{0{,}85}{R}\Delta t$

ii) ELSE IF $R \leq 1,25$ the solution is accepted and the next time increment evolves as:

ii-a) if $R \leq 0,5$ then $\Delta t^{\star} = 1,5\Delta t$
ii-b) if $0,5 < R \leq 0,8$ then $\Delta t^{\star} = 1,25\Delta t$
ii-c) if $0,8 < R \leq 1,25$ then $\Delta t^{\star} = \Delta t/R$.

Thermomechanical analysis of hollow brick wall: In this example we consider a thermomechanical coupling in the cellular units placed within a brick wall under sustained fire conditions. The mechanical and thermal properties of the brick material and interface mortar layers are chosen as given in Table 3.

In the first analysis we assume that the geometry and the loading allow for exploiting the periodicity conditions. This implies that the analysis can be carried out on a single cellular unit (with the size $570 \times 200 \times 200\,mm^3$) isolated from the whole structure at the level of interface (of the thickness equal to $10\,mm$) with neighboring units, by applying the corresponding boundary conditions which assure periodicity. More precisely, for the typical unit assembly in a brick wall (see Figure 8) with only partial overlapping of successive layers, the same periodicity conditions are enforced only over half of the brick. The chosen finite element

Table 3. Mechanical and thermal properties of the brick and mortar interface

Property	Brick	Mortar joint
density	1.870	2.100
heat capacity	836 $J.kg^{-1}.K^{-1}$	$950 J.kg^{-1}.K^{-1}$
conductivity (parallel to flakes)	0.55 $W.m^{-1}.K^{-1}$	1.15 $W.m^{-1}.K^{-1}$
conductivity (perpendicular to flakes)	0.35 $W.m^{-1}.K^{-1}$	-
thermal expansion coefficient at θ_{ref}	7.10^{-6} K^{-1}	1.10^{-5} K^{-1}
Young modulus	12 GPa	15 GPa
Poisson ratio	0.2	0.25
σ_y at θ_{ref}	14.5 MPa	-
fracture energy	80 $J.m^{-2}$	-

mesh (see Figure 8) consists of three vertical layers of flat shell elements which brings the total number of these element to 384 for the entire brick. The model of the interface joints is constructed with elastic solid elements covering the cells of the brick placed only at the top, as imposed by periodicity boundary conditions.

Both mechanical and thermal loading is applied in this case. The mechanical loading is supposed to represent the dead load on the brick chosen as a compressive loading of 1.3 MPa, which is introduced directly at the level of each element as the initial compressive loading in the bricks, remaining constant afterwards. The thermal loading is then applied, in terms of the uniform temperature field applied only at the brick facet exposed to fire. The time evolution of this temperature field is given as

$$\theta(t) = \theta_0 + 345 \log(8t + 1) \tag{29}$$

where θ_0 is the initial temperature and t the time in minutes.

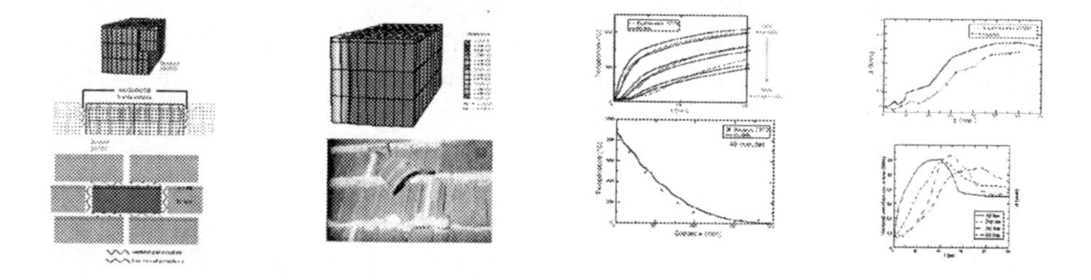

Figure 8. Unit assembly in a brick wall and analysis with periodic BC; experimental and numerical results for damage and temperature profile.

First, the results are presented in terms of temperature field. Figure 8 shows the evolution of the temperature in three different cells. The experimental results are provided for thermocouples inside the cells. Therefore, we compare these values with the temperatures of the two surfaces on both sides of each cell obtained by the finite element analysis. The comparison shows that we are able to capture quite well the temperature evolutions even far from the exposed face of the wall. This result is reiterated by a temperature profile at 48 minutes after the beginning of fire, also shown in Figure 8. The key point in order to obtain such good result is the introduction of radiative exchanges in the heat transfer model. Finally, we show in Figure 8) the comparison between experimental and numerical results pertaining to the horizontal displacement of the wall built with ten rows of bricks, clearly indicating a significant bending deformation. This bending is due to the temperature gradient through the wall. We conclude that the wall stiffness is quite well predicted by our numerical model, despite a slight overestimate of the displacement near the boundary, which is quite likely the consequence of the uncertainty in boundary conditions. With respect to computed mechanics fields, in Figure 8) we also show the evolution of the sum of vertical reactions at selected nodes. Each curve corresponds to a line of nodes parallel to the exposed face of the wall and positive values are used for compression (with prestressed initial value due to mechanical constant loading).

We have improved this result even further by means of a global analysis on the entire wall structure (e.g. without periodic boundary conditions), by using close to 100000 degrees of freedom; The obtained results remain confidential and are not presented herein.

4. Multi-scale in Time and Load of Short Duration: Impact of Aeroplane on Massive Structures

In this section we present our recent work on trying to incorporate the significant advances in nonlinear analysis within the design procedure of a complex engineering structure under equally complex loading. The case in point is the impact by aeroplane on a massive structures, such as a nuclear power plant (see Figure 9) or military installations.

The traditional design procedures for this kind of problem are not applicable to complex structure and even less so to complex, non-proportional loading case (such as the one produced by frictional contact of the aeroplane with the structure). Namely, a number of empirical expressions, which have been proposed and used currently, are based on rather simplified interpretations of experimental results of projectile tests perforating different plate-like structures. There exists a very large dispersion in these experimental results on a single structural component and more importantly their lack of pertinence to complex structures, which makes them

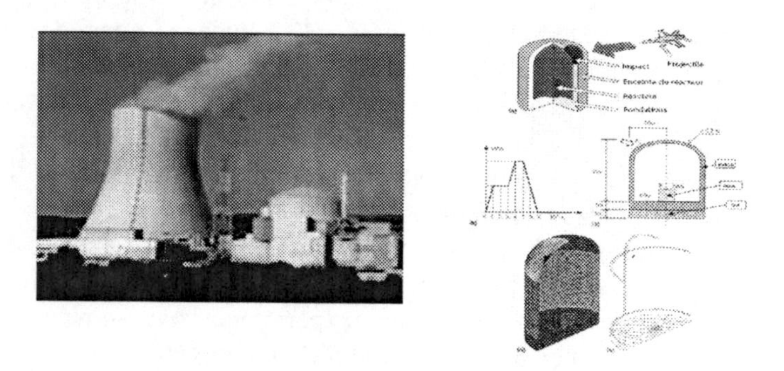

Figure 9. Nuclear power plant: massive structure built to sustain aeroplane impact.

practically inapplicable to the case of complex structures under complex loading. Therefore, a novel approach to nonlinear analysis of this problem is sought, which is capable of taking into account the final goal of engineering design for such a structure.

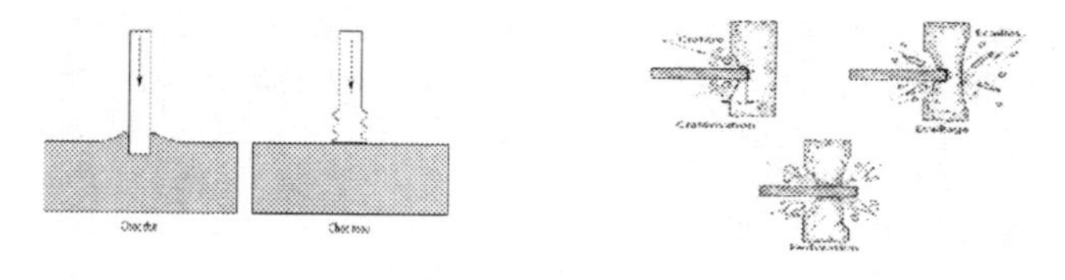

Figure 10. Hard vs. soft impact and damage modes of interest for this study.

The main difficulty in dealing with this class of impact problems is related to the type of impact one should take into account; In that respect, we distinguish between the hard impact, where the projectile is much more stiff than the target, and the soft impact, where the target is at least as stiff as the projectile (see Figure 10). Another difficulty is related to a great diversity of damage modes, including perforation or spalling, which have to be accounted for (see Figure 10). In order to deal with this kind of complexity, and at the same time keep a reasonable efficiency, we split the analysis in the local and global phase. The local phase will deal with the analysis of a single structural component (e.g. an impacted area of a nuclear power plant or a concrete slab in a military installation, see Figure 11, whereas the global analysis will deal with the complete (complex) structure. Moreover, the later stage of global analysis will incorporate directly the result of the local analysis in the manner which is both sufficiently representative of the local analysis results and which provides a much higher efficiency of the computations in the global phase, so that one can carry our all the parametric studied which are needed in the design phase (see Fig. 11).

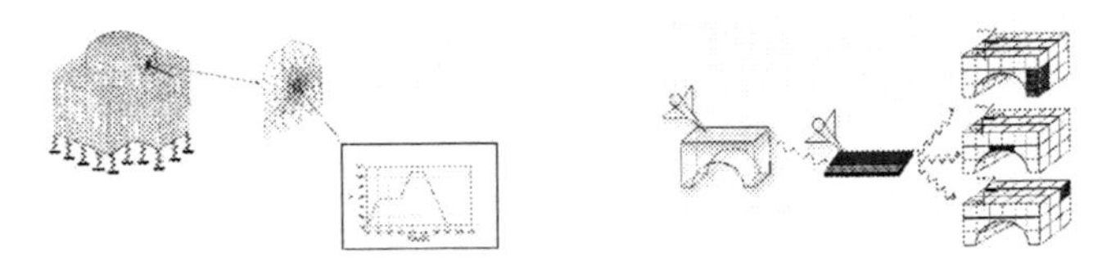

Figure 11. Design procedure based on mesh transfer between local phase and global phase.

The local phase of the analysis is carried out first, by using a very refined model of a sigle component corresponding to the impacted zone. The analysis of this kind considers the finite element model of the projectile (of the finite element model of an aeroplane), which can account for the large plastic deformations developing at impact and frictional contact with potentially large frictional sliding. For that reason, an explicit, central difference scheme is used for such an analysis according to

$$\begin{aligned} \mathbf{d}_{n+1} &= \mathbf{d}_n + \Delta t\, \mathbf{v}_n + \Delta t^2\, \mathbf{a}_n \\ \mathbf{M}\mathbf{a}_{n+1} &= \mathbf{r}_{contact,n+1} - \hat{\mathbf{f}}^{int}(\mathbf{d}_{n+1}, [\boldsymbol{\epsilon}^{vp}, \boldsymbol{\mathcal{D}}, \ldots]_{GNP}) \\ \mathbf{v}_{n+1} &= \mathbf{v}_n + \tfrac{\Delta t}{2}(\mathbf{a}_n + \mathbf{a}_{n+1}) \end{aligned} \tag{30}$$

where $\mathbf{d}$, $\mathbf{v}$ and $\mathbf{a}$ are, respectively, displacements, velocities and accelerations and $\mathbf{M}$ is the mass matrix, which is taken of a diagonal form in order to enhance the computation efficiency. We indicated in the last expression that the main source of nonlinearity, besides impact and frictional contact by the aeroplane with contact forces denoted as $\mathbf{r}_{contact}$, are the plastic and damage deformations of the impacted component of the structure, which are all contained in term $\mathbf{f}^{int}$.

For the purpose of structural design, we ought to provide a very reliable representation of the damage in the impacted structural component, where the latter is built of reinforced concrete. Several damage mechanism have to be represented: the first is concrete damage in tension leading to cracking, the second one is the damage in compression with the original feature that the concrete initially hardens due to compaction with an increase of stress until the ultimate value where the concrete will break, and the last one is the dependence of compressive and tensile strength of concrete with respect to the rate of deformation (see Figure 12). The fracture of concrete in tension is described by a criterion depending directly upon the principal values of tensile elastic strains in a quite similar manner as for the model presented in the previous section, including the modification for different fracture energies for cracking in tension and cracking in compression. The criterion for the concrete damage in compression is constructed as a modified form

(Herve et al., 2005) of the Gurson plasticity criterion, defined with

$$\phi(\sigma_{ij}, \sigma_m, f^*) = \frac{3J_2}{\sigma_m^2} + 2q_1 f^* \cosh(q_2 \frac{I_1}{2\sigma_m}) - (1 + (q_3 f^*)^2) \leq 0 \qquad (31)$$

where σ_{ij} are the components of the nominal stress tensor, σ_m is the reference value of the stress in the matrix, f^* is the concrete porosity and q_1, q_2, q_3 are the chosen coefficients. The graphic illustration of this criterion for both compressive and tensile stress is provided in Figure 12.

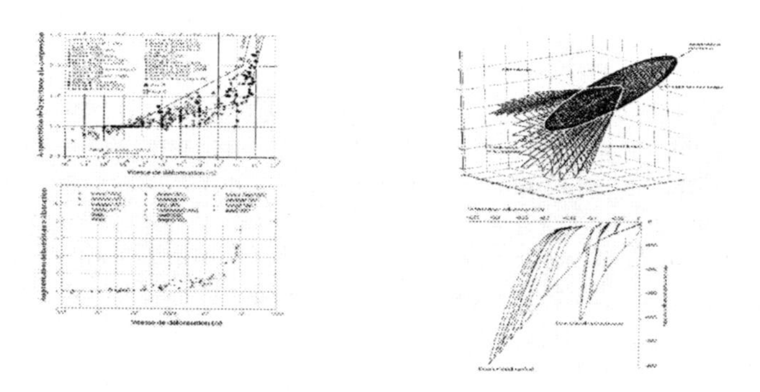

Figure 12. Constitutive model of concrete for dynamic analysis: dependence of compressive and tensile strength on rate of deformation, yield-damage criterion and hardening introduced by compaction.

Porosity is made dependent on equivalent plastic deformation, and the same parameter defines the threshold for element erosion, defining the stage where the element is damaged to that extent that it ought to be completely removed from the mesh. The evolution equations for plastic deformation in compression and damage deformation in tension are chosen as rate dependent, in order to account for the effects of rate of deformation, which is quite pronounced for this class of problems (see Figure 12).

The numerical implementation of the proposed model is incorporated within the proposed computational framework in (30). However, contrary to explicit scheme computations for the global momentum balance equations, the integration of the evolution equations for internal variables, such as plastic strains, damage compliance and hardening/softening variables, is carried out by an implicit scheme. The latter provides at each step the admissible values of stress with respect to the chosen criteria, and leads to a very robust numerical implementation. For more details on model numerical implementation we refer to (Herve et al., 2005).

Sandia's Laboratory tests simulations: In order to check the reliability of the proposed constitutive model we carried out simulations of impact tests performed

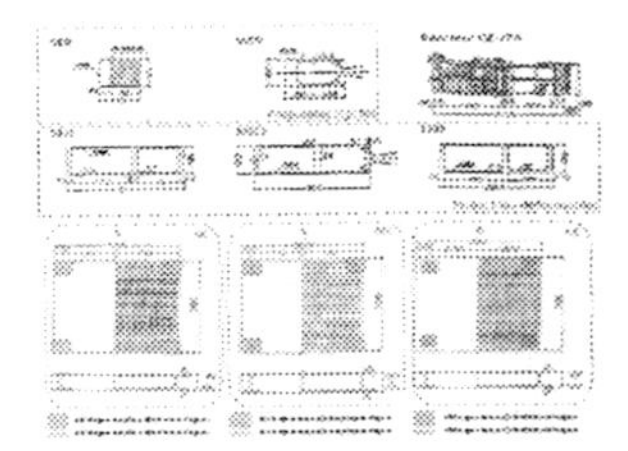
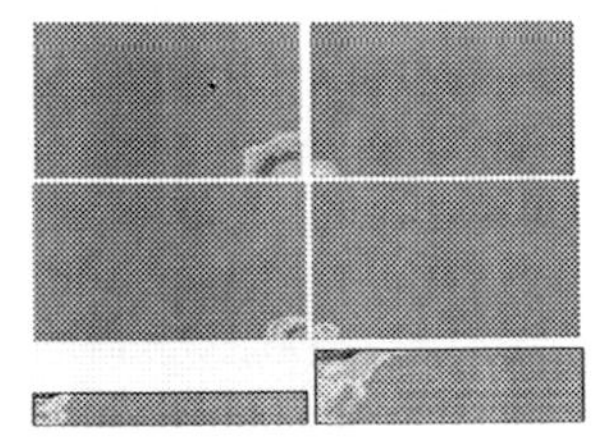

Figure 13. Experimental results of slab perforation obtained at Sandia US National Lab. and results of our numerical simulations.

in Sandia Laboratory by Sugano and co-workers. Following the test program, we simulate impact on the reinforced concrete slabs by different size missiles equivalent to aircraft engine. In particular, we performed simulations for Large size Equivalent and Deformable missile (LED), Medium size Equivalent and Deformable missile (MED) and Small size Equivalent Rigid missile (SER); See Figure 13 for details. The missiles are used for impacting different concrete slabs (see also Figure 13), with a particular choice made for two of them as summarized in Table 4. In numerical simulations the missiles where modelled by using the shell finite element models, except for SER missile with the mesh of 3D solid elements. The concrete slab are modelled with under-integrated 3D solid elements with the proposed constitutive model, and the reinforcement is modelled with truss-bar elastoplastic elements.

Table 4. Simulated impact tests - missile and slab characteristics and experimental results.

No.	Missile type	Velocity (m/s)	Slab thick. (m)	Reinf. ratio	Slab type	Perfor.	Spal.	Penetr.
S10	SER	141	0.15	0.4	Small # 1	No	Yes	Yes
L5	LED	214	1.60	0.4	Large # 3	No	Some	Yes

The qualitative results observed in experiments are presented in Table 4; we note that penetration means that the missile penetrated the slab without having gone through it, perforation means that missile went through, scabbing means that the impact generated a scab on the rear face of the slab.

Very similar results are obtained in our numerical simulations and presented in Figure 13 for a slabs S10 and L5, respectively. The computations where stopped when the velocity of the missile stopped decreasing, which implies that the missile was finally stopped by the concrete. We observed a fair amount of damage of a thin slab in this numerical simulations, not enough to allow for the missile penetration but only the spalling which occurs at the rear face. For the thick slab, the damage remains localized in the impact area with a large undamaged volume separating the damage on the front face and damage on the rear face. The latter implies the presence of only some small cracks. On the basis of results obtained in our numerical simulations, we can conclude that the proposed model is capable of providing a very good correlation with the tests.

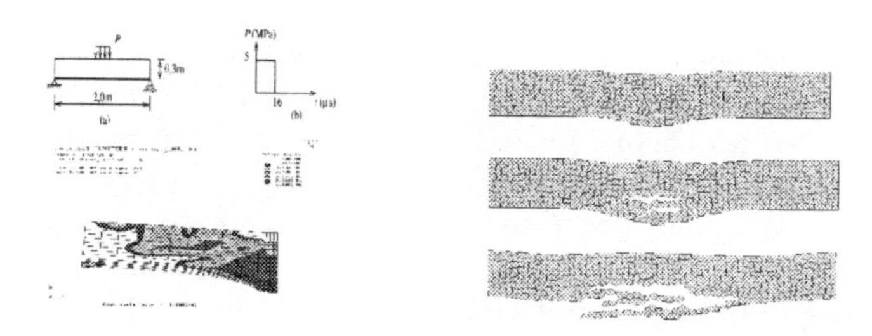

Figure 14. Numerical simulation of 3-point bending testing: continuum model with element erosion can remove completely damaged elements, and discrete model can also represent spalling as well as the motion of any detached piece of specimen.

We note in passing that yet another kind of result can be of interest for local analysis of structural component regarding the spalling phenomena, where a completely damaged piece can detach from the main structure and fly away. If the mass of that piece is too large or if its velocity is too high, it is very likely to have the damage of the interior equipment or storage material, even if the integrity of the structural component has been preserved under impact. The spalling phenomena of this kind can be easily illustrated on a 3-point bending test in dynamics, which is performed by dropping a weight from a certain height, producing at impact a compressive wave travelling downwards from the upper surface. When this wave reaches the lower surface of the specimen, it will reflect, double in size and turn into a tensile wave. That is a moment when the damage can be introduced in tension sensitive material, with pieces which can detach and freely move away from the main structure.

In the numerical simulation of this kind of test we can reproduce some or all of these results, depending upon the kind of model which we are using; See Figure 14. Our damage-plasticity continuum model is capable of predicting the extent of the damage zone associated with a large value of damage variable, or even eliminate all the elements which are extensively damaged by activating the erosion criterion. However, with such a continuum model we can eliminate but can no longer follow the motion of the pieces which are now detached from the structural component. For that reason, we have developed (Ibrahimbegovic and Delaplace, 2003) a discrete model based on Vornoi cell representation of the specimen, where the cohesive forces between the adjacent cells are represented by geometrically exact Reissner beam model (Ibrahimbegovic and Taylor, 2002). This particular feature of the beam model with its capability for representing overall large motion (accompanied by small strains) is crucial for the present application, where the detached piece is represented by several Vornoi cells with preserved cohesive forces; See Figure 14. A special attention is payed (Delaplace and Ibrahimbegovic, 2005) to implementing the appropriate time-integration schemes capable of controlling the high frequency content of motion (Ibrahimbegovic and Mamouri, 2002) and thus minimizing the risk of spurious stress oscillation introduced by brittle fracture.

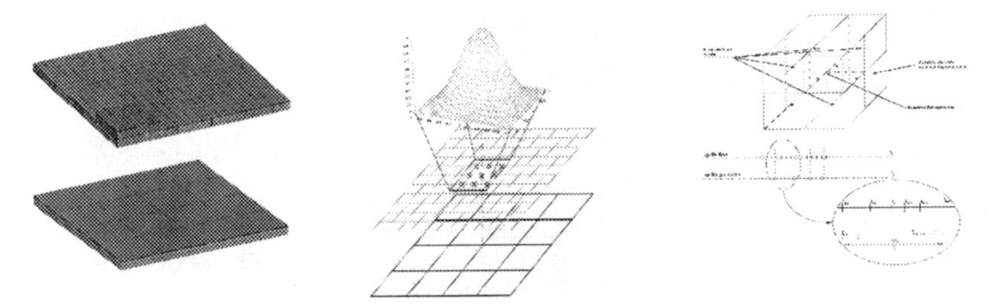

Figure 15. Coarse and ne mesh representation of plate-like structural component and space-time eld transfer between two meshes based on diffuse approximation.

Having completed the local analysis phase of aeroplane impact on a single structural components (or impacted zone), we further turn to the global phase. The main goal of this phase is to check the integrity of the whole structural assembly, and perform in a very efficient manner any eventual parametric study where different design possibilities are proposed, with each one integrating the structural component which is considered in the local phase. Therefore, the global analysis can integrate directly the results obtained in the local analysis phase, but with two possible simplifications. The first one concerns the replacement of fine mesh used in the local analysis phase by a coarse mesh which is chosen of the same grading as the mesh used for the rest of the structural assembly. We can easily avoid in this manner any potential risk that the waves propagating across the interface between the impacted zone and the rest of the structure remain trapped within the refined mesh zone. The second simplification consists of replacing the true loading on impacted structural component which stems from time consuming computation of frictional contact and impact of the aeroplane with the equivalent nodal loading applied at the nodes of the coarse mesh of the structural component which is used in the global analysis phase. This task is very much like the construction of a phenomenological model of material obtained from a detailed representation of the material microstructure.

The problem of this kind belongs to the class of field transfer problems. The main difficulty we have to deal with herein concerns the best choice between two possible manners to accomplish this task. In order to fix these ideas we further consider a simple problem where a single structural component, a slab represented by a fine mesh of 3D solid elements, is impacted by a small plate (see Figure 15). This nonlinear impact problem is solved by central difference scheme by using typically small time steps which are imposed by the small size of the finite elements in the fine mesh. We well further seek to recover as close as possible the results of the nonlinear impact analysis on the coarse mesh by using again the central difference scheme, driven no longer with contact, but with an equivalent loading applied on the coarse mesh.

The first transfer problem concerns the space coordinates, or in other words, the transfer between the fine and the coarse mesh. The field transfer is done for any point of the coarse mesh, denoted as $\mathbf{x}$, where we need the best possible coarse mesh representation of the results obtained on the fine mesh. The coarse mesh representation is constructed in accordance with moving least square approximation of the fine mesh transfer. More precisely, for the displacement component on the coarse mesh $d_i^{coarse}(\mathbf{x})$ we assume that one can use the projection which is based on the nodal values of the corresponding displacement component obtained in the neighborhood of $\mathbf{x}$ on the fine mesh, $d_i^{fine}(\mathbf{x}_j)$

$$d_i^{coarse}(\mathbf{x}) := \Pi(d_i^{fine}(\mathbf{x}_j^{fine})) \; ; \; \forall \mathbf{x}_j \in \mathcal{N}(\mathbf{x}) \tag{32}$$

The projection of this kind should be in accordance with the moving least square approximation which is written as

$$d_i^{coarse} = \begin{bmatrix} 1 & x_1 & x_2 & x_3 \end{bmatrix}^T \begin{bmatrix} a_0 \\ a_1 \\ a_2 \\ a_3 \end{bmatrix} = \mathbf{p}(\mathbf{x})^T \mathbf{a} \tag{33}$$

where $\mathbf{a}$ are the approximation parameters to be determined from the least square fit. The latter can formally be formulated as the following minimization problem

$$\min_a J(\mathbf{a}) \; ; \; J(\mathbf{a}) = \frac{1}{2} \sum_{j \in \mathcal{N}(\mathbf{x})} W(\mathbf{x}, \mathbf{x}_j) \left[\mathbf{p}(\mathbf{x}_j)^T \mathbf{a} - d_i^{fine}(\mathbf{x}_j) \right]^2 \tag{34}$$

where $W(\mathbf{x}, \mathbf{x}_j)$ are the bell-shaped weighting functions (see Figure 15), which are chosen in order to limit the influence of the points placed farther away from the point $\mathbf{x}$ as well as to provide a continuous approximation when moving across domain of influence of the neighboring nodes on the coarse mesh. The latter is true (Villon, 2002) only when taking more than a minimum of four points in the neighborhood $\mathbf{x}_j \in \mathcal{N}(\mathbf{x})$ $j = 1, \ldots, m \geq 4$, or otherwise the weighting functions will play no role. The cost function in the minimization problem in (34) can also be stated in matrix notation as

$$J(\mathbf{a}) = \frac{1}{2}\mathbf{a}^T\mathbf{PWP}^T\mathbf{a} - \mathbf{a}^T\mathbf{PW}\tilde{\mathbf{d}}^{fine} + \frac{1}{2}\tilde{\mathbf{d}}^{fine,T}\mathbf{W}\tilde{\mathbf{d}}^{fine} \tag{35}$$

where the weighting factors, interpolation polynomial and fine mesh displacement values are stored as

$$\mathbf{W} = \begin{bmatrix} W(\mathbf{x},\mathbf{x}_1) & \ldots & 0 \\ & \ddots & \\ & & \ddots \\ 0 & \ldots & W(\mathbf{x},\mathbf{x}_m) \end{bmatrix} ; \mathbf{P}^T = \begin{bmatrix} \mathbf{p}(\mathbf{x}_1)^T \\ \vdots \\ \mathbf{p}(\mathbf{x}_m)^T \end{bmatrix} ; \tilde{\mathbf{d}}^{fine} = \begin{bmatrix} d_i^{fine}(\mathbf{x}_1) \\ \vdots \\ d_i^{fine}(\mathbf{x}_m) \end{bmatrix} \tag{36}$$

The Kuhn-Tucker optimality condition of this minimization problem leads to the optimal value of approximation parameters

$$\mathbf{a} = [\mathbf{PWP}^T]^{-1}\mathbf{PW}\tilde{\mathbf{d}}^{fine} \tag{37}$$

which defines completely the chosen approximation in (32).

Having clarified the field transfer in space between the fine and the coarse mesh, we now consider the transfer of evolution problem in time, where the time steps used on the selected coarse mesh are likely to be much bigger than those used previously on the fine mesh. The main difficulty in that respect pertains to two possible manners of performing such a field transfer. In order to further elaborate on this idea, we consider a single time step of the coarse mesh computation which corresponds to a number of small steps of the fine mesh; See Figure 15. One possible way for the field transfer is by first carrying out the computations on the fine mesh by using the central difference scheme in (30) and then use the projection (32) of such a result to the coarse mesh. This kind of field transfer can formally be written as

$$\begin{array}{l} \mathbf{d}_{n+1}^{coarse}(\mathbf{x}) := \Pi(\mathbf{d}_{n+1}^{fine}(\mathbf{x}_j^{fine})) \; ; \; \mathbf{v}_{n+1}^{coarse}(\mathbf{x}) := \Pi(\mathbf{v}_{n+1}^{fine}(\mathbf{x}_j^{fine})) \; ; \; \forall \mathbf{x}_j \in \mathcal{N}(\mathbf{x}) \\ \mathbf{a}_{n+1}^{coarse}(\mathbf{x}) := \Pi(\mathbf{a}_{n+1}^{fine}(\mathbf{x}_j^{fine})) \; ; \; \mathbf{f}_{n+1}^{coarse}(\mathbf{x}) := \Pi(\mathbf{r}_{contact,n+1}^{fine}(\mathbf{x}_j^{fine})) \; ; \end{array} \tag{38}$$

where $\mathbf{d}_{n+1}^{fine}$, $\mathbf{v}_{n+1}^{fine}$ and $\mathbf{a}_{n+1}^{fine}$ are, respectively, the computed vectors of displacements, velocities and acceleration at time t_{n+1}, while $\mathbf{r}_{contact,n+1}^{fine}$ are nodal forces computed from contact problem on the fine mesh. The last expression states that any time we need the corresponding values on the coarse mesh they can be obtained by projection, which completely eliminates the need for central difference scheme computations on the coarse mesh.

There is an alternative way to carry out the transfer, which implies the need for computations on the coarse mesh as well. Namely, we first start with the transfer of the results obtained on the fine mesh, and then follow up with the computation carried out on the coarse mesh. The latter can formally be written according to

$$\begin{aligned}
&\mathbf{d}_{n+1}^{coarse}(\mathbf{x}) = \Pi(\mathbf{d}_n^{fine}(\mathbf{x}_j^{fine})) + \Delta t\, \Pi(\mathbf{v}_n^{fine}(\mathbf{x}_j^{fine})) + \Delta t^2\, \Pi(\mathbf{a}_n^{fine}(\mathbf{x}_j^{fine}))\\
&\mathbf{M}\mathbf{a}_{n+1}^{coarse}(\mathbf{x}) = \mathbf{g}_{n+1}(\mathbf{x}) - \hat{\mathbf{f}}^{int}(\mathbf{d}_{n+1}^{coarse}(\mathbf{x}), \ldots)\\
&\mathbf{v}_{n+1}^{coarse}(\mathbf{x}) = \mathbf{v}_n^{coarse}(\mathbf{x}) + \tfrac{\Delta t}{2}(\Pi(\mathbf{a}_n^{fine}(\mathbf{x}_j^{fine})) + \mathbf{a}_{n+1}^{coarse}(\mathbf{x}))
\end{aligned} \tag{39}$$

From standpoint of computational efficiency, there is one crucial difference between computational procedure on fine and on coarse mesh: the former is driven by impact and frictional contact and the latter is driven by equivalent nodal loading $\mathbf{g}_{n+1}$. Such an equivalent nodal loading is obtained as the solution of the minimization problem seeking to render the results of two field transfert procedures as close as possible

$$\min_{\mathbf{g}_{n+1}} J(\mathbf{g}_{n+1}) \tag{40}$$

with the explicit form of the cost function defined as

$$\begin{aligned}
J(\mathbf{g}_{n+1}) \ = & \tfrac{1}{2}\int_{\Omega^{coarse}} \{(\mathbf{d}_{n+1}^{coarse}(\mathbf{x}) - \Pi(\mathbf{d}_{n+1}^{fine}(\mathbf{x})))^T(\mathbf{d}_{n+1}^{coarse}(\mathbf{x}) - \Pi(\mathbf{d}_{n+1}^{fine}(\mathbf{x})))\\
& +(\mathbf{v}_{n+1}^{coarse}(\mathbf{x}) - \Pi(\mathbf{v}_{n+1}^{fine}(\mathbf{x})))^T(\mathbf{v}_{n+1}^{coarse}(\mathbf{x}) - \Pi(\mathbf{v}_{n+1}^{fine}(\mathbf{x})))\\
& +(\mathbf{a}_{n+1}^{coarse}(\mathbf{x}) - \Pi(\mathbf{a}_{n+1}^{fine}(\mathbf{x})))^T(\mathbf{a}_{n+1}^{coarse}(\mathbf{x}) - \Pi(\mathbf{a}_{n+1}^{fine}(\mathbf{x})))\}\, dV
\end{aligned} \tag{41}$$

The dependence of such a cost function on the equivalent load vector on the coarse mesh $\mathbf{g}_{n+1}$ is defined through (39).

In seeking to enforce further result correspondance of these two types of field transfer, we can also require that the work of the equivalent loads on the coarse mesh over any time step of the coarse mesh computations remains as close as possible to the work of contact forces on the fine mesh. The latter can formally be written as

$$C(\mathbf{g}_{n+1}) \ = \int_{\Omega^{coarse}} \{\tfrac{t_{n+1}-t_n}{2}(\mathbf{g}_{n+1}^T\mathbf{v}_{n+1}^{coarse} + \mathbf{g}_n^T\mathbf{v}_n^T) - W_{contact}^{fine}\}\, dV \tag{42}$$

where $W_{contact}^{fine}$ is the work of contact forces which is computed on the fine mesh during the same time step. One can thus modify the optimization problem in (40) by adding the last condition as the constraint

$$\min_{g_{n+1}} \max_{\lambda_{n+1}} L(\mathbf{g}_{n+1}, \lambda_{n+1})\ ;\ L(\mathbf{g}_{n+1}, \lambda_{n+1}) = J(\mathbf{g}_{n+1}) + \lambda_{n+1} C(\mathbf{g}_{n+1}) \tag{43}$$

Four different methods for field transfer are developed on the basis of the minimization problems presented herein. The first one considers the direct transfer of the nodal values computed on the fine mesh and the second one will add the work conservation constraint. The third method seeks to improve upon the computation of the cost function by using the patch-like computations. Finally, the fourth method is using the direct transfer of the values at the Gauss numerical points, leading in general to the highest precision of results.

We have carried out the computations with these four methods for the impacted thick plate component in Figure 15. The results on representing the upper and lower bounds are obtained by carrying out the computations of the impact problem on the fine and on the coarse mesh, respectively. The first set of computations, performed for the case of the elastic plate (see Figure 16), have shown that all the methods can give fairly good results, since a reasonably good representation of the fundamental vibration modes is the only condition which should be fulfilled. The second computations with four methods is carried out for elastoplastic plate (see Figure 17). In this case, the fourth method clearly shows far superior results, which is the consequence of the highest level of consistency we impose for this kind of field transfer.

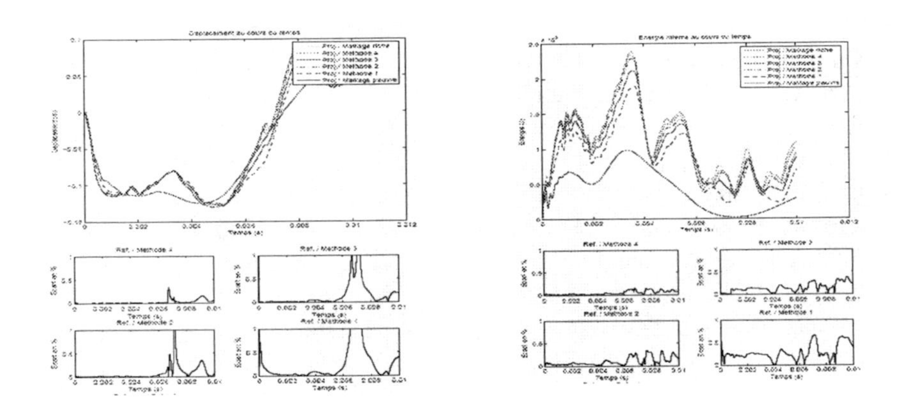

Figure 16. Computed displacement and energy with ne and coarse mesh and four different projection methods - elastic case.

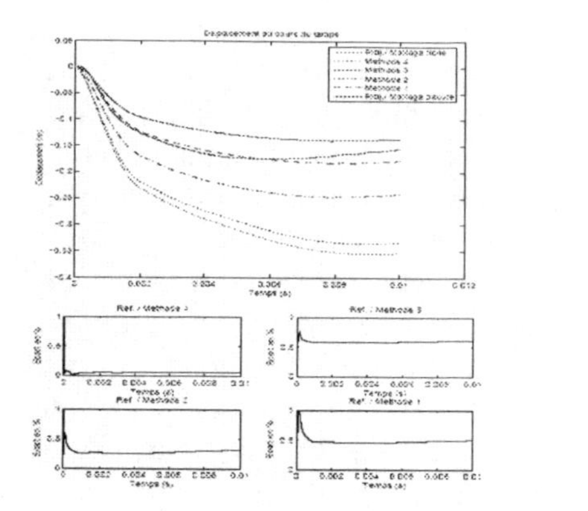

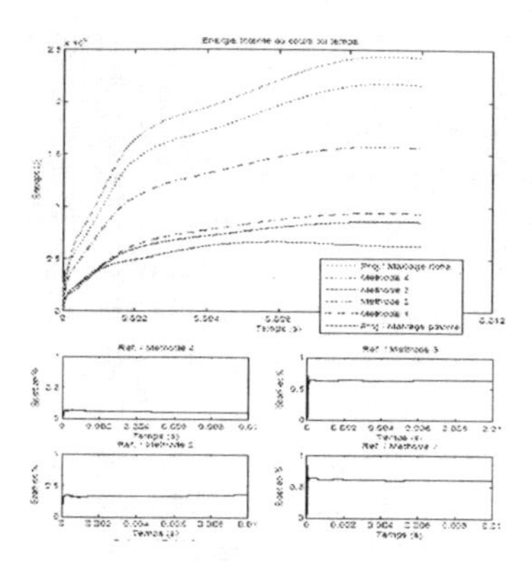

Figure 17. Computed displacement and energy with fine and coarse mesh and four different projection methods - elastoplastic case.

5. Concluding Remarks

We have presented in the foregoing a multi-scale modelling approach for predicting the inelastic behavior and ultimate limit load of complex massive structures, where the micro-scale is defined as the scale which allows the most reliable interpretation of the inelastic mechanisms. We have also discussed the problems where the reliable loading interpretation requires dealing with a coupled problem. An operator split type solution procedure for this kind of coupled problem is proposed, by taking into account the multi-scale approach in time. We considered herein two extreme cases: the first is a thermomechanical coupling problem of fire resistance with a very long duration of elevated temperature, and the second is a very short duration of aeroplane impact on a massive structure.

We have shown that the proposed procedures for nonlinear inelastic analysis of complex structures have reached a fair level of reliability, where they can be used to supplement the time consuming and costly experimental procedure on structures. We have also shown how to reduce the amount of information coming from refined models used for nonlinear analysis of a particular structural component, in view of preparing the reduced model for parametric studies and a novel design procedure, which should verify the integrity of not only a single component but also of the whole structural assembly.

Current works deal with the development of computer models capable of providing additional information from the structural testing under heterogeneous stress field, and accounting for heterogeneities in material properties (Ibrahimbegovic and Markovic, 2003). The model reduction procedures in dynamics, as those proposed earlier in (Ibrahimbegovic et al., 1990a), are also studied to be used in dynamics (Markovic et al., 2005) and in earthquake engineering design (Davenne et al., 2003). Model and mesh adaptivity issues, where the choice between refined and coarse model is made, geared towards the ultimate load computation of a complex massive structures also belong to our current research activities of this kind (Brancherie et al., 2005), (Bohinc et al., 2005). The procedures for optimal design of complex structures in presence of inelastic nonlinear behavior are also being developed (Ibrahimbegovic et al., 2005b). Finally, we are currently seeking a deeper understanding of the physical aspects which would allow to construct the key ingredient of probabilistic description of inelastic nonlinear phenomena and thus feed the proposed computational procedure for probabilistic studies of nonlinear behavior of complex structures, presently available mostly for linear problems (Matthies, 2004).

Acknowledgements

The research results summarized herein are obtained in several research projects carried out at "Laboratoire de Mécanique et Technologie de Cachan", with participation of a couple of senior colleagues, Dr. Luc Davenne and Prof. Pierre Villon, and four of my former doctoral students at "Ecole Normale Supérieure de Cachan": Dr. Delphine Brancherie, supported by the French Ministry of Research and CNRS under ACI 2159; Dr. Jean-Baptiste Colliat, supported by the French Ministry of Research and CTTB; Dr. Norberto Dominguez, supported by the French-Mexican Cooperation Program and EDF/AMA and Dr. Guillaume Hervé, supported by the French Ministry of Research and CEA/DAM.

References

Bohinc, U., Ibrahimbegovic, A., and Brank, B. (2005) Robust plate finite elements for adaptive modeling of structures, In J. Korelc (ed.), *Congress of Slovenain Society of Mechanics*, Ljubljana, Slovenia, Univerza v Ljubljani.

Brancherie, D. and Ibrahimbegovic, A. (2004) Macro-scale model of dissipative phenomena produced at micro-scale: theoretical formulation and numerical implementation, *European J. Finite Elements* **13**, 461–475.

Brancherie, D. and Ibrahimbegovic, A. (2005) Fracture of massive structures: a novel approach to constructing an anisotropic damage model combining continuum-type hardening and discrete-type softening models, *International Journal for Numerical Methods in Engineering* p. in press.

Brancherie, D., Villon, P., Ibrahimbegovic, A., Rassineux, A., and Breitkopf, P. (2005) Transfer operator based on diffuse interpolation and energy conservation for damage materials, In K. Bathe (ed.), *3rd MIT Conference in Computational Fluid and Solid Mechanics*, Amsterdam, Netherlands, Elsevier.

Colliat, J.-B., Davenne, L., and Ibrahimbegovic, A. (2002) Modeling of failure of masonry wall under in-plane loading, *French J. Civil Engineering* **4**, 593–606.

Colliat, J.-B., Ibrahimbegovic, A., and Davenne, L. (2005a) Heat conduction and radiation heat exchange in cellular structures using flat shell elements, *Communications in Numerical Methods in Engineering* p. in press.

Colliat, J.-B., Ibrahimbegovic, A., and Davenne, L. (2005b) Saint-Venant multi-surface plasticity model in strain space and in stress resultants, *Engineering Computations* **41**, 536–557.

Davenne, L., Ragueneau, F., Mazars, J., and Ibrahimbegovic, A. (2003) Efficient Approach to Earthquake Engineering Analysis, *Computers and Structures* **81**, 1223–1239.

Delaplace, A. and Ibrahimbegovic, A. (2005) Time-integration schemes for dynamic fracture problem using the discrete model, *International Journal for Numerical Methods in Engineering* p. in press.

Dominguez, N., Brancherie, D., Davenne, L., and Ibrahimbegovic, A. (2005) Prediction of crack pattern distribution in Reinforced Concrete by coupling a strong discontinuity model of concrete cracking and a bond-slip of reinforcement model, *Engineering Computations* **41**, 558–582.

Herve, G., Gatuingt, F., and Ibrahimbegovic, A. (2005) On numerical implementation of a coupled rate dependent damage-plasticity constitutive model for concrete in application to high-rate dynamics, *Engineering Computations* **22**, 583–604.

Ibrahimbegovic, A. (2003) On the geometrically exact Formulation of Structural Mechanics and Its Applications to Dynamics, Control and Optimization, *Comptes Rendus de l'Academie des Sciences. Part II : Mcanique* **331**, 383–394.

Ibrahimbegovic, A. (2006) *Nonlinear mechanics of deformable solids: Theoretical formulation and finite element implementation (in French)*, Paris, Herves-Science, Lavoisier.

Ibrahimbegovic, A. and Brancherie, D. (2003) Combined hardening and softening constitutive model of plasticity: precoursor to shear slip line failure, *Computational Mechanics* **31**, 88–100.

Ibrahimbegovic, A., Chen, H., Wilson, E., and Taylor, R. (1990)a Ritz method for dynamic analysis of large discrete linear systems with non-proportional damping, *Earthquake Engineering and Structural Dynamics* **19**, 877–889.

Ibrahimbegovic, A. and Chorfi, L. (2002) Covariant principal axis formulation of associated coupled thermoplasticity at finite strains and its numerical implementation, *International Journal of Solids and Structures* **39**, 499–528.

Ibrahimbegovic, A., Colliat, J.-B., and Davenne, L. (2005)a Thermomechanical coupling in folded plates and non-smooth shells, *Computer Methods in Applied Mechanics and Engineering* **194**, 2686–2707.

Ibrahimbegovic, A. and Delaplace, A. (2003) Microscale and mesoscale discrete models for dynamics fracture of structures built of brittle material, *Computers and Structures* **81**, 1255–1265.

Ibrahimbegovic, A. and Frey, F. (1994) Stress Resultant Geometrically Nonlinear Shell Theory With Drilling Rotations. Part III: Linearized Kinematics, *International Journal for Numerical Methods in Engineering* **37**, 3659–3683.

Ibrahimbegovic, A., Gresovnik, I., Markovic, D., Melnyk, S., and Rodic, T. (2005b) Shape optimization of two-phase inelastic material with microstructure, *Engineering Computations* **22**, 605–645.

Ibrahimbegovic, A., Knopf-Lenoir, C., Kucerova, A., and Villon, P. (2004) Optimal design and optimal control of structures undergoing finite rotations and elastic deformations, *International Journal for Numerical Methods in Engineering* **61**, 2428–2460.

Ibrahimbegovic, A. and Kozar, I. (1995) Nonlinear Wilson's brick element for finite elastic deformation of three-dimensional solids, *Communications in Numerical Methods in Engineering* **11**, 655–664.

Ibrahimbegovic, A. and Mamouri, S. (2002) Energy conserving/decaying implicit time-stepping scheme for nonlinear dynamics of three-dimensional beams undergoing finite rotation, *Computer Methods in Applied Mechanics and Engineering* **191**, 4241–4258.

Ibrahimbegovic, A. and Markovic, D. (2003) Strong coupling methods in multi-phase and multi-scale modeling of inelastic behavior of heterogeneous structures, *Computer Methods in Applied Mechanics and Engineering* **192**, 3089–3107.

Ibrahimbegovic, A., Markovic, D., and Gatuingt, F. (2003) Constitutive Model of Coupled Damage-Plasticity and Its Finite Element Implementation, *Revue eurpenne des lments finis* **12**, 381–405.

Ibrahimbegovic, A. and Taylor, R. (2002) On the role of frame-invariance in structural mechanics models at finite rotations, *Computer Methods in Applied Mechanics and Engineering* **191**, 5159–5176.

Ibrahimbegovic, A., Taylor, R., and Wilson, E. (1990b) A robust quadrilateral membrane finite element with drilling degrees of freedom, *International Journal for Numerical Methods in Engineering* **30**, 445–457.

Ibrahimbegovic, A. and Wilson, E. (1991a) A modified method of incompatible modes, *Communications in Numerical Methods in Engineering* **7**, 187–194.

Ibrahimbegovic, A. and Wilson, E. (1991b) Unified Computational Model for Static and Dynamic Frictional Contact Analysis, *International Journal for Numerical Methods in Engineering* **34**, 233–247.

Lemaitre, J. and Chaboche, J.-L. (1988) *Mcanique des matriaux solides*, Paris, Dunod.

Markovic, D. and Ibrahimbegovic, A. (2004) On micro-macro interface conditions for micro-scale based FEM for inelastic behavior of heterogeneous materials, *Computer Methods in Applied Mechanics and Engineering* **193**, 5503–5523.

Markovic, D. and Ibrahimbegovic, A. (2005) Complementary energy based FE modeling of coupled elasto-plastic and damage behavoir for continuum microstructure computations, *Computer Methods in Applied Mechanics and Engineering* **-**, in press.

Markovic, D., Park, K.-C., and Ibrahimbegovic, A. (2005) Reduction of substructural interface degrees-of-freedom in flexibility based component mode sythesis, *International Journal for Numerical Methods in Engineering* **-**, in press.

Matthies, H. (2004) Computational aspects of probability in nonlinear mechanics, In A. Ibrahimbegovic and B. Brank (eds.), *Multi-physicss and Multi-scale Computer Models in Nonlinear Analysis and Optimal Design of Engineering Structures Under Extreme Conditions,*, Amsterdam, Netherlands, IOS Press.

Villon, P. (2002) Transfert de champs plastiquement admissibles, *Comptes Rendus de l'Academie des Sciences. Part II : Mcanique* **330**, 313–318.

ON RELIABLE FINITE ELEMENT METHODS FOR EXTREME LOADING CONDITIONS

K.-J. Bathe *(kjb@mit.edu)*
Massachusetts Institute of Technology, Cambridge, MA 02139, U.S.A.

Abstract. In this paper we focus on the analysis of solids and structures when these are subjected to extreme conditions of loading resulting in large deformations and possibly failure. The analysis should be conducted with finite element methods that are as reliable as possible and effective. The requirement of reliability is important in any finite element analysis but is particularly important in simulations involving extreme loadings since physical test data are frequently not available, or only available for some similar conditions. To then reach a high level of confidence in the computed solutions requires that reliable finite element procedures be used.

While in this paper a large field of analysis is covered, the presentation is narrow because it only focuses on our research results, mostly published in the last decade (since 1995), and only on some of our contributions. Hence, this paper is not written to fully survey the field.

Key words: finite element methods, structures, fluids, extreme loading

1. Introduction

Finite element methods are now widely used in engineering analysis and we can expect a continued growth in the use of these methods. Finite element programs are extensively employed for linear and nonlinear analyses, and the simulations of highly nonlinear events are of much interest[1-3]. In particular, the simulations may be used to postulate accidents and natural disasters, and thus can be used to study how a structure will perform in such severe events. Then, if necessary, remedies in the design of a structure can be undertaken, and equally important, warning devices might be installed, that all lead to greater public safety.

An important point is that for large civil and mechanical engineering structures (long-span bridges, off-shore plants, storage tanks, power plants, chemical plants, underground storage caverns, to name just a few), physical tests can only be performed to a limited extent. Parts of the structures can

A. Ibrahimbegovic and I. Kozar (eds.),Extreme Man-Made and Natural Hazards in Dynamics of Structures, 71–102.

be tested (like connections between pipes) but a complete structure may only be tested when completely assembled in the field and sometimes that is not possible either. Hence, the simulations of such structures subjected to severe loadings probe into, and try to predict, the future based on scarce actual physical tests. It is then very important to use reliable finite element methods in order to have the highest possible confidence in the computed results[3].

The objective in this paper is to survey some finite element analysis techniques with a particular focus on the reliability of the methods. Of course, any simulation starts with the selection of a mathematical model, and this model must be chosen judiciously. Once an appropriate mathematical model has been selected, for the questions asked, the finite element analysis is performed.

In this paper, we consider the solution of problems involving 2D and 3D solids, plates and shells, fluid-structure interactions, and general multi-physics events. The simulation of structures subjected to extreme loading conditions frequently comprises multi-physics events that involve fluids and severe thermal effects. For all these analyses, we need to use effective and reliable finite elements, efficient methods for the solution of the equations, for steady-state (static) and transient conditions, and finally, as far as possible, appropriate error measures.

While a large analysis field is covered in this paper, we shall focus only on our research accomplishments since 1995, and that also only partially. Indeed, throughout our research endeavors since the 1970's, the aim was always to only develop reliable and effective finite element procedures[3].

Therefore, this paper is not intended to be a survey paper, neither to give all of our achievements nor to give adequate acknowledgement of the many important works of other researchers. All we focus on is to survey some valuable analysis methods for structures in extreme loading conditions.

2. On the Selection of the Mathematical Model

The first step of any structural analysis involves the selection of an appropriate mathematical model[3,4]. This model needs to be selected based on the geometry, material properties, the loading and boundary conditions (the effect of the 'rest of the universe' on the structure) and, most importantly, the questions asked by the analyst. If fluid-structure interactions are important, then the mathematical model should also include the fluid and the loading and boundary conditions thereon.

The purpose of the analysis is of course to answer certain questions regarding the stiffness, strength and possible failure of the structure under consideration. In the case of extreme loading conditions, we would like to ascertain the behavior of the structure in postulated accidents, either man-made or imposed by nature, or in natural disasters. Hence, when studying the behavior of the structure, we would like to predict the future not only when the structure is operating in normal conditions, which requires a linear analysis, but also when the structure is subjected to extreme conditions of loading, which requires a highly nonlinear analysis.

The mathematical model should naturally be as simple as possible to answer the engineering questions but not be too simple (an observation attributed to A. Einstein) because the answers may then be erroneous and misleading. For most structures, it is necessary to choose a mathematical model that can only be solved using numerical methods and finite element procedures are widely used.

It is clear that the finite element solution of the mathematical model will contain all the assumptions of the mathematical model and hence cannot predict any response not contained in this model. Selecting the appropriate mathematical model is therefore most important. But it is also clear that the analysis can only give insight into the physical behavior of a structure, that is, nature, because it is impossible to reproduce nature exactly.

In this paper we only consider deterministic analyses. If non-deterministic simulations need to be carried out, then still, the considerations given here are all valid because the deterministic procedures are basic methods used in those analyses as well[1].

A fundamental question must always be whether the mathematical model used is appropriate. This question can be addressed by the process of hierarchical modeling[3-4]. In this process of mathematical modeling and finite element solution, it is best to use one finite element program for the solution of the different linear and nonlinear mathematical models, because the assumptions in the finite element procedures are then the same in all solutions. In the case of extreme conditions of loading, this finite element program need be used for linear and highly nonlinear structural and multi-physics problems, including fluid flows, severe temperature conditions, and the full mechanical interactions. In this paper we are focusing on ADINA for such analyses.

3. The Finite Element Solution of Solids and Structures

With the mathematical model chosen, the finite element procedures are used to solve the model. It is important that in this phase of the analysis well-founded and reliable procedures be used. By reliability of a finite element procedure we mean that in the solution of a well-posed mathematical model, the procedures always, for a reasonable finite element mesh, give a reasonable solution. And if the mesh is reasonably fine, an accurate solution of the chosen mathematical model is obtained[3].

It is sometimes argued that since the geometry, material conditions, loadings are not known to great accuracy, there is no need to solve the mathematical model accurately. This is a reasonable argument provided the mathematical model is solved to 'sufficient' accuracy *and* a control on the level of accuracy is available. Such a control on the accuracy of the finite element solution of the mathematical model is difficult to achieve and reliable finite element procedures are best used.

The reliability of a finite element procedure means in particular that when some geometric or material properties are changed in the mathematical model, then for a given finite element mesh the accuracy of the finite element solution does not drastically decrease. Hence, pure displacement-based finite element methods are not reliable when considering almost incompressible materials (like a rubber material, or a steel in large strains). As well-known, these methods 'lock', and for such analyses, well-founded mixed methods need be used[3].

To exemplify what can go wrong in a finite element solution when unreliable finite element procedures are used, we refer to ref. 3, page 474, where the computed frequencies of a cantilever bracket are given. If reduced integration is used, then phantom frequencies (that are totally non-physical) are predicted. Such ghost frequencies may also be predicted when some hour-glass control with reduced integration is employed.

Similar situations also arise in the analysis of plates and shells, when reduced integration is employed, but in shell analyses, in addition, the use of flat shell elements can result in unreliable solutions, see Section 3.2. And similar conditions can also arise in the analysis of fluids and the interactions with structures, see Section 4.

3.1. SOLIDS

The use of well-formulated mixed methods has greatly enhanced the reliable analysis of solids and structures[3].

Considering the analysis of solids, in large strains, the condition of an almost incompressible material response is frequently reached. This situation is of course encountered in simulations involving rubber-like materials (that already are almost incompressible in small strain conditions) but also in simulations of many inelastic conditions, like elasto-plasticity and visco-plasticity. In these analyses, the displacement/pressure (u/p) finite elements are very effective and if the appropriate pressure interpolations are chosen, the elements are also optimal[3]. An optimal element uses for a given displacement interpolation the highest pressure interpolation that satisfies the inf-sup condition[3-6]

$$\inf_{q_h \in Q_h} \sup_{\mathbf{v}_h \in V_h} \frac{(q_h, \text{div } \mathbf{v}_h)}{\|q_h\|_0 \|\mathbf{v}_h\|_1} \geq \beta > 0 \tag{1}$$

where V_h is the finite element displacement space, Q_h is the finite element pressure space, and β is a constant independent of h (the element size).

Reference 3 gives a table of u/p elements that satisfy the above inf-sup condition. These elements can of course also be used for analyses that do not involve the incompressibility condition (but are computationally slightly more costly than the pure displacement-based elements).

For 2D solutions, for example, the 9/3 element (9 nodes for the displacement interpolations, and 3 degrees of freedom for the pressure interpolation) is an effective element. But, in practice, a 4-node element is desirable and frequently the 4/1 element (4 nodes for the displacement interpolations and a constant pressure) is used. While the 4/1 element does not satisfy the above inf-sup condition, it can be quite effective when used with care. A 4-node element that does satisfy the inf-sup condition was presented by Pantuso and Bathe[7,8]. Of course, equivalent elements exist for 3D solutions.

In any simulation, it is desirable to obtain, as the last step of the analysis, some indication regarding the accuracy of the finite element solution when measured on the exact solution of the mathematical model. We briefly address this issue in Section 3.4.

3.2. PLATES AND SHELLS

In the analysis of plates and shells, the situation is more complex than encountered in the analysis of solids. Here, for a given mesh, an optimal element would give the same error irrespective of the thickness of the plate or shell, and for any plate and shell geometry, boundary conditions and admissible loading used[3,5]. As well known, all displacement-based shell elements formulated using the Reissner-Mindlin kinematic assumption do not satisfy this condition in the analysis of general shells, and for this reason research has focused on the development of mixed elements, based in essence on the general formulation:

Find $\boldsymbol{u}_h \in V_h$ and $\boldsymbol{\xi}_h \in E_h$

$$a(\mathbf{u}_h, \mathbf{v}_h) + b(\boldsymbol{\xi}_h, \mathbf{v}_h) = f(\mathbf{v}_h) \quad \forall\, \mathbf{v}_h \in V_h \tag{2}$$

$$b(\boldsymbol{\eta}_h, \mathbf{u}_h) - t^2 c(\boldsymbol{\eta}_h, \boldsymbol{\xi}_h) = \mathbf{0} \quad \forall\, \boldsymbol{\eta}_h \in E_h \tag{3}$$

where $a(\cdot,\cdot)$, $b(\cdot,\cdot)$, and $c(\updownarrow,\square)$, are bilinear forms, $f(\cdot)$ is a linear form, t is the thickness of the shell, and V_h , E_h are the finite element displacement and strain spaces.

Mixed elements, however, then should satisfy the consistency condition, the ellipticity condition, and ideally the inf-sup condition[5,6, 9-12]

$$\sup_{\mathbf{v}_h \in V_h} \frac{b(\boldsymbol{\eta}_h, \mathbf{v}_h)}{\|\mathbf{v}_h\|_V} \geq c \sup_{\mathbf{v} \in V} \frac{b(\boldsymbol{\eta}_h, \mathbf{v})}{\|\mathbf{v}\|_V} \quad \forall \, \boldsymbol{\eta}_h \in E_h \tag{4}$$

where V is the complete (continuous) displacement space, and c is a constant independent of t and h.

If an element satisfies these conditions, the discretization is optimal. However, the first mixed elements proposed were not tested for these conditions, and only relatively lately a more rigorous testing has been proposed [13, 14].

Of course, the resulting test problems can (and should) also be used to evaluate any already earlier-proposed shell element, even if it seems that the element is effective. For example, when so evaluating flat shell elements formulated by superimposing membrane and bending actions, a non-convergence behavior may be found, see ref. 12.

We have proposed some time ago the MITC plate and shell elements, see refs. 3, 15 and the references therein, and have also given some further recent MITC developments for shells [16, 17]. For the analysis of plates, the inf-sup condition can be re-written into a simpler form and we were able to check that the MITC plate elements are optimal (or close thereto)[9,10] (see Section 7.1).

However, for shell elements the inf-sup condition needs to be evaluated as given above and in general only numerical tests seem possible because V is the complete space in which the shell problem is posed [11,12]. Then it can be equally effective to instead solve well-designed test problems [12, 13, 18-20]. In these tests an appropriate norm to measure the error must be used – a norm that ideally is applicable to all shell problems – and the s-norm proposed by Hiller and Bathe is effective[12, 20].

The difficulty in formulating a general shell element lies in that the element should ideally be optimal in the analysis of membrane-dominated shells, bending-dominated shells, and mixed- behavior shells. For the membrane-dominated case, actually, the displacement-based shell elements perform satisfactorily but for the bending-dominated and mixed cases, the displacement-based elements 'lock' and a mixed formulation need be used – which satisfies consistency, ellipticity, and ideally the inf-sup condition. The inf-sup condition can be by-passed, but then nonphysical numerical factors enter the element formulation.

At present, there seems no shell element formulated based on the Reissner-Mindlin assumption that is 'proven analytically' to satisfy all three conditions. However, well-designed numerical tests can be performed to identify whether the conditions seem to be satisfied and the MITC shell elements have performed well in such tests. Indeed, the MITC4 shell element (the 4-node element[3]) has shown excellent convergence and is used abundantly in a number of finite element codes (see Section 7.2).

Considering convergence studies of the general shell elements used in engineering practice, a basic step was to identify the underlying shell mathematical model used [21]. The different terms in the variational formulation of this shell model can thus also be studied, and it is possible to identify how well the specific terms are approximated in the finite element solution [22].

The MITC interpolations can of course also be employed in the formulation of 3D-shell elements and here too the underlying shell model was identified [23].

3.3. NONLINEAR ANALYSIS OF SOLIDS AND STRUCTURES, INCLUDING CONTACT CONDITIONS

The above considerations naturally also hold for geometric and material nonlinear analyses. The large deformation analysis of solids and structures has now been firmly established [3, 24], although of course improvements are sought, specifically in establishing more comprehensive material models. A particular area of interest is to increase the accuracy of the response predictions when considering inelastic orthotropic metals, where the anisotropy may exist initially or be induced by the response[25-27]. The elastic response may be anisotropic, the yielding may be anisotropic and the directions and magnitude of anisotropy may change during the response.

The development of increasingly more comprehensive material models will clearly continue for some time and will also involve the molecular modeling of materials and coupling of these models to finite element discretizations.

A field that has still undergone, in recent years, considerable developments, is the more accurate analysis of contact problems, in particular when higher-order elements are used. Such higher-order elements are typically the 10-node or 11-node tetrahedral elements generated in free-form meshing. The consistent solution algorithm proposed in ref. 28 is quite valuable in that the patch test can be satisfied exactly[28-30]. The algorithm is also used effectively in gluing different meshes in a multi-scale analysis (see Section 7.4), and a mathematical analysis has given insight into the performance of the discretization scheme[30].

The basic approach is that along the contactor surface, the tractions are interpolated and the gap and slip between the contactor and target are evaluated from the nodal positions and displacements. Let λ be the contact pressure and $g(s)$ be the gap at the position s along the contact surface Γ_C, then the normal contact conditions are

$$g \geq 0, \quad \lambda \geq 0, \quad g\lambda = 0 \tag{5}$$

where the last equation in (5) is the complementary condition. We use a constraint function $w_n(g, \lambda)$ to turn the inequality constraints of contact into the equality constraint

$$w_n(g, \lambda) = 0 \tag{6}$$

which gives in variational form

$$\int_{\Gamma_C} w_n(g, \lambda)\, \delta\lambda \, \mathrm{d}\Gamma_C = 0 \tag{7}$$

Whether the gap is open or closed on the surface is automatically contained in the formulation using the constraint function. For frictional contact conditions another constraint function is used.

The scheme published in ref. 28 should be used with appropriate interpolations for the contact pressure – for given geometry and displacement interpolations – and appropriate numerical integration schemes to evaluate the integrals enforcing the contact conditions.

The inf-sup condition for the contact discretization is given by

$$\sup_{\mathbf{v}_h \in V_h} \frac{\int_{\Gamma_C} \mu_h g(\mathbf{v}_h)\, \mathrm{d}\Gamma_C}{\|\mathbf{v}_h\|_1} \geq \beta \sup_{\mathbf{v} \in V} \frac{\int_{\Gamma_C} \mu_h g(\mathbf{v})\, \mathrm{d}\Gamma_C}{\|\mathbf{v}\|_1} \qquad \forall \mu_h \in M_h \tag{8}$$

where M_h is the space of contact tractions and • is a constant, greater than zero. This condition can be satisfied as discussed in refs. 28, 30.

For some example solutions involving contact conditions, see Sections 7.3, 7.8, 7.10, 7.11.

While we considered so far the analysis of solids and structures, of course similar considerations also hold when developing discretization schemes for fluids, including the interactions with structures, see Section 4 below, and in these cases nonlinearities are usually present.

3.4. MEASURING THE FINITE ELEMENT SOLUTION ERRORS

Once a finite element solution of a mathematical model has been obtained, ideally, we could assess the solution error. Much research has focused on the 'a posteriori' assessment of these errors[31, 32]. However, the problem is formidable since the error between the numerical solution and the exact solution of the mathematical model shall be established when the exact solution is unknown.

A review of techniques currently available to establish this error has been published in ref. 31. It was concluded that no technique is currently available that establishes lower and upper bounds, proven to closely bracket the exact solution, when considering general analysis and an acceptable computational effort. The simple recovery-based error 'estimators' are still, in many regards, the most attractive and can of course be used in general linear and nonlinear analyses. However, they only give an indication of the error and need be used with care.

Considering recovery-based estimators, higher-order accuracy points or approximations thereof are frequently used[33] but then the error estimator may only be applicable in linear analysis. In our experience, the stress bands proposed by Sussman and Bathe[34,35] (see also Refs. 3 and 31) are quite effective to obtain error estimates in general analyses (see Section 7.5). Of course, in practical engineering analysis, frequently very fine meshes are used, simply to ensure that an accurate solution has been obtained. Then no error assessment is deemed necessary.

The recovery-based error estimators can be used directly to estimate the error in different regions of the domain analyzed. However, in some analyses, it may be of interest to control the error in only a specific quantity, like the bending moment at a section of the structure. Then we may need to use a fine mesh only in certain regions of the complete analysis domain. A typical example is the analysis of fluid flows with structural interactions. It may not be necessary to solve for the fluid flow very accurately in the complete fluid domain in order to only predict the stresses, accurately, in the structure at a certain location. In these cases, the concept of goal-oriented error estimation can be very effective and has considerable potential for further developments[31, 36,37].

4. The Finite Element Solution of Fluids and Interactions with Structures

The analyses of fluids and fluid structure interactions (FSI) have obtained in recent years much attention because the dynamic behavior of structures can be much influenced by surrounding fluids. If the fluid can be idealized as inviscid and undergoing only small motions, the analysis is much simpler than when actual flow and Euler or Navier-Stokes fluid assumptions are necessary. However, many analysis cases can now be modeled effectively, and when free surfaces or interactions with structures undergoing large deformations are considered, an arbitrary Lagrangian-Eulerian (ALE) formulation is widely used.

4.1. ACOUSTIC FLUIDS AND INTERACTIONS WITH STRUCTURES

The first models to describe an acoustic fluid were simply extensions of solid analysis discretizations, assuming a large bulk modulus and a small shear modulus (corresponding to the fluid viscosity). However, these fluid models are not effective because they 'lock' and even when formulated in a mixed formulation need be implemented to prevent loss of mass. In addition, the formulations then contain many zero energy modes, or modes of very small energy[38-41].

A clearly more effective approach is to use a potential formulation[3] (see Section 7.6) and such formulation extended also for actual flow has been used very successfully to solve large and complex fluid-structure interactions[42].

4.2. EULER AND NAVIER-STOKES FLUIDS AND FLUID STRUCTURE INTERACTIONS

The solution of fluid flow structure interactions (FSI) requires effective finite element / finite volume techniques to model the fluid including high Péclet and Reynolds number conditions, effective finite element methods for the structure, and the proper coupling of the discretizations[43-47].

Since in engineering practice, frequently rather coarse meshes are much desirable for the fluid flow (and indeed may have to be used because the 3D fluid mesh would otherwise result into too many degrees of freedom), we have concentrated our development efforts on establishing finite element discretization schemes that are stable even when coarse meshes are used for very high (element) Péclet and Reynolds numbers and show optimal accuracy[48-51]. The basic approach in the development is to use flow-condition-based interpolations (FCBI) in the convective terms of the fluid and to use element control volumes (like in the finite volume method) in order to assure local mass and momentum conservation[51-56].

To present the basic approach of the FCBI schemes, consider the solution of the following Navier-Stokes equations:

Find the velocity $\mathbf{v}(\mathbf{x}) \in V$ and the pressure $p(\mathbf{x}) \in P$ in the domain Ω such that

$$\nabla \cdot \mathbf{v} = 0, \quad \mathbf{x} \in \Omega \tag{9}$$

$$\nabla \cdot (\mathbf{v}\mathbf{v} - \boldsymbol{\tau}) = \mathbf{0}, \quad \mathbf{x} \in \Omega \tag{10}$$

$$\boldsymbol{\tau} = \boldsymbol{\tau}(\mathbf{v}, p) = -p\mathbf{I} + \frac{1}{\mathrm{Re}}\left\{\nabla\mathbf{v} + (\nabla\mathbf{v})^{\mathrm{T}}\right\} \tag{11}$$

where we assume that the problem is well-posed in the Hilbert spaces V and P, $\boldsymbol{\tau}$ is the stress tensor and Re is the Reynolds number. Equations (9) and (10) are subject to appropriate boundary conditions.

In the FCBI approach, we use for the solution a Petrov-Galerkin variational formulation with subspaces U_h, V_h, and W_h of V, and P_h and Q_h of P. The formulation for the numerical solution is:

Find $\mathbf{u}_h \in U_h$, $\mathbf{v}_h \in V_h$ and $p_h \in P_h$ such that for all $w \in W_h$ and $q \in Q_h$

$$\int_\Omega w\nabla \cdot \left(\mathbf{u}_h\mathbf{v}_h - \boldsymbol{\tau}(\mathbf{u}_h, p_h)\right) \mathrm{d}\Omega = \mathbf{0} \tag{12}$$

$$\int_\Omega q\nabla \cdot \mathbf{u}_h \,\mathrm{d}\Omega = 0 \tag{13}$$

The trial functions in U_h and P_h are the usual functions of finite element interpolations for velocity and pressure, respectively. These are selected to satisfy the inf-sup condition of incompressible analysis[3]. An important point is that the trial functions in V_h are different from the functions U_h and are defined using the flow conditions in order to stabilize the convection term. The weight functions in the spaces W_h and Q_h are step functions, which enforce the local conservation of momentum and mass, respectively.

The resulting FCBI elements do not require a tuning of upwind parameters, of course satisfy the property of local and global mass and momentum conservation, pass the inf-sup test of incompressible analysis and appropriate patch tests on distorted meshes, and the interpolations can be used to establish a consistent Jacobian for the iterations in the incremental step by step solutions. The elements can be used with first and second-order accuracy.

We have developed FCBI elements for incompressible, slightly compressible and low-speed compressible flows. Each of these flow categories are abundantly used in engineering practice with various turbulence models. For some solutions, see Sections 7.7, 7.8, 7.9.

In addition, of course, there is the category of high-speed compressible flows and here we use the established and widely used Roe schemes.

These fluid flow models can all be coupled with structural models, in large deformations, with contact conditions, and with piezo-electric, thermal and electro-magnetic effects[3, 47, 57].

5. An Extension of the Finite Element Method – the Method of Finite Spheres

In engineering practice, usually a major effort is necessary in a finite element analysis to establish an adequate mesh of elements. To mesh a part of a motor car may take months, when the actual computer runs, including the plotting of the results, may take only a few days.

Much research has lately been focused on the development of meshfree discretization schemes for which no mesh is used for the interpolations and numerical integrations. However, while some techniques are referred to as meshless, they are actually not truly meshless methods because a mesh is still needed for the numerical integration.

In our research we have focused on the development of the method of finite spheres (MFS) which is one of the most effective truly meshless techniques available[58-63]. In this technique no mesh is used, and the complete domain needs simply be covered by the spheres (or disks in 2D solutions). The nodal unknowns are located at the nodes representing the centers of the spheres. Of course, the accuracy of the numerical solution (compared to the exact solution of the mathematical model) depends on the number of spheres used. The main advantage is that the spheres (disks) overlap (whereas the classical finite elements abut to each other) and hence highly distorted or sliver elements do not exist.

The MFS can simply be understood to be a specific finite element method and, of course, the elements (disks or spheres) of the MFS can be coupled to traditional finite elements [63].

So far we have tested the MFS in the analysis of solids, and have found that while the technique of course gives much more flexibility in placing nodal unknowns in the analysis, the cost of the numerical integrations is high. The traditional finite elements are much less costly in the required numerical integrations. However, further research should increase the effectiveness of the MFS and there is good potential in the technique, in particular when coupled to traditional finite elements[63,64]. Then the spheres would only be employed in those regions where highly distorted traditional finite elements would otherwise be used, due to the geometric complexities or the deformations that have taken place in a large deformation solution.

Of course, in principle, meshless methods can be employed also for the analysis of shells, for fluid flow simulations and FSI analyses, but the reliability issues, for example the basic phenomenon of ‘locking’, need be addressed as well, just like in the classical finite element discretizations[3, 59].

6. The Solution of the Algebraic Finite Element Equations

Consider that an appropriate finite element model has been established, for a static/steady-state or transient/dynamic analysis. The next step is to solve the governing finite element equilibrium equations.

For static (steady-state) and implicit dynamic (transient) solutions, direct sparse Gauss elimination methods are effective up to model sizes of about half a million equations[65]. For larger models, (iterative) algebraic multi-grid methods are much more effective. For ill-conditioned systems, combinations of these techniques can be used. Since in fluid flow analyses, the number of algebraic equations to be solved is usually large, the solutions are generally obtained using an algebraic multi-grid procedure, sweeping through the momentum, continuity, turbulence, mass transfer, etc., equations.

Considering transient structural response, a widely-used scheme of time integration is the Newmark method trapezoidal rule. However, if large deformations over long-time durations need be solved, then the scheme given in ref. 66 is much more effective. In this time integration, we consider each time step Δt to consist of two equal sub-steps, each solved implicitly. The first sub-step is solved using the trapezoidal rule with the usual assumptions

$$ {}^{t+\Delta t/2}\dot{\mathbf{U}} = {}^{t}\dot{\mathbf{U}} + \left[\frac{\Delta t}{4}\right]\left({}^{t}\ddot{\mathbf{U}} + {}^{t+\Delta t/2}\ddot{\mathbf{U}}\right) \tag{14} $$

$$ {}^{t+\Delta t/2}\mathbf{U} = {}^{t}\mathbf{U} + \left[\frac{\Delta t}{4}\right]\left({}^{t}\dot{\mathbf{U}} + {}^{t+\Delta t/2}\dot{\mathbf{U}}\right) \tag{15} $$

Then the second sub-step is solved using the three-point Euler backward method with the governing equations

$$ {}^{t+\Delta t}\dot{\mathbf{U}} = c_1\,{}^{t}\mathbf{U} + c_2\,{}^{t+\Delta t/2}\mathbf{U} + c_3\,{}^{t+\Delta t}\mathbf{U} \tag{16} $$

$$ {}^{t+\Delta t}\ddot{\mathbf{U}} = c_1\,{}^{t}\dot{\mathbf{U}} + c_2\,{}^{t+\Delta t/2}\dot{\mathbf{U}} + c_3\,{}^{t+\Delta t}\dot{\mathbf{U}} \tag{17} $$

where $c_1 = 1/\Delta t$, $c_2 = -4/\Delta t$, $c_3 = 3/\Delta t$.

This scheme is in essence a fully implicit second-order accurate Runge-Kutta method and requires per step about twice the computational effort as the trapezoidal rule. However, the accuracy per time step is significantly increased, and in particular the method remains stable when the trapezoidal rule fails to give the solution (see Section 7.12) This scheme is an option in ADINA for nonlinear dynamics, and is also employed in the ADINA FSI solutions. The Navier-Stokes and structural finite element equations are fully coupled in FSI solutions, and are either solved iteratively or directly using this time integration scheme.

For explicit dynamic solutions of structural response, the usual central difference method is effectively used. However, we employ this method with the same finite elements that are also used in static or implicit dynamic analyses (see Sections 7.10, 7.11). Hence the only difference to an implicit solution is that the time integrator is the central difference method and that the solution scheme can only be used with a lumped mass matrix. Of course, the time step needs to satisfy the stability limit[3] and therefore many time steps are frequently used. This is clearly the method of choice for fast transient analyses.

Since the element discretizations are the same in explicit and implicit dynamic solutions, restarts from explicit to implicit solutions, and vice versa, are directly possible. This option can be of use when an initial fast transient response is followed by a slow transient, almost static response. For example, in the analysis of metal forming problems, the initial response might be well calculated using the explicit solution scheme, and the spring-back might be best calculated using the implicit scheme.

A key point is that explicit dynamic solutions can be obtained very efficiently using distributed memory processing (DMP) environments. The scalability is excellent, so that a hundred processors, or more, can effectively be used. This possibility renders explicit time integration very attractive, so that in some cases even solutions that normally would be obtained using implicit integration, because a relatively long time scale governs the response, might be more effectively calculated with the explicit scheme.

7. Illustrative Solutions using ADINA

The objective in this section is to briefly give some solutions that demonstrate the capabilities mentioned above. The solutions have all been obtained when studying the finite element methods mentioned above or using the ADINA program for the analysis of solids, structures, fluids and multi-physics problems[67]. Additional solutions can be found on the ADINA web site[67] and, for example, in ref. 68.

7.1. EVALUATION OF PLATE BENDING ELEMENTS

In engineering practice mostly shell elements based on the Reissner-Mindlin kinematic assumption are employed for the analysis of plates, since plates are just a special case of a shell. If a shell element (used in a plate bending analysis) is based on the Reissner-Mindlin kinematic assumption and satisfies the inf-sup condition for plate bending, then it does not 'lock', meaning that the convergence curves for any (admissable) plate bending problem have the optimal slope and do not shift as the thickness of the plate decreases.

Figure 1 shows a square plate, fixed around its edges. The plate is loaded with uniform pressure. Uniform meshes of the MITC4 and MITC9 shell elements have been used and the solutions were calculated for decreasing element sizes (h) and decreasing thickness (t) of the plate.

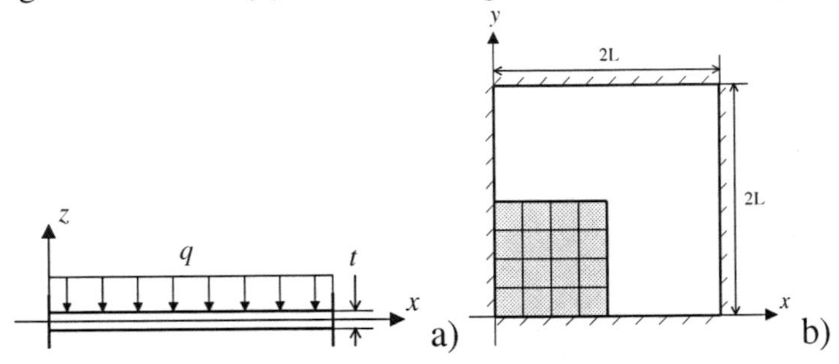

Figure 1. Clamped plate under uniform load: a) Side view of plate, b) Plan view showing also quarter of plate represented by shell elements.

Figure 2 shows the convergence curves obtained using the s-norm[20]. We notice that the calculated convergence curves have the optimal slope and the curves do not shift as the thickness of the plate decreases.

We show these results because it is important to test element formulations in this way – that is, the performance of a formulation in terms of convergence curves should be evaluated as the thickness of the plate decreases.

7.2. EVALUATION OF SHELL ELEMENTS

The thorough evaluation of shell elements is much more difficult than the evaluation of plate bending elements, because a shell element should perform well in membrane-dominated, bending-dominated and mixed problem solutions, and for any curvature shell. Hence, well-formulated linear static test problems need be selected from each of these categories, and also an appropriate norm need be used in the convergence calculations. Dynamic solutions are, of course, also of value but really only after the shell element has displayed excellent behavior in linear static solutions.

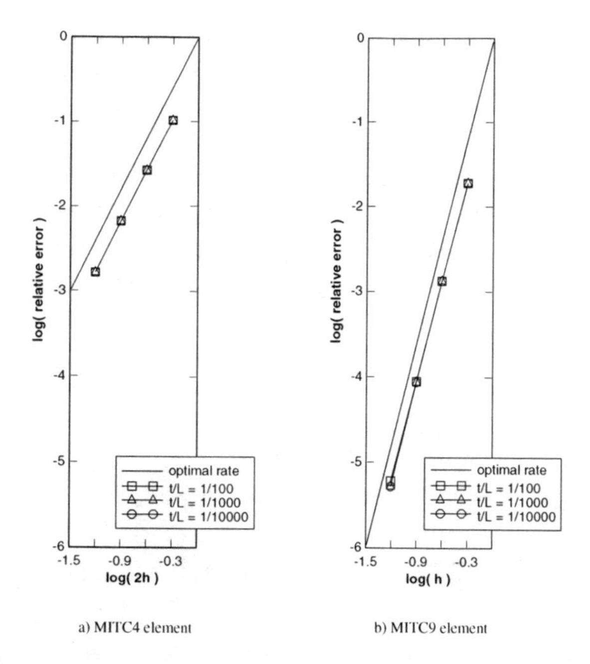

Figure 2. Clamped plate under uniform load: convergence curves using square of the s-norm.

The objective in the evaluation of a shell element must then be to identify how well the element performs in the solutions of these static analysis test problems, as the shell thickness decreases. The meshing used need to also take into account boundary layers. Ideally, in each of the test problems, the convergence curves have the optimal slope and do not shift as the shell thickness decreases.

While a number of different problems need to be solved for a full evaluation of an element, we demonstrate the task by considering the axisymmetric hyperboloid shell problem shown in Figure 3[12]. If the shell is clamped at both ends, the problem is membrane-dominated, and if the shell is free at both ends, the problem is bending-dominated. These are difficult test problems to solve because the shell is doubly-curved.

Figure 4 shows the convergence curves obtained when using the s-norm[20]. We see that the MITC4 element performs very well, independent of the shell thickness, and does not lock.

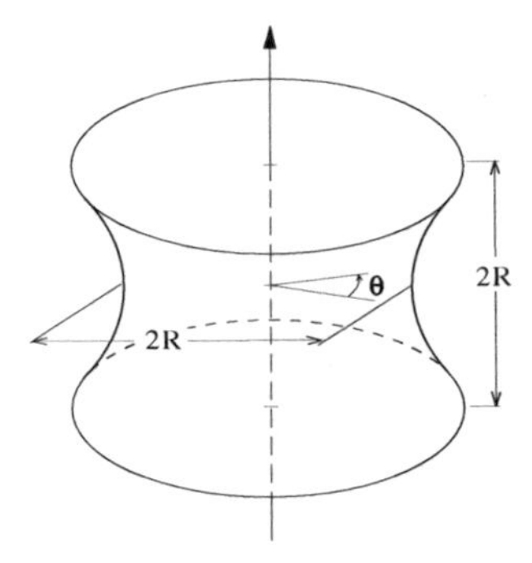

Figure 3. The axisymmetric hyperboloid shell problem. The shell is subjected to the pressure $p = p_0 \cos(2\theta)$; for the model solved using symmetry conditions see ref. 12

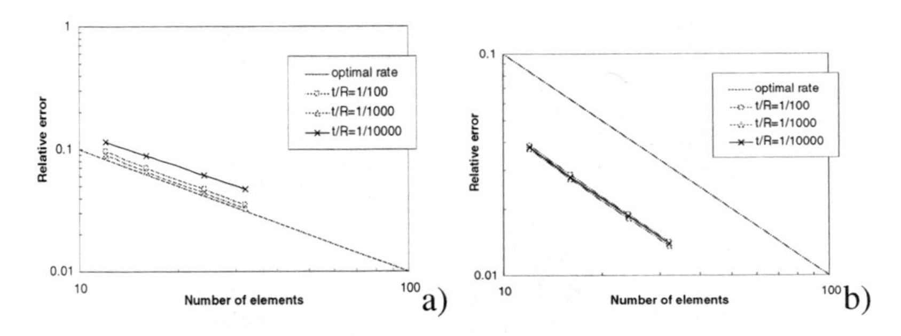

Figure 4. The axisymmetric hyperboloid shell problem: convergence curves using MITC4 element and s-norm[12, 20] a) Clamped ends, b) Free ends.

7.3. CRUSH ANALYSIS OF MOTOR CARS

A difficult nonlinear problem involving shell analysis capabilities, multiple contact conditions, large deformations with elasto-plasticity and fracture, is the roof crush analysis of motor cars. 'Crushing' a motor car roof is a slow physical process taking seconds in contrast to 'crashing' a motor car against another object, which is an event of milliseconds.

While explicit dynamic simulations are widely used to evaluate the crash behavior of motor cars, the crushing of an automobile roof is more appropriately – and more effectively – simulated using implicit dynamic, or static, solution techniques[65].

Figure 5 shows a typical finite element model of a car, and Figure 6 gives computed results for another car. The ADINA solution was obtained using an implicit dynamic (slow motion) analysis with the actual physical crushing speed of 0.022 mph. In the explicit solutions not using ADINA, quite different results were obtained for different crushing speeds, and it is seen that when the speed of crushing is close to the laboratory test speed, the explicit solution is unstable. In this explicit code, the elements used are not stable in static analysis (they do not satisfy the conditions mentioned in Sections 3.2 and 7.2).

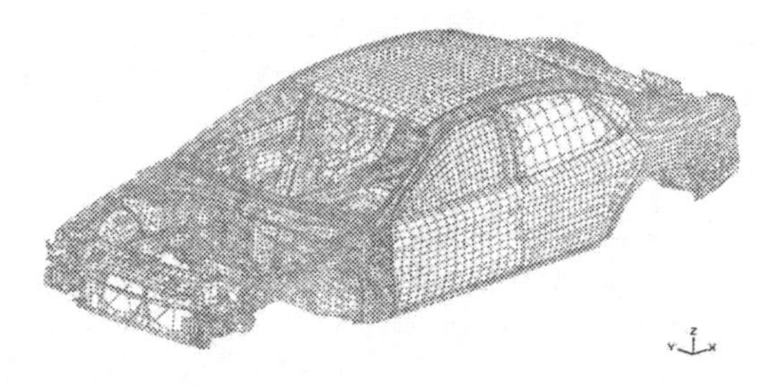

Figure 5. Finite element model of a car.

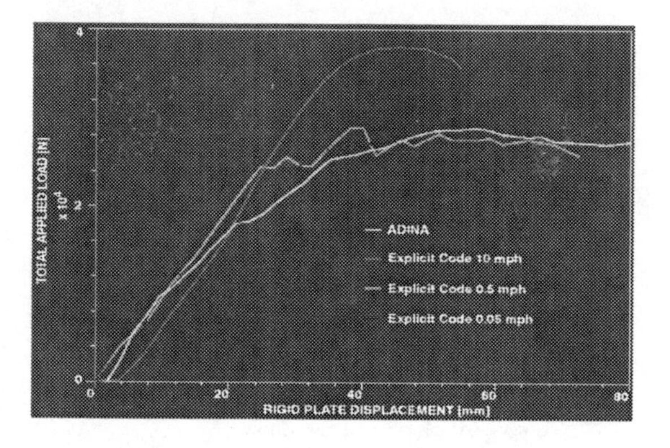

Figure 6. Results of a roof crush simulation for a car.

7.4. GLUING OF DISSIMILAR MESHES FOR MULTI-SCALE ANALYSIS

In a multi-scale analysis, it can be effective to create a fine mesh for a certain small region and then have successively coarser meshes away from that region. Then it may also be effective to mesh some regions with free-form tetrahedral elements while other regions are meshed with brick elements. It is often a challenge to connect these regions with dissimilar meshes together.

However, a powerful gluing feature makes it easy to connect regions with dissimilar meshes. This feature is illustrated in the simple 2D and 3D examples shown in Figures 7 and 8. The theory for this analysis feature is given in ref. 28.

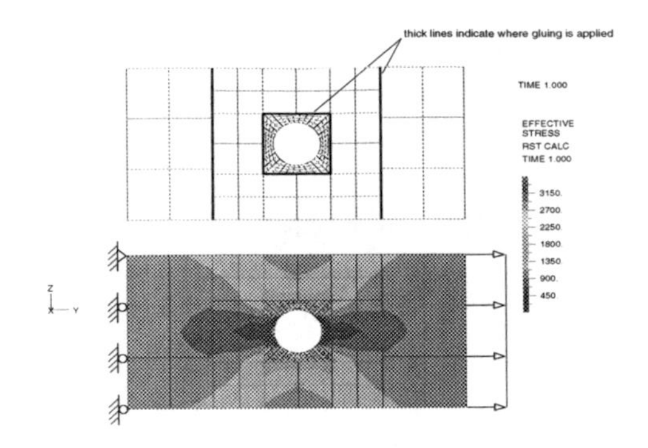

Figure 7. Gluing of dissimilar meshes, 2D.

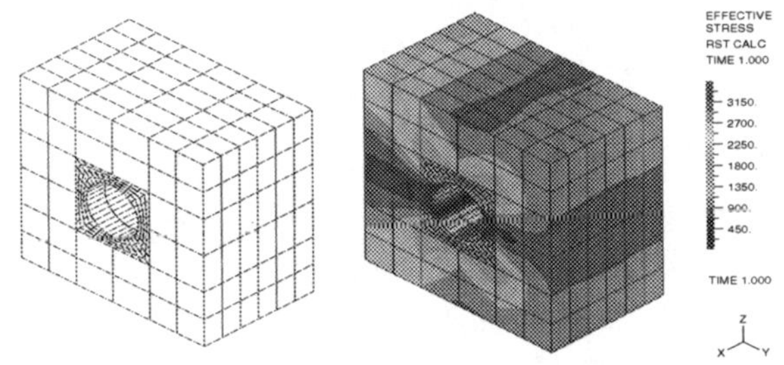

Figure 8. Gluing of dissimilar meshes, 3D.

7.5. USING AN ERROR ESTIMATOR

In this example, we demonstrate the use of the error estimation based on refs. 34, 35 and available in ADINA, see also ref. 31. Figures 9 and 10 show results obtained in the study of a cantilever structure with the left end fixed, in linear, nonlinear and FSI solutions. Of particular interest is the stress solution around the elliptical hole in the middle of the cantilever. In the linear analysis, the structure is simply subjected to the uniform pressure p. In the nonlinear analysis, this pressure is increased to $6p$ and causes large deformations. Finally, in the fully coupled FSI solution, the steady-state force effects of a Navier-Stokes fluid flow around the structure (of magnitude about $2p$) are considered in addition to the pressure of $6p$.

In each case, a coarse mesh solution (for the mesh used see Figure 9), the error estimation for the longitudinal (bending) stress using the coarse mesh and the 'exact' error have been calculated. The exact errors have been obtained by comparing the coarse mesh solutions with very fine mesh solutions (that in practice would of course not be computed). The error estimation is seen in these cases to be conservative and not far from the exact error.

While this error estimation can be used for general stress and thermal analyses of solids and shells, including contact conditions and FSI, a word of caution is necessary: As with all existing practical error estimation techniques, there is no proof that the error estimate is always accurate and a conservative prediction. Hence the presently available error estimation procedures are primarily useful to estimate whether the mesh is fine enough, and if a refinement is necessary where such refinement should be concentrated.

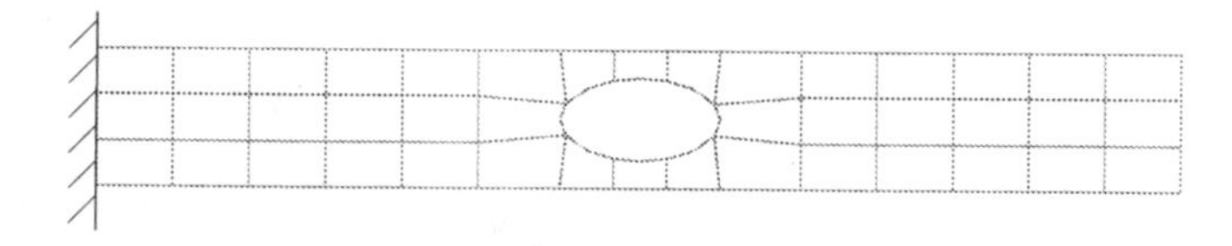

Figure 9. Coarse mesh of cantilever, used in all solutions, with boundary conditions; 9-node elements.

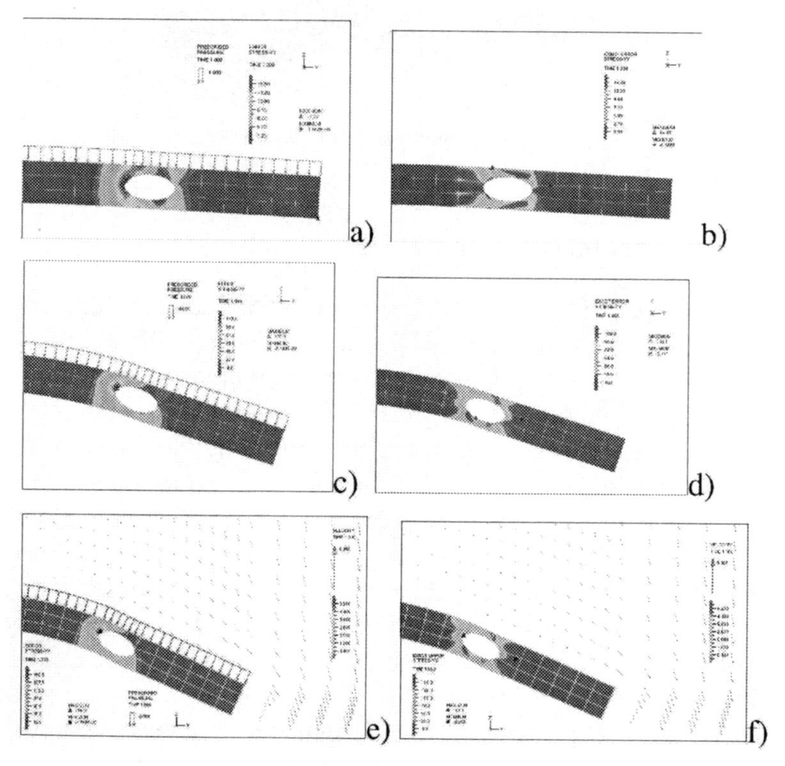

Figure 10. Results in analysis of cantilever structure: a) Linear analysis, estimated error in region of interest shown, b) Linear analysis, exact error in region of interest shown, c) Nonlinear analysis, estimated error in region of interest shown, d) Nonlinear analysis, exact error in region of interest shown, e) FSI analysis, estimated error in region of interest shown, f) FSI analysis, exact error in region of interest shown.

7.6. FREQUENCY SOLUTION OF FLUID STRUCTURE SYSTEM

In many dynamic analyses of fully coupled fluid structure systems, the fluid can be assumed to be an acoustic fluid. In such cases, it can be effective to perform a frequency solution and mode superposition analysis for the dynamic response.

The major expense is then in solving for the frequencies and mode shapes, which requires the solution of the quadratic eigenvalue problem[3]

$$\left(\mathbf{K}-\lambda\mathbf{C}-\lambda^2\mathbf{M}\right)\boldsymbol{\varphi}=\mathbf{0} \tag{18}$$

The Lanczos method can be used efficiently for the solution of this eigenvalue problem.

Figures 11 and 12 show two vibration modes of a reactor vessel with piping analyzed using ADINA. The model has about 600,000 degrees of freedom and the 100 lowest frequencies and corresponding mode shapes of the quadratic eigenvalue problem were computed in about half an hour on an IBM Linux machine with four processors.

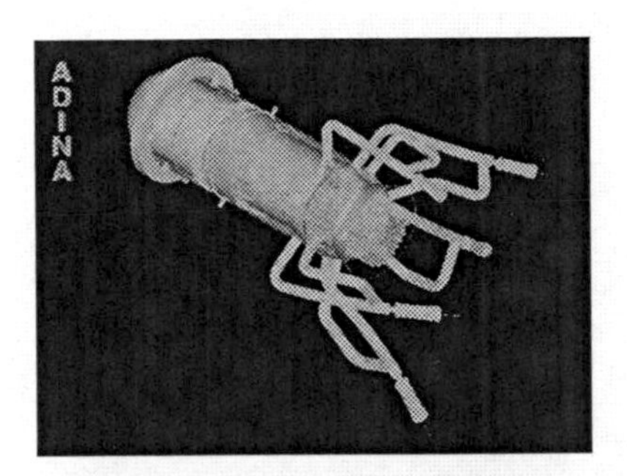

Figure 11. Low vibration mode of a reactor vessel, frequency 1.649 Hz.

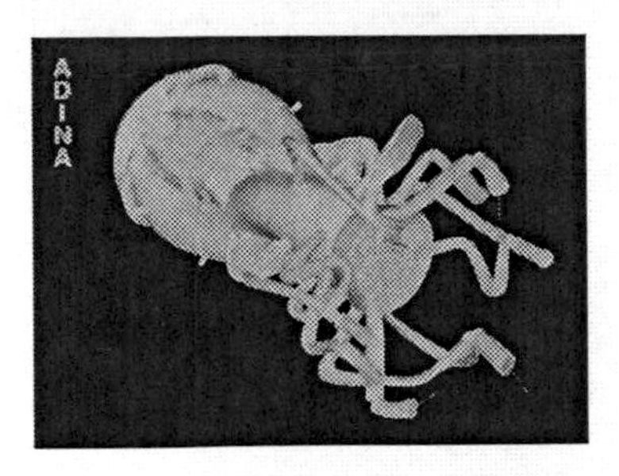

Figure 12. *Higher vibration mode of a reactor vessel, frequency 39.98 Hz.*

7.7 SOLUTION OF LARGE FINITE ELEMENT FLUID SYSTEMS

Today's CFD and FSI solutions require generally the solution of large finite element systems that involve millions of degrees of freedom.

Here we give two examples of solutions of large fluid flow models using an efficient algebraic multi-grid solver. Figure 13 shows the problems solved and Figure 14 gives the solution times (clock times) and memory used. It is important to note that the solution times and the memory used increase approximately linearly with the number of degrees of freedom. Also, in both problem solutions, 8 million equations are solved in less than 2 hours on the single processor PC.

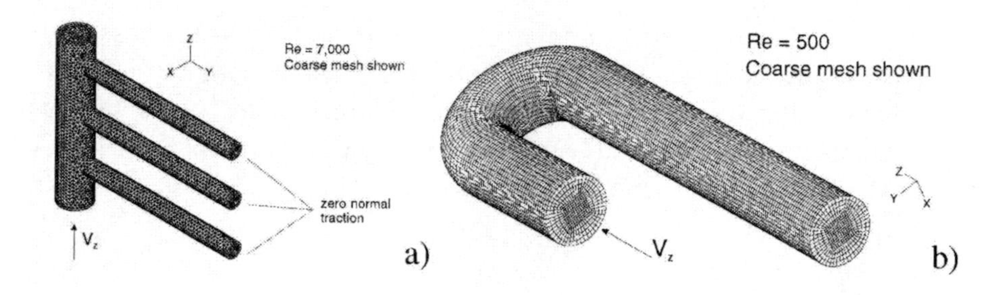

Figure 13. Models to demonstrate the solution of large finite element fluid systems: a) Manifold problem, b) Turnaround duct problem.

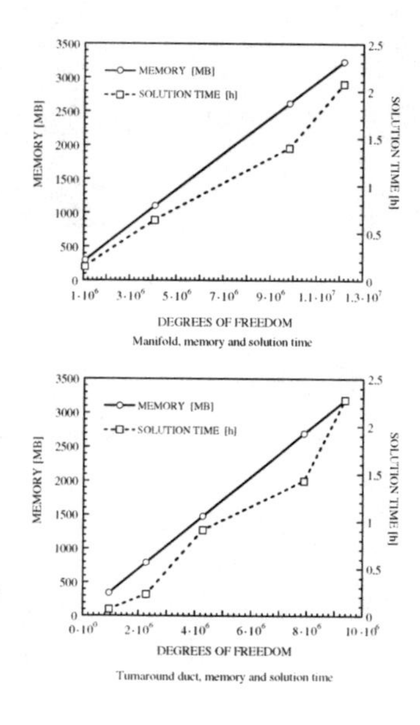

Figure 14. Solution times (clock times) and memory usage for the manifold and turnaround duct problems; single processor PC used, 3.2 GHz.

7.8. ANALYSIS OF OIL TANK

There is much interest in simulating accurately the sloshing of fluids in large diameter tanks. The fluid need be modeled as a Navier-Stokes fluid and an Arbitrary-Lagrangian-Eulerian formulation is effectively used for the fully coupled FSI solution.

Figure 15 shows a typical flexible tank filled with oil. A pontoon floats on the oil surface to prevent the oil from contacting the air. The tank is subjected to a horizontal ground motion of magnitude 1 meter and frequency 0.125 Hz.

Figure 15 also gives the ADINA model used for the analysis; here the fluid mesh (8 – node FCBI elements) and the structural meshes of the tank and pontoon (MITC4 shell elements) are shown separately.

Figure 16 shows some results of the analysis, namely snap-shots of the oil sloshing in the tank and stresses in the tank wall.

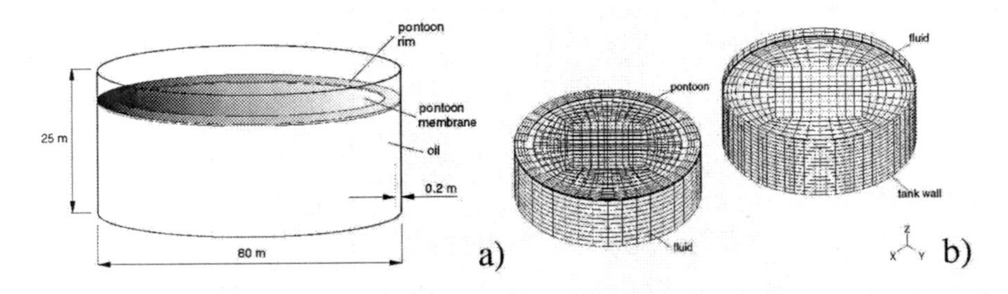

Figure 15. Analysis of oil tank: a) Schematic of tank, b) The finite element model of oil tank with pontoon.

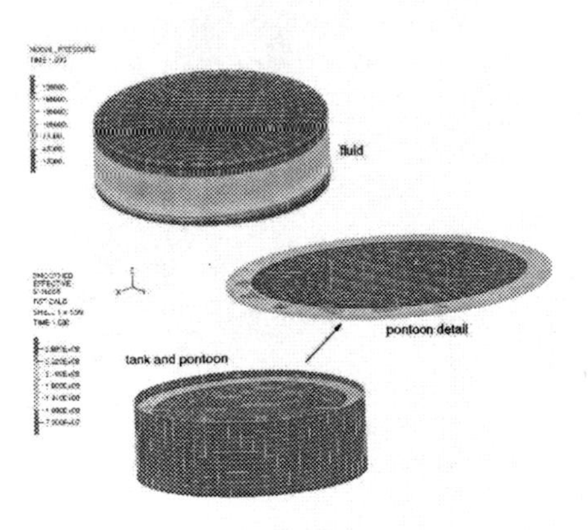

Figure 16. A snap-shot of the response of the tank.

7.9. ANALYSIS OF TURBULENT FLOW IN AN EXHAUST MANIFOLD

The simulation of turbulent flow in exhaust manifolds is of much interest. Here we consider a Volvo Penta 6-cylinder diesel engine. ADINA was used with the shear stress transport (SST) turbulence model to solve for the turbulent flow in the manifold.

Figure 17 shows the manifold considered and indicates the calculated fluid flow. Figure 18 gives a detail of the manifold with the calculated pressure contours. The Reynolds number at the inlets of the manifold is approximately 13,000, and the model was solved with about 3½ million unknowns. With the fluid flow also a thermo-mechanical analysis of the structure is directly possible.

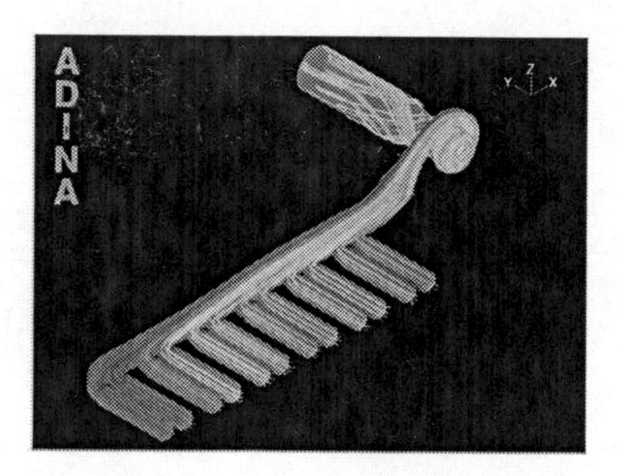

Figure 17. Exhaust manifold, also showing particle traces.

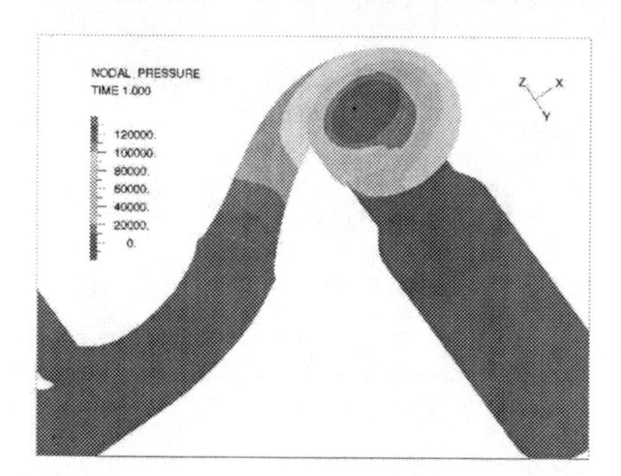

Figure 18. Exhaust manifold: pressure contours.

7.9. CRASH ANALYSIS OF CYLINDERS

For very fast transient analyses, the use of explicit time integration can be much more effective than using implicit time integration. Figures 19 and 20 show the analysis of two cylinders in impact. The response has been calculated using the central difference method of explicit time integration with MITC4 shell elements used to model the cylinders. This type of analysis involving crash, contact, large deformations with fracture within milliseconds is clearly most effectively carried out using explicit time integration.

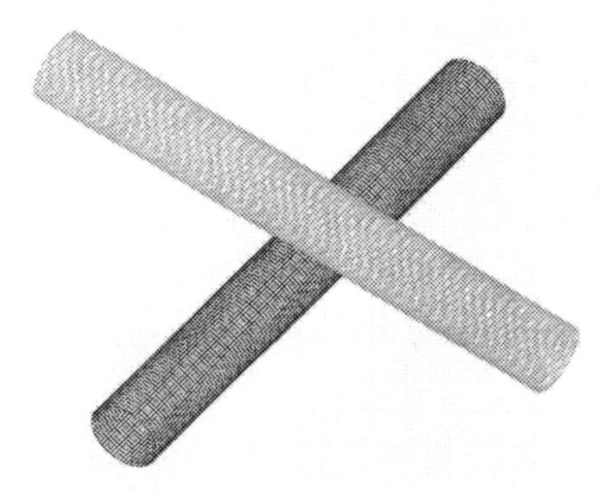

Figure 19. Crash analysis of cylinders. The top cylinder crashes onto the bottom cylinder.

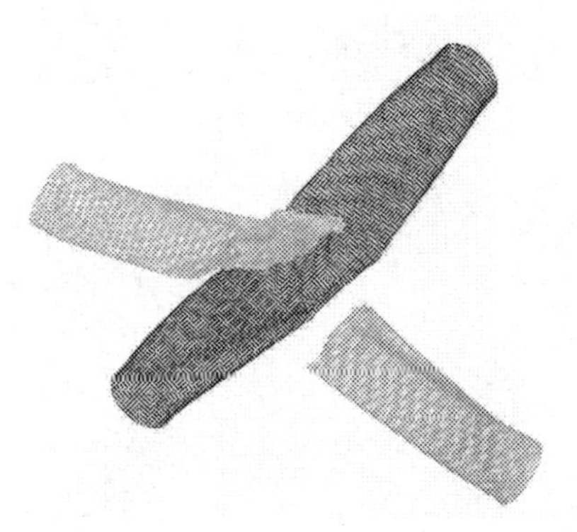

Figure 20. Response of cylinders at time = 0.040 s.

7.10. EXPLICIT AND IMPLICIT SOLUTIONS OF A METAL FORMING PROBLEM

The forming of the S-rail shown in Figure 21 is a widely-used verification problem of metal forming procedures. This problem can be solved using explicit or implicit time integration in a finite element program.

7.11. EXPLICIT AND IMPLICIT SOLUTIONS OF A METAL FORMING PROBLEM

The forming of the S-rail shown in Figure 21 is a widely-used verification problem of metal forming procedures. This problem can be solved using explicit or implicit time integration in a finite element program.

Figure 22 shows the results obtained using ADINA in implicit integration (the trapezoidal rule is used) and in explicit integration (the central difference method is used). The same mesh of MITC4 shell elements was employed. As seen, the final deformations and stresses are very similar using the two analysis techniques. Hence the only reason for using one or the other solution procedure in ADINA is that one technique may be computationally much more effective. In this case, the solution times are quite comparable and hence either the implicit or the explicit time integration might be used.

In this analysis ADINA was used within the NX Nastran environment and the plots were obtained using the Femap program.

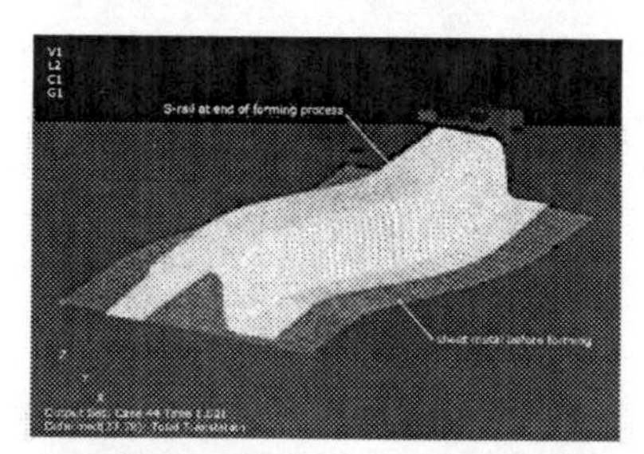

Figure 21 Forming of an S-rail.

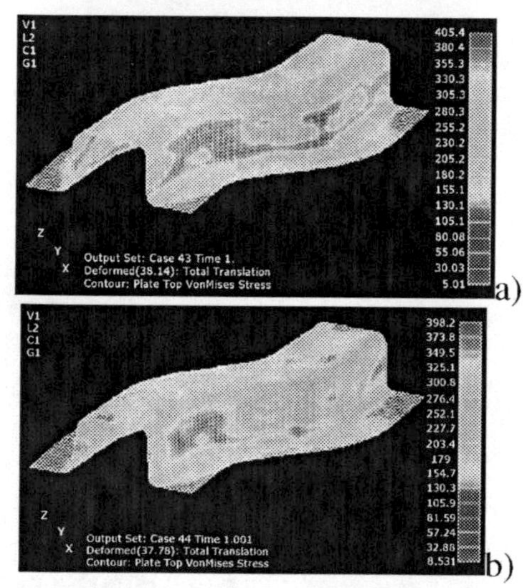

Figure 22. Results obtained in forming of the S-rail: a) Results using implicit integration, b) Results using explicit integration.

7.12 THE NONLINEAR DYNAMIC LONG DURATION SOLUTION OF A ROTATING PLATE

The Newmark method trapezoidal rule of time integration can become unstable in large deformation, long time duration analyses, and in such cases the time integration scheme given in equations (14) to (17) can be much more effective.

Figure 23 shows a plate, free to rotate, which is subjected to a twisting moment. Once the moment is removed, the plate should continue to rotate at constant angular speed.

Figure 24 gives the kinetic energy of the plate as a function of time. If the trapezoidal rule of time integration is used, the solution becomes unstable after a few rotations, although Newton-Raphson equilibrium iterations to a tight convergence tolerance are performed in each time step. On the other hand, the scheme of equations (14) to (17) with a much larger time step gives a stable and accurate solution.

More experiences with this time integration scheme are given in ref. 66.

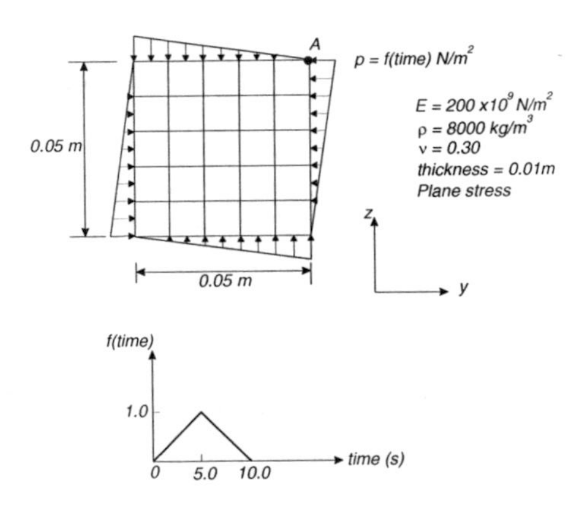

Figure 23. Rotating plate problem

8. Concluding Remarks

The objective in this paper was to present some developments regarding the analysis of structures when subjected to severe loading conditions. The required simulations will frequently result in highly nonlinear analyses, involving multi-physics conditions with fluid-structure interactions and thermal effects.

The focus in this paper was on the need to use reliable finite element methods for such simulations. Since test data for the envisioned scenarios will be scarce, it is important to have as high a confidence as possible in the computed results without much experimental verification. Such confidence is, however, only possible if reliable finite element methods are used. Furthermore, it can be effective to employ a single program system to perform the analyses, in hierarchical modeling involving linear to highly nonlinear solutions.

In this paper various solutions obtained with ADINA have been presented that show the applicability and versatility of the program in the study of linear and highly nonlinear problems, including multi-physics conditions.

While many different and complex problems can at present be solved, there are still major challenges in developing more effective and more comprehensive analysis procedures. Eight key challenges have been summarized in ref. 1, see the Preface of the 2005 Volume.

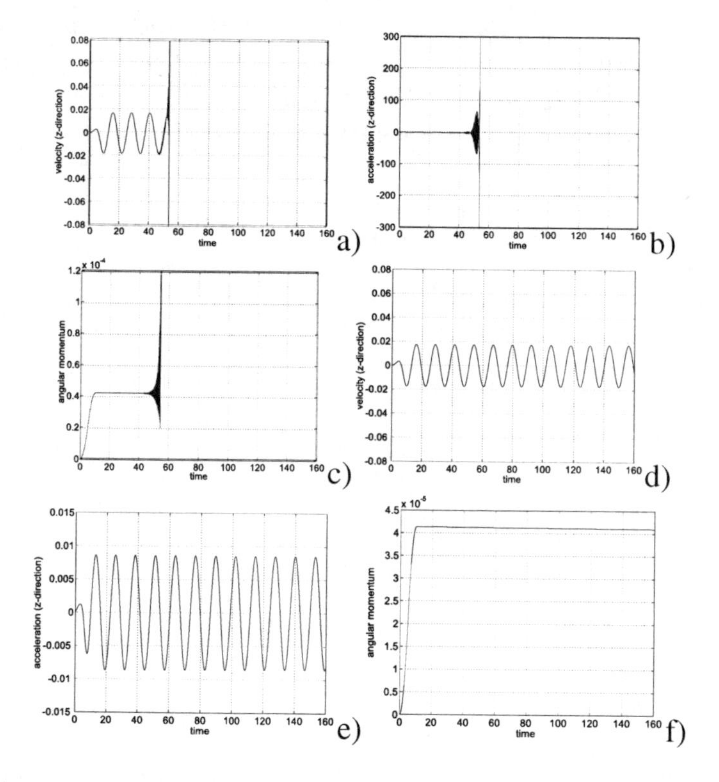

Figure 24. Results obtained in the analysis of the rotating plate problem: a) Velocity at point A using the trapezoidal rule; $\Delta t = 0.02$ s, b) Acceleration at point A using the trapezoidal rule; $\Delta t = 0.02$ s, c) Angular momentum using the trapezoidal rule; $\Delta t = 0.02$ s, d) Velocity at point A using the scheme of Equations (14) to (17); $\Delta t = 0.4$ s, e) Acceleration at point A using the scheme of Equations (14) to (17); $\Delta t = 0.4$ s, f) Angular momentum using the scheme of Equations (14) to (17); $\Delta t = 0.4$ s

References

1. K. J. Bathe, (ed.), Computational Fluid and Solid Mechanics, (Elsevier, 2001); *Computational Fluid and Solid Mechanics 2003,* (Elsevier, 2003); *Computational Fluid and Solid Mechanics 2005*, (Elsevier, 2005). Proceedings of the First to Third MIT Conferences on Computational Fluid and Solid Mechanics.
2. O.C. Zienkiewicz and R.L. Taylor, *The Finite Element Method*, (Butterworth-Heinemann, 2005).
3. K. J. Bathe, *Finite Element Procedures*, (Prentice Hall, 1996).
4. M. L. Bucalem and K. J. Bathe, *The Mechanics of Solids and Structures – Hierarchical Modeling and the Finite Element Solution*, (Springer, to appear).
5 K. J. Bathe, The inf-sup condition and its evaluation for mixed finite element methods, *Computers & Structures*, **79**:243-252, 971, (2001).
6. F. Brezzi and M. Fortin, *Mixed and Hybrid Finite Element Methods*, Springer Verlag, New York, 1991.
7. D. Pantuso and K. J. Bathe, A four-node quadrilateral mixed-interpolated element for solids and fluids, *Mathematical Models and Methods in Applied Sciences*, **5**(8):1113-1128, (1995).
8. D. Pantuso and K. J. Bathe, On the stability of mixed finite elements in large strain analysis of incompressible solids, *Finite Elements in Analysis and Design*, **28**:83-104, (1997).
9. A. Iosilevich, K. J. Bathe and F. Brezzi, Numerical inf-sup analysis of MITC plate bending elements, *Proceedings, 1996 American Mathematical Society Seminar on Plates and Shells, Québec, Canada*, (1996).
10. A. Iosilevich, K. J. Bathe and F. Brezzi, On evaluating the inf-sup condition for plate bending elements, *Int. Journal for Numerical Methods in Engineering*, **40**:3639-3663, (1997).
11. K. J. Bathe, A. Iosilevich and D. Chapelle, An inf-sup test for shell finite elements, *Computers & Structures*, **75**: 439-456, (2000).
12. D. Chapelle and K.J. Bathe, *The Finite Element Analysis of Shells – Fundamentals*, (Springer, 2003).
13. D. Chapelle and K. J. Bathe, Fundamental considerations for the finite element analysis of shell structures, *Computers & Structures*, **66**(1):19-36, (1998).
14. K. J. Bathe, A. Iosilevich and D. Chapelle, An evaluation of the MITC shell elements, *Computers & Structures*, **75**: 1-30, (2000).
15. M. L. Bucalem and K. J. Bathe, Finite element analysis of shell structures, *Archives of Computational Methods in Engineering*, **4**:3-61, (1997).
16. K. J. Bathe, P. S. Lee and J. F. Hiller, Towards improving the MITC9 shell element, *Computers & Structures*, **81**: 477-489, (2003).
17. P. S. Lee and K. J. Bathe, Development of MITC isotropic triangular shell finite elements, *Computers & Structures*, **82**:945-962, (2004).
18. P. S. Lee and K. J. Bathe, On the asymptotic behavior of shell structures and the evaluation in finite element solutions, *Computers & Structures*, **80**:235-255, (2002).
19. K. J. Bathe, D. Chapelle and P. S. Lee, A shell problem 'highly-sensitive' to thickness changes, *Int. Journal for Numerical Methods in Engineering*, **57**:1039-1052, (2003).
20. J. F. Hiller and K. J. Bathe, Measuring convergence of mixed finite element discretizations: an application to shell structures, *Computers & Structures*, **81**:639-654, (2003).
21. D. Chapelle and K. J. Bathe, The mathematical shell model underlying general shell elements, *Int. J. for Numerical Methods in Engineering*, **48**:289-313, (2000).

P. S. Lee and K. J. Bathe, Insight into finite element shell discretizations by use of the basic shell mathematical model, *Computers & Structures*, **83**:69-90, (2005).
23. D. Chapelle, A. Ferent and K. J. Bathe, 3D-shell elements and their underlying mathematical model, *Mathematical Models & Methods in Applied Sciences*, **14**:105-142, (2004).
24. M. Kojic and K.J. Bathe, *Inelastic Analysis of Solids and Structures*, (Springer, 2005).
25. G. Gabriel and K. J. Bathe, Some computational issues in large strain elasto-plastic analysis, *Computers & Structures*, **56**(2/3):249-267, (1995).
26. K. J. Bathe and F. J. Montans, On modeling mixed hardening in computational plasticity, *Computers & Structures*, **82**:535-539, (2004).
27. F. J. Montans and K. J. Bathe, Computational issues in large strain elasto-plasticity: an algorithm for mixed hardening and plastic spin, *Int. J. for Numerical Methods in Eng.*, **63**:159-196, (2005).
28. N. Elabbasi and K. J. Bathe, Stability and patch test performance of contact discretizations and a new solution algorithm, *Computers & Structures*, **79**:1473-1486, (2001).
29. N. Elabbasi, J. W. Hong and K. J. Bathe, The reliable solution of contact problems in engineering design, *Int. J. of Mechanics and Materials in Design*, **1**:3-16, (2004).
30. K. J. Bathe and F. Brezzi, Stability of finite element mixed interpolations for contact problems, *Proceedings della Accademia Nazionale dei Lincei, s. 9*, **12**:159-166, (2001).
31 T. Grätsch and K. J. Bathe, *A posteriori* error estimation techniques in practical finite element analysis, *Computers & Structures*, **83**:235-265, (2005).
32 M. Ainsworth and J.T. Oden, A posteriori Error Estimation in Finite Element Analysis, J. Wiley & Sons, 2000.
33. J. F. Hiller and K. J. Bathe, Higher-order-accuracy points in isoparametric finite element analysis and an application to error assessment, *Computers & Structures*, **79**:1275-1285, (2001).
34. T. Sussman and K. J. Bathe, Studies of Finite Element Procedures – On Mesh Selection, *Computers & Structures*, Vol. 21, pp. 257-264, 1985.
35. T. Sussman and K. J. Bathe, Studies of Finite Element Procedures – Stress Band Plots and the Evaluation of Finite Element Meshes, *Engineering Computations*, Vol. 3, pp. 178-191, 1986
36. T. Grätsch and K. J. Bathe, Influence functions and goal-oriented error estimation for finite element analysis of shell structures, *Int. J. for Numerical Methods in Eng.*, **63**:709-736, (2005).
37. T. Grätsch and K. J. Bathe, Goal-oriented error estimation in the analysis of fluid flows with structural interactions, *Comp. Meth. in Applied Mech. and Eng.*, in press.
38. K. J. Bathe, C. Nitikitpaiboon and X. Wang, A mixed displacement-based finite element formulation for acoustic fluid-structure interaction, *Computers & Structures*, **56**:(2/3), 225-237, (1995).
39. X. Wang and K. J. Bathe, On mixed elements for acoustic fluid-structure interactions, *Mathematical Models & Methods in Applied Sciences*, **7**(3):329-343, (1997).
40. X. Wang and K. J. Bathe, Displacement/pressure based mixed finite element formulations for acoustic fluid-structure interaction problems, *Int. Journal for Numerical Methods in Engineering*, **40**:2001-2017, (1997).
41. W. Bao, X. Wang, and K. J. Bathe, On the inf-sup condition of mixed finite element formulations for acoustic fluids, *Mathematical Models & Methods in Applied Sciences*, **11**(5):883-901, (2001).
42. T. Sussman and J. Sundqvist, Fluid-structure interaction analysis with a subsonic potential-based fluid formulation, *Computers & Structures*, **81**:949-962, (2003).

43. K. J. Bathe, Simulation of structural and fluid flow response in engineering practice, *Computer Modeling and Simulation in Engineering*, **1**:47-77, (1996).
44. K. J. Bathe, H. Zhang and S. Ji, Finite element analysis of fluid flows fully coupled with structural interactions, *Computers & Structures*, **72**:1-16, (1999).
45. S. Rugonyi and K. J. Bathe, On the finite element analysis of fluid flows fully coupled with structural interactions, *Computer Modeling in Engineering & Sciences*, **2**:195-212, (2001).
46. S. Rugonyi and K. J. Bathe, An evaluation of the Lyapunov characteristic exponent of chaotic continuous systems, *Int. Journal for Numerical Methods in Engineering*, **56**:145-163, (2003).
47. K. J. Bathe and H. Zhang, Finite element developments for general fluid flows with structural interactions, *Int. Journal for Numerical Methods in Engineering*, **60**:213-232, (2004).
48. K. J. Bathe, D. Hendriana, F. Brezzi and G. Sangalli, Inf-sup testing of upwind methods, *Int. J. for Numerical Methods in Engineering*, **48**:745-760, (2000).
49. Y. Guo and K. J. Bathe, A numerical study of a natural convection flow in a cavity, *Int. J. for Numerical Methods in Fluids*, **40**:1045-1057, (2002).
50. D. Hendriana and K. J. Bathe, On upwind methods for parabolic finite elements in incompressible flows, *Int. J. for Numerical Methods in Engineering*, **47**:317-340, (2000).
51. K. J. Bathe and H. Zhang, A Flow-Condition-Based Interpolation finite element procedure for incompressible fluid flows, *Computers & Structures*, **80**:1267-1277, (2002).
52. K. J. Bathe and J. P. Pontaza, A Flow-Condition-Based Interpolation mixed finite element procedure for high Reynolds number fluid flows, *Mathematical Models and Methods in Applied Sciences*, **12**(4): 525-539, (2002).
53. H. Kohno and K. J. Bathe, Insight into the Flow-Condition-Based Interpolation finite element approach: Solution of steady-state advection-diffusion problems, *Int. J. for Numerical Methods in Eng.*, **63**:197-217, (2005).
54. H. Kohno and K. J. Bathe, A Flow-Condition-Based Interpolation finite element procedure for triangular grids, *Int. J. Num. Meth. in Fluids*, **49**:849-875, (2005).
55. H. Kohno and K. J. Bathe, A 9-node quadrilateral FCBI element for incompressible Navier-Stokes flows, *Comm. in Num. Methods in Eng.*, (in press).
56. B. Banijamali and K. J. Bathe, The CIP Method embedded in finite element discretizations of incompressible fluid flows, submitted.
57. P. Gaudenzi and K. J. Bathe, An iterative finite element procedure for the analysis of piezoelectric continua, *J. of Intelligent Material Systems and Structures*, **6**(2): 266-273, (1995).
58. S. De and K. J. Bathe, The method of finite spheres, *Computational Mechanics*, **25**:329-345, (2000).
59. S. De and K. J. Bathe, Displacement/pressure mixed interpolation in the method of finite spheres, *Int. J. for Numerical Methods in Engineering*, **51**:275-292, (2001).
60. S. De and K. J. Bathe, Towards an efficient meshless computational technique: the method of finite spheres, *Engineering Computations*, **18**:170-192, (2001).
61. S. De and K. J. Bathe, The method of finite spheres with improved numerical integration, *Computers & Structures*, **79**:2183-2196, (2001).
62. S. De, J. W. Hong and K. J. Bathe, On the method of finite spheres in applications: Towards the use with ADINA and in a surgical simulator, *Computational Mechanics*, **31**:27-37, (2003).

63. J. W. Hong and K. J. Bathe, Coupling and enrichment schemes for finite element and finite sphere discretizations, *Computers & Structures*, **83**:1386-1395, (2005).
64. M. Macri and S. De, Towards an automatic discretization scheme for the method of finite spheres and its coupling with the finite element method, *Computers & Structures*, **83**:1429-1447, (2005).
65. K. J. Bathe, J. Walczak, O. Guillermin, P. A. Bouzinov and H. Chen, Advances in crush analysis, *Computers & Structures*, **72**:31-47, (1999).
K. J. Bathe and M. M. I. Baig, On a composite implicit time integration procedure for nonlinear dynamics, *Computers & Structures*, **83**:2513 – 2534, (2005).
67. K. J. Bathe, ADINA System, *Encyclopaedia of Mathematics*, **11**:33-35, (1997); http://www.adina.com.
68 J. Tedesco and J. Walczak, Guest Editors of Special Issue, *Computers and Structures*, **81**:455-1109,(2003).

Part II: PROBABILITY ASPECTS, DESIGN AND UNCERTAINTY

QUANTIFYING UNCERTAINTY: MODERN COMPUTATIONAL REPRESENTATION OF PROBABILITY AND APPLICATIONS

Hermann G. Matthies
Institute of Scientific Computing
Technische Universität Braunschweig
Brunswick, Germany
(`wire@tu-bs.de, http://www.wire.tu-bs.de`*)*

Abstract. Uncertainty estimation arises at least implicitly in any kind o f modelling of the real world. A recent development is to try and actually quantify the uncertainty in probabilistic terms. Here the emphasis is on uncertain systems, where the randomness is assumed spatial. Traditional computational approaches usually use some form of perturbation or Monte Carlo simulation. This is contrasted here with more recent methods based on stochastic Galerkin approximations. Also some approaches to an adaptive uncertainty quantification are pointed out.

Key words: uncertainty quantification, spatially stochastic systems, stochastic elliptic partial differential equations, stochastic Galerkin methods, Karhunen-Loève expansion, Wiener's polynomial chaos, white noise analysis, sparse Smolyak quadrature, Monte Carlo methods, stochastic finite elements

1. Introduction

Quantification of uncertainty means to be in some way able to attach a measure to something which may be very vague. This will be done here by expressing uncertainty through probabilistic models. Probabilistic or stochastic mechanics—e.g. [76]—deals with mechanical systems, which are either subject to random external influences—a random or uncertain environment, or are themselves uncertain, or both, cf. e.g. the reports [68, 91, 92, 101, 48]. Not considered will be probabilistic methods to solve deterministic problems in mechanics, an area often just labelled as "Monte Carlo methods", e.g. [89]. That term will be used here as well, although in a more general meaning.

One may further make a distinction as to whether the mechanical system is modelled as finite-dimensional—i.e. its time-evolution may be de-

A. Ibrahimbegovic and I. Kozar (eds.),Extreme Man-Made and Natural Hazards in Dynamics of Structures, 105–130.

scribed by an ordinary differential equation (ODE) in time—or infinite-dimensional, e.g. a partial differential equation (PDE) with spatial differential operators is used as a mathematical model. In the second case the numerical treatment involves also a discretisation of the spatial characteristics of the mechanical system. Often the specific procedure is not important—but in case something specific has to be taken, it is assumed here that the mechanical system has been discretised by finite elements, e.g. [113]. This then leads to an area often referred to as "stochastic finite elements" (SFEM), e.g. [56, 68, 101], or "probabilistic/random finite elements", e.g. [61]. This is the area which will be dealt with in more detail in the following, and the reports [68, 91, 92, 101, 48, 72] contain more comprehensive surveys.

As the field is very large, this lecture tries to cover mainly new developments not yet contained in text books, and necessarily reflects to some extent the preferences and work of the author. This manuscript builds on the material contained in [72], and only gives a very brief recapitulation necessary to introduce basic notions and some of the notation.

2. Uncertainty in the Model and Input

In making scientific predictions, uncertainty enters in many guises and at different places [68]. One of the basic problems facing anyone doing a scientific prediction is coming to terms with the uncertainty inherent in the predicition. Here we go even further in our wish not only to express this uncertainty somehow verbally, but also to be able to quantify it.

A model which we make from the physical world usually first—at least implicitly—defines the part to be considered, which is often termed as the "system". Everything else is the "rest of the world", and interacts with the system in form of exterior excitation, action, loading etc., different terms being used in different areas of science. The assumption here is basically that the interaction with the rest of the world is essentially only one-way, i.e. that the re-action of the system on the rest of the world can be neglected in this context. Obviously there is uncertainty in this step, but as this is on a pre-mathematical level, it is usually difficult to be quantified.

The next step consist of making a mathematical model of those features of the system and the outside action which we deem important in the chosen context. Obviously again here is uncertainty involved, and again it is difficult to be quantified. But let us remark that at this stage it is possible to choose a mathematical model which expressively contains elements to model some of the uncertainties involved quantitatively.

There are different kinds of uncertainty, the most important distinction being between *aleatoric* uncertainty, which is regarded as inherent in the phenomenon and can not be reduced, and *epistemic* uncertainty, which results from our incomplete knowledge and could in principle be reduced, though this may be impractical, not possible in the framework of available time and resources, or many similar reasons.

So, not by any lack of ability of the modeller, there may be elements in the mathematical setup which can not be predictited with certainty.

In an abstract setting, we may describe the situation as follows: We have a mathematical model B, which maps the action F of the rest of the world (the input) onto the system into the reaction U (the output) of the system:

$$\mathsf{B} : \mathsf{F} \to \mathsf{U}. \tag{1}$$

Usually the mathematical model will look just the other way around, i.e. we describe which action $\mathsf{f} \in \mathsf{F}$ is caused by which reaction $\mathsf{u} \in \mathsf{U}$, and the mapping in Eq. (1) has to be inverted $\mathsf{B} = \mathsf{A}^{-1}$ (an equation has to be solved) to obtain the response of the system:

$$\mathsf{Au} = \mathsf{f}. \tag{2}$$

In this setup, there may be uncertainty in the mathematical model B, and/or the action f. And what we would like to know is how this uncertainty is reflected in the response u.

To be more concrete, let us look at an example which will be used throughout the sequel: We look at the problem of predicting the grounwater flow through some aquifer. This is the first step in modelling, we have chosen our system. To make the following presentation simple, we will model this as flow through a porous medium based on Darcy's law. After this second step, the governing equations are:

$$\begin{aligned}
&\text{for} \quad x \in \mathcal{G} \quad \text{and} \quad t \in [0, T]: \\
c\,\partial_t u(t,x) - \nabla \cdot (\kappa(x) \nabla u(t,x)) &= f(t,x) \\
u(t,x) &= h(t,x) \quad \text{for} \quad x \in \Gamma_d \subseteq \partial\mathcal{G} \\
\text{and} \quad \kappa(x)\,\partial_n u(t,x) &= g(t,x) \quad \text{for} \quad x \in \Gamma_n \subseteq \partial\mathcal{G}, \\
u(0,x) &= u_0(x) \quad \text{for} \quad x \in \mathcal{G}.
\end{aligned} \tag{3}$$

Here ∇ is the nabla operator w.r.t. the spatial variables x, and ∂_t and ∂_n are the partial derivatives w.r.t time t resp. the outside normal n of the spatial domain $\mathcal{G}$ occupied by the aquifer. The time interval of interest is $[0, T]$, u is the hydraulic head of the flow, and mass conservation together with Darcy's constitutive law with conductivity κ and specific storage c lead to Eq. (3).

This is the diffusion equation, and certainly it also models many other phenomena. Many things usually are uncertain in Eq. (3). This may be the sinks or sources f, or the boundary data g and h (the action from the rest of the world—uncertainty in the input F), or the domain $\mathcal{G}$, the extent of the boundary parts Γ_d and Γ_n, or the conductivity κ or the storage coefficient c (uncertainty in the system B), or the initial data u_0.

Compared to the abstract setting Eq. (1), the space for the u's is U, f, g, and h are in F, and the partial differential equation in Eq. (3) is the inverse $\mathsf{A} = \mathsf{B}^{-1}$ in Eq. (2) of the mapping B. The uncertainty in F is often aleatoric, it may even be modelled as random process (e.g. rainflow, discharge from a river, etc.). On the other hand, the uncertainty in the system B may often be epistemic (e.g. we can not measure the conductivity at all locations).

For the sake of simplicity, we shall look only at stationary states of Eq. (3), and henceforth assume that all time derivatives are zero, and the action F is independent of time. Furthermore, for simplicity assume that $\Gamma_n = \emptyset$, and $h = 0$.

2.1. REPRESENTING UNCERTAINTY

The description of uncertainty [72] can be done in different ways [75, 68, 27, 8, 81, 9], some of the most common methods are:

- *worst-case scenarios* just trying to provide bounds, the mathematical tools beeing *interval analysis, convex models*, etc., e.g. [10, 75, 26].
- methods based on *fuzzy set theory*, linguistically often identified as beeing concerned with *possibility*, e.g. [112, 27, 65].
- *evidence theory*, which tries to create upper and lower bounds on the *likelihood* of events, these being linguistically termed as *plausibility* and *belief*, e.g. [95, 106, 59].
- *probabilistic* or *stochastic* models, which offer mathematically the richest structure. This is the kind of description which will be assumed here, e.g. [1, 3, 25, 29, 36, 39, 40, 43, 60, 62, 84, 97, 105].

Purely from the linguistic terms appearing in this short—and incomplete—description, one may glean that this is by no means an easy subject, nor that there is common agreement on which of these approaches may be best, even just in a certain setting. One may summerise the situatuion as stating that *there is uncertainty on how to best describe uncertainty.*

As it is the method which allows the mathematically most detailled description (but also likely requires the most extensive description), and as it seems to be the one mostly chosen in practice, the probabilistic or stochastic approach will be assumed in the sequel. Such a description is usually accomplished by representing certain quantities in the model as *random variables*, *stochastic processes*, or *random fields*. Elementary probability and random variables assumed to be known, in the following a short recap on the aspects of stochastic processes and random fields in mechanics which are important for the computational approaches will be given, more and especially more references may be found in [72].

2.2. STOCHASTIC MODELS AND TECHNIQUES

It will be assumed that the reader has some elementary knowledge of probability and stochastic processes, e.g. as contained in [84, 25, 35, 60, 62], cf. also [64, 74]. The text book [84] also has an introductory description of in particular linear time-invariant systems with (stationary) stochastic input, an area quite well known and hence not addressed here. In particular we would like to draw on the introductory material contained in [72].

The main objective of the probabilistic aspects considered here—where they are not only a computational device, e.g. [89], but part of the model, e.g. [64, 74]—is to make the uncertainty present in the model mathematically tractable and quantifiable [75, 27]. The first step herein is to represent the uncertainty involved by means of *probabilistic* methods. Therefore we shall quickly just summerise the ways random variables, and especially stochastic processes and random fields are described.

2.2.1. *Stochastic Processes and Random Fields*

In what follows, it will be useful to consider a set Ω of random elementary events, together with a class of subsets $\mathfrak{A}$ of Ω—technically a σ-algebra e.g. [60, 62, 25, 35]—to which a real number in the interval $[0,1]$ may be assigned, the *probability* of occurrence—mathematically a measure P. A $\mathcal{V}$-valued random variable r is then a function relating to each $\omega \in \Omega$ an element $\mathsf{r}(\omega) \in \mathcal{V}$, where $\mathcal{V}$ is a vector space, usually the real numbers $\mathbb{R}$—some technicalities have to be added, such that r is measurable. Viewing random variables and other random elements as functions on Ω will help to make a connection later to other well-known approximation processes. For later reference, it is also good to view the space of all such $\mathcal{V}$-valued random variables as linear combinations of elements like $r(\omega)\boldsymbol{v}$, where r is a real-valued (ordinary) random variable, and $\boldsymbol{v} \in \mathcal{V}$, i.e. as elements of the tensor product $L_0(\Omega) \otimes \mathcal{V}$, where $L_0(\Omega)$ is the space of real valued random variables.

Such a real-valued random variable r is considered completely specified by its distribution function [60, 62, 84, 25, 35]

$$\forall s \in \mathbb{R}: \quad F_{\mathsf{r}}(s) = \Pr\{\mathsf{r}(\omega) \leq s\} := \int_{\{\mathsf{r}(\omega)\leq s\}} dP(\omega). \tag{4}$$

The simplest characterisation of a random variable involves its mean, or average, or expected value $\mathbf{E}(\mathsf{r})$, and the splitting into the mean $\bar{\mathsf{r}}$ and the fluctuating part $\tilde{\mathsf{r}}$,

$$\mathsf{r}(\omega) = \bar{\mathsf{r}} + \tilde{\mathsf{r}}, \quad \text{with} \quad \mathbf{E}(\tilde{\mathsf{r}}) = 0, \tag{5}$$

$$\bar{\mathsf{r}} := \mathbf{E}(\mathsf{r}) := \int_\Omega \mathsf{r}(\omega)\, dP(\omega), \tag{6}$$

as well as the covariance

$$C_{\mathsf{r}} := \mathbf{E}(\tilde{\mathsf{r}} \otimes \tilde{\mathsf{r}}) = \mathbf{E}\left(\tilde{\mathsf{r}}^{\otimes 2}\right), \tag{7}$$

or higher moments $\mathbf{E}\left(\tilde{\mathsf{r}}^{\otimes k}\right)$ of order k. If the cross-covariance $C_{\mathsf{r}_{1,2}} = \mathbf{E}(\tilde{\mathsf{r}}_1 \otimes \tilde{\mathsf{r}}_2)$ of two random variables r_1 and r_2 vanishes, they are called uncorrelated.

Assume that part of the known input data $\mathsf{f} \in \mathsf{F}$ for the system B under consideration is such a random variable. Then the output resp. state of the system $\mathsf{u} \in \mathsf{U}$ will contain also random components, i.e. is a random variable. One is interested then in the characterisation of $\mathsf{u}(\omega)$, say with the moments (mean, variance, etc.) $\mathbf{E}\left(\mathsf{u}^{\otimes k}\right) = \int_\Omega \mathsf{u}^{\otimes k}\, dP(\omega)$, or exceedance probabilities $\Pr(\{\mathsf{u} \geq \alpha\}) := \int_\Omega \chi_{\{\mathsf{u}\geq\alpha\}}\, dP(\omega)$.

If the vector space $\mathcal{V}$ is a space of functions $\mathcal{F}(\mathcal{T})$ on some time interval $\mathcal{T} = [0, T]$—often the interval $(-\infty, +\infty)$ is considered—we say that the random variable $\mathsf{s}(\omega, t) = \mathsf{s}_t(\omega) \in L_0(\omega) \otimes \mathcal{F}(\mathcal{T})$ is a stochastic process. The so-called second order information are then the mean function $\bar{\mathsf{s}}(t) \in \mathcal{F}(\mathcal{T})$, with the fluctuating part $\tilde{\mathsf{s}}(t, \omega)$, which has zero mean, and the covariance function

$$C_{\mathsf{s}}(t_1, t_2) := \mathbf{E}(\tilde{\mathsf{s}}_{t_1} \otimes \tilde{\mathsf{s}}_{t_2}) \in \mathcal{F}(\mathcal{T}) \otimes \mathcal{F}(\mathcal{T}). \tag{8}$$

In case the random properties of $\mathsf{s}(t, \omega)$ are independent of t, the process is called strictly stationary or strict sense stationary (SSS) [84]; and in case of only the mean $\bar{\mathsf{s}}$ being independent of t, i.e. constant, and the covariance a function of the time lag, i.e. $C_{\mathsf{s}}(t_1, t_2) = c_{\mathsf{s}}(t_1 - t_2)$, the process is called weakly stationary or weak sense stationary (WSS). In this case the process is often characterised by its spectrum $S_{\mathsf{s}}(\nu)$, the Fourier transform of the time-lag covariance function.

It is well known that in this case the process may be synthesised from its spectrum, e.g. [96]:

$$\mathsf{s}(t, \omega) = \bar{\mathsf{s}}(t) + \sum_{k=-\infty}^{\infty} \varsigma_k(\omega) \sqrt{S_{\mathsf{s}}(\nu_k)} \exp(i 2\pi \nu_k t), \tag{9}$$

where the $\varsigma_k(\omega)$ are uncorrelated random variables of unit variance and vanishing mean [62], the series converging in $L_2(\Omega \times \mathcal{T}) = L_2(\Omega) \otimes L_2(\mathcal{T})$.

In applications, the input may be represented as in Eq. (9), and from that a description of the output in a similar representation derived. This then offers the basis to compute the quantities of interest alluded to before.

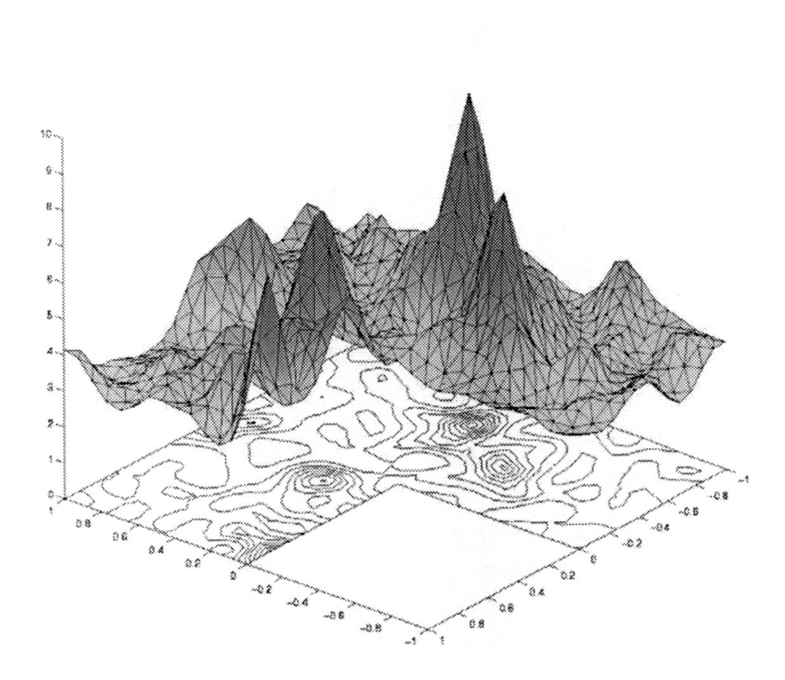

Figure 1. Realisation of κ.

In case the space $\mathcal{V}$ is a space of functions $\mathcal{F}(\mathcal{G})$ on a spatial domain, we speak of a random field. The description of random fields [1, 104, 105, 17, 33, 96], where we assign a random variable $\mathsf{r}(x, \omega)$ to each point $x \in \mathcal{G}$ in some domain $\mathcal{G}$ in space, is in many ways very similar to the case of stochastic processes. The only difference is that the "index set" has changed from the interval $\mathcal{T}$ to the domain in space $\mathcal{G}$. The decomposition into mean and fluctuating part with zero mean of Eq. (5) is as before, and so is the definition of the covariance function Eq. (8). What was called stationary before is now called homogeneous. A realisation—the spatial field $\mathsf{r}(\cdot, \omega) \in \mathcal{F}(\mathcal{G})$ for one specific ω—of the positive non-Gaussian random field $\kappa(x, \omega)$ representing the conductivity in Eq. (3) on an L-shaped domain $\mathcal{G} \subset \mathbb{R}^2$ is shown in Fig. 1.

In case the domain $\mathcal{G}$ is a product of intervals, a multi-dimensional spectrum may be computed, but most space-domains of interest are not as simple. This necessitates the generalisation in the next section 2.2.2 of the Fourier expansion, which will then be valid also for non-stationary or non-homogeneous stochastic processes or random fields.

2.2.2. *Karhunen-Loève Expansion*

What is needed is an effective computational representation of random fields, so that the desired output fucntionals (e.g. moments, exceedance probabilities) may be efficiently computed.

What is desired is a decomposition similar to the spectral synthesis Eq. (9) into products of deterministic functions on the domain—in Eq. (9) the exponentials—and functions only dependent on ω, i.e. simple random variables. This is a linear combination—just like Eq. (9)—of elements from $L_0(\Omega) \otimes \mathcal{F}(\mathcal{G})$. It is furnished by the *Karhunen-Loève expansion* [62], also known as *proper orthogonal decomposition*, and may also be seen as a *singular value decomposition* [38] of the random field. Given the covariance function $C_{\mathsf{r}}(x_1, x_2) \in \mathcal{F}(\mathcal{G}) \otimes \mathcal{F}(\mathcal{G})$—defined analogous to Eq. (8)—one considers the *Fredholm eigenproblem* [20, 88]:

$$\int_{\mathcal{G}} C_{\mathsf{r}}(x_1, x_2)\phi_k(x_2)\, dx_2 = \varrho_k^2 \phi_k(x_1), \quad x_1 \in \mathcal{G}. \tag{10}$$

This equation can usually not be solved analytically, but standard numerical methods [2] (e.g. based on the FEM) may be used to discretise and resolve this Fredholm integral equation.

As the covariance function is—via its definition Eq. (8)—symmetric and positive semi-definite, Eq. (10) has positive decreasingly ordered eigenvalues $\{\varrho_k^2\}_{k=1,\infty}$ with only accumulation point zero, and a complete and orthonormal—in $L_2(\mathcal{G})$ —set of eigenfunctions $\{\phi_k\}$. This yields the spectral decomposition

$$C_{\mathsf{r}}(x_1, x_2) = \sum_{k=1}^{\infty} \varrho_k^2\, \phi_k(x_1) \otimes \phi_k(x_2). \tag{11}$$

This gives us similarly to Eq. (9) a possibility to synthesise the random field through its Karhunen-Loève expansion (KLE). Two such truncated expansions for the field in Fig. 1 are shown in Fig. 2:

$$\mathsf{r}(x, \omega) \approx \bar{\mathsf{r}}(x) + \sum_{k=1}^{M} \varrho_k\, \rho_k(\omega) \phi_k(x), \tag{12}$$

where again $\rho_k(\omega)$ are uncorrelated random variables with unit variance and zero mean $\mathbf{E}\,(\rho_k \rho_m) = \delta_{km}$. This is a reflection of the orthonormality of the eigenfunctions, where the inner product in $L_2(\Omega)$ is given by

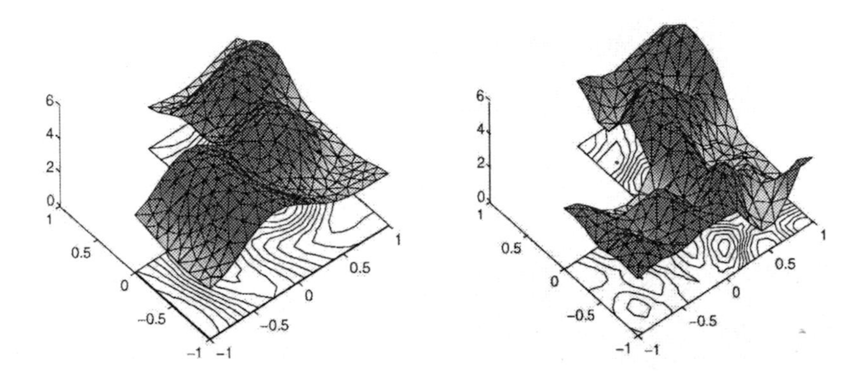

Figure 2. Realisations of κ with 20 and 50 KL-Terms.

$\langle \rho_k | \rho_m \rangle := \mathbf{E}(\rho_k \rho_m)$; uncorrelated is the stochastic pendant to orthogonal. This expansion converges without further assumptions in $L_2(\Omega) \otimes L_2(\mathcal{G}) \cong L_2(\Omega, L_2(\mathcal{G}))$ as $M \to \infty$.

In case the stochastic process $\mathsf{s}(t, \omega)$ or the random field $\mathsf{r}(x, \omega)$ is *Gaussian* [60, 62, 84, 48], the random variables $\varsigma_k(\omega)$ in Eq. (9), and the random variables $\rho_k(\omega)$ in Eq. (12) are also Gaussian—a linear combination of Gaussian variables is again Gaussian. As they are uncorrelated, they are also independent [62], and the synthesis is particularly simple. Non-Gaussian fields may be synthesised as functions of Gaussian fields, e.g. [39, 32, 48, 55] and the references therein. One simple method is the so-called "translation"

$$\mathsf{r}(x, \omega) = \Phi(x, \gamma(x, \omega)), \tag{13}$$

where $\gamma(x, \omega)$ is a Gaussian field, and Φ some appropriate function.

A specific possibility is explored in the next section 2.2.3, cf. [104, 96, 90].

2.2.3. *Wiener's Polynomial Chaos Expansion*

As suggested by Wiener [107], any random variable $\mathsf{r}(\omega)$—subject to some restrictions of a technical nature—may be represented as a series of polynomials in uncorrelated and hence independent Gaussian variables $\boldsymbol{\theta} = (\theta_1, \ldots, \theta_\jmath, \ldots)$ [72, 60, 33, 39, 42, 66, 43, 68, 32, 69, 40, 108, 48, 73], the *polynomial chaos expansion* (PCE):

$$\mathsf{r}(\omega) = \sum_{\alpha \in \mathcal{J}} r^{(\alpha)} H_\alpha(\boldsymbol{\theta}(\omega)). \tag{14}$$

The word *chaos* was used here by N. Wiener [107], and has nothing to do with modern usage of the word in mathematics, where it characterises the unpredictable behaviour of dynamical systems. Here $\mathcal{J} := \{\alpha \,|\, \alpha = (\alpha_1, \ldots, \alpha_\jmath, \ldots),\ \alpha_\jmath \in \mathbb{N}_0\}$ is the set of infinite multi-indices, i.e. sequences of non-negative integers, only finitely many of which are non-zero [72, 42, 66, 69, 48, 73]. The $r^{(\alpha)}$ are the generalised Fourier coefficients for the orthogonal basis of Hermite polynomials:

$$H_\alpha(\boldsymbol{\theta}(\omega)) = \prod_{\jmath=1}^{\infty} h_{\alpha_\jmath}(\theta_\jmath(\omega)), \tag{15}$$

where the $h_\ell(\xi)$ are the usual Hermite polynomials, and $\boldsymbol{\theta}(\omega) := (\theta_1(\omega), \ldots)$ is an infinite sequence of normalised independent Gaussian random variables. Orthogonality is reflected in $\langle H_\alpha | H_\beta \rangle := \mathbf{E}\,(H_\alpha(\boldsymbol{\theta}) H_\beta(\boldsymbol{\theta})) = \alpha!\, \delta_{\alpha,\beta}$, where $\alpha! := \prod_{\jmath=1}^{\infty} \alpha_\jmath!$.

This then gives a convenient representation in *independent* identically distributed (**iid**) Gaussian variables of the stochastic process resp. random field:

$$\mathsf{r}(x,\omega) = \bar{\mathsf{r}}(x) + \sum_{k=1}^{\infty} \sum_{\alpha \in \mathcal{J}} \varrho_k\, r_k^{(\alpha)} H_\alpha(\boldsymbol{\theta}(\omega)) \phi_k(x). \tag{16}$$

One should point out once more that the approach described here is *one possibility* to discretise random fields in the context of the stochastic finite element method, see the reports [68, 96, 91, 92, 101, 48] and the references therein for other possibilities.

One important part which has been achieved should not go unmentioned though: Effectively we have transferred everything from the measure space $(\Omega, \mathfrak{A}, P)$ to the space of Gaussian random variables $\Theta = \{\boldsymbol{\theta}\}$ with a Gaussian measure, and this Gaussian measure is a product measure, allowing convenient integration and the use of Fubini's theorem.

2.3. SPARSE REPRESENTATION

The representation in Eq. (16) is one in tensor products, and this is computationally very significant, as it allows great savings. Although one has a convenient computational representation, it may still contain too many terms which only play a minor rôle. Tensor products are a key ingredient for so-called sparse approximations, where we appproximate an object of dimension $M \times N$ with many fewer terms than $O(MN)$.

2.3.1. *Optimal Approximation*

The first possibility is offered by the optimality porperty of the KLE: The approximation in Eq. (12) has the smallest error in $L_2(\Omega) \otimes L_2(\mathcal{G})$ of all possible approximations in M uncorrelated random variables. If we neglect those terms with small singular value ϱ_k, the error in variance may directly be estimated as $\sum_{k>M} \varrho_k^2$, and a first and most important reduction is accomplished. Going a step further, in the sum over α, all terms with a small product $\varrho_k^2 (r_k^{(\alpha)})^2 \alpha!$ may again be neglected, as they contribute only little to the variance, and hence to the variablility of the input.

2.3.2. *Call for Adaptivity*

These considerations only concern the input, but we are interested in functionals (so-called statistics) of the output variables, and hence it would be better to be able to control them. Hence we need error estimates, and the influence of each term on these error estimates. This is furnished by adaptive procedures. These are well established by now in several fields. As we shall put here the computation of the output quantities in the context of Galerkin procedures in section 3.2, many methods developed in the field of partial differential equations can be transferred here.

Hence it will be important to find efficient computational procedures to compute output quantities in the form as in Eq. (12), And in particular, it will be necessary to get good estimates on the size of the coefficients, and then be able to retain only those terms in the expansion which are necessary to reach a certain desired accuracy in the result. This whole procedure will have to be performed adaptively, in as much as which expansion terms in the spatial domain are retained, and also as regards the question which expansion terms in the "stochastic dimension" to retain. This covers the question which particular dimensions are to participate, and what polynomial degrees are necessary in this particular dimension to achieve a desired accuracy. Let us also point out here that there is nothing special about Hermite polynomials—being merely a convenient choice, other functions certainly can and will be used.

3. Uncertainty Propagation

To be able to compute the expansion coefficients for the output quantitites, it has to be investigated how the uncertainty is transmitted through the system. Here we give only a synapsis, more can be found in [72], and the references therein. The multitude of computational approaches can only be touched upon, subsequently we will concentrate on just one approach, the stochastic Galerkin methods in section 3.2.

3.1. COMPUTATIONAL TECHNIQUES OVERVIEW

Stochastic systems can be interpreted mathematically in several ways. At the moment we concentrate on randomness in space. More information, references, and reviews on stochastic finite elements can be found in [68, 91, 101, 48]. A somewhat specialised field is the area of structural reliability, e.g. see [3, 23, 41, 64, 74].

3.1.1. *Monte Carlo Simulation and Other Direct Integration Methods*

Both Monte Carlo Simulation [16, 83, 93] and its faster cousin Quasi Monte Carlo [16, 77] are best viewed as integration techniques. When solving for the response of a stochastic system, the ultimate goal is usually the computation of response statistics, i.e. functionals of the solution, or more precisely expected values of some functions of the response, such as

$$\Psi_u(x) = \mathbf{E}\left(\Psi(x, \omega, u(x, \omega))\right) = \int_\Omega \Psi(x, \omega, u(x, \omega))\, dP(\omega). \tag{17}$$

Such an integral may numerically be approximated by a weighted sum of samples of the integrand.

Monte Carlo (MC) methods can be used directly for this, but they require a high computational effort [16]. Variance reduction techniques are employed to lower this effort somewhat. Quasi Monte Carlo (QMC) methods [16, 77] may reduce the computational effort considerably without requiring much regularity. But often we have high regularity in the stochastic variables, and this is not exploited by MC and QMC methods. Monte Carlo chooses the samples randomly according to the underlying probability measure, and the weights all equal, e.g. [16]. Quasi Monte Carlo, e.g. [16, 77] chooses deterministic low-discrepancy evaluation points and still all weights equal. Common Quadrature rules choose special evaluation points and corresponding weights, .e.g. [98, 30, 78, 79, 86, 87].

Monte Carlo is very robust, and in principle almost anything can be tackled by it. But on the other hand it is very slow, and this is the main disadvantage. Its other main advantage is that it is not affected by the dimension of the integration domain as most other methods are. Sparse (Smolyak) quadrature methods are an efficient alternative. These have initially been described in [98], and have found increasing attention in recent years, e.g. [78, 87, 30].

As integrals over the probability space are also part of most other approaches, these integration methods are usually found inside many other techniques.

If the purpose of the computation of the expectation is to calculate the probability of rare events—like in the evaluation of reliability—there are special methods like FORM or SORM specifically tailored for this task, e.g. [3, 23, 41, 64, 74].

3.1.2. *Perturbation Techniques*

Alternatives to Monte Carlo (e.g. [83, 24]) methods, which compute the first moments of the solution have been developed in the field of stochastic mechanics—cf. [61], for example perturbation methods, e.g. [56], or methods based on Neumann-series, e.g. [33, 5].

The perturbation approach—e.g. see [56, 68]– does not really start from a probabilistic consideration, but from expecting the stochastic response to be a small variation about the mean. Assuming the stochastic parameters to vary a little bit around their mean, a formal perturbation expansion—usually only up to first or second order—is performed, which gives the variation of the response about the mean. By taking expected values one gets approximate expressions for the most basic response quantities, the mean and the covariance.

As may be expected, the perturbation approach usually only works well when the stochastic variations are not too large. It becomes important in another guise though, namely in its use as an error estimator. Here it is then known as the *adjoint* or *dual* method. It may also be used in a stochastic context, see [52] and the references therein, also to the usual use as error estimator.

3.1.3. *Response Surfaces*

Another possibility [58] in stochastic computations is to evaluate the desired response in some—maybe well chosen—points of the stochastic parameter space, and then to fit these response points with some analytic expression—often a low order polynomial. This then is the response surface, which may be taken as a substitute of the true response. All subsequent computations can then be done with this response surface, e.g. [3, 41, 64, 74]. Current fitting techniques usually only work well for low order and not too many dimensions of the stochastic parameter space.

3.1.4. *Galerkin Methods*

Following [33], stochastic Galerkin methods have been applied to various linear problems, e.g. [31, 32, 85, 70, 49, 45, 108, 109]. Recently, non-linear problems with stochastic loads have been tackled, e.g. [109], and some first results of both a theoretical and numerical nature for non-linear stochastic operators are in [50, 53].

These Galerkin methods allow us to have an explicit functional relationship between the independent random variables and the solution—in contrast with usual Monte Carlo approaches—so that subsequent evaluations of functionals (statistics like the mean, covariance, or probabilities of exceedance) are very cheap. This may be seen as a way to systematically calculate more and more accurate "response surfaces" [58]. They are described in more detail in the next section 3.2.

3.2. STOCHASTIC GALERKIN METHODS

For the sake of simplicity, we focus on stationary systems, like example Eq. (3) without the time dependence and the time derivatives, and we have stochastic coefficients, hence the equations are labelled as stochastic partial differential equations (SPDEs). In the mathematical formulation, and also to make sure that the numerical methods will work well, we strive to have similar overall properties as in the deterministic system. The deterministic system is well-posed in the sense of Hadamard, i.e. the solution depends continuously on the right hand side f and g in the norm of L_2, and on the coefficient of diffusivity κ, but in the norm of L_∞. For this to hold, it is necessary that the operator representing Eq. (3) is continuous and continuously invertible. The simplest way of achieving the same kind of properties in the stochastic case is to require that:

$$0 < \kappa_- < \kappa(x,\omega) < \kappa_+ < \infty. \tag{18}$$

All other conditions being extended also naturally into the stochastic domain, it is possible to show that the problem is well-posed [73].

The so-called strong form in Eq. (3) is not a good starting point for the Galerkin approach [33], and as in the purely deterministic problem [113, 100, 18, 99] a variational formulation is needed [73]: Look for the solution u in a space $\mathcal{V} \otimes \mathcal{S}$ of functions which have a spatial dependence (the variables from $\mathcal{V}$), and a stochastic dependence (the variables from $\mathcal{S}$), such that for all $w \in \mathcal{V} \otimes \mathcal{S}$:

$$\mathbf{b}(u,w) := \int_\Omega \int_\mathcal{G} (\nabla w(x,\omega))^T \kappa(x)(\nabla u(x,\omega))\, dx\, dP(\omega) = \langle\!\langle \mathcal{A}(u), w \rangle\!\rangle =$$

$$\langle\!\langle f, w \rangle\!\rangle := \int_\Omega \int_\mathcal{G} f(x,\omega) w(x,\omega)\, dx\, dP(\omega). \tag{19}$$

Assume that the spatial part of the SPDE has been approximated by a Galerkin method—here we use the finite element method (FEM). In some sense an arbitrary spatial discretisation could be used, but as we deal with Galerkin methods in the stochastic domain, assuming this also in the spatial domain gives a certain measure of unity to the presentation. What is used in the discretisation amounts to the finite element method in space—it does not matter which variant—and a spectral or pure p-method in the stochastic dimension [33].

3.2.1. *Spatial Discretisation*

Performing a Galerkin approximation in the spatial part amounts to taking only a finite-dimensional subspace $\mathcal{V}_N \subset \mathcal{V}$. Let $\{s_1(x), \ldots, s_N(x)\}$ be a basis in $\mathcal{V}_N$, we then approximate the solution by [18, 100, 113]

$$u(x, \omega) = \sum_{k=1}^{N} s_k(x) u_k(\omega) = \boldsymbol{s}(x)\boldsymbol{u}(\omega), \tag{20}$$

where the $\{u_k(\omega)\}$ now are random variables in $\mathcal{S}$, and for concise notation we have set $\boldsymbol{s}(x) = [s_1(x), \ldots, s_N(x)]$ and $\boldsymbol{u}(\omega) = [u_1(\omega), \ldots, u_N(\omega)]^T$.

Inserting this ansatz into Eq. (19), and applying the spatial Galerkin conditions, we then require that for all $\varphi \in \mathcal{S}$:

$$\int_\Omega \varphi(\omega)\boldsymbol{K}(\omega)\boldsymbol{u}(\omega)\, dP(\omega) = \int_\Omega \varphi(\omega)\boldsymbol{f}(\omega)\, dP(\omega), \tag{21}$$

where $\boldsymbol{K}(\omega) = (K_{\imath\jmath}(\omega)) = \left(\int_R (\nabla s_\imath(x))^T \kappa(x, \omega)(\nabla s_\jmath(x))\, dx\right)$ is the stiffness matrix, and the RHS is $\boldsymbol{f}(\omega) = [f_1(\omega), \ldots, f_N(\omega)]^T$ with $f_\jmath(\omega) = \int_R s_\jmath(x) f(x, \omega)\, dx$. The variational Eq. (21) will be written as $\boldsymbol{K}(\omega)\boldsymbol{u}(\omega) = \boldsymbol{f}(\omega)$, understood in a weak sense. This involves the variable $\omega \in \Omega$, and is still computationally intractable, as in general we need infinitely many coordinates to parametrise Ω.

3.2.2. *Stochastic Discretisation*

In section 2.2.3 we used the PCE to represent a stochastic process. Here we want to extend the Galerkin idea, and for that reason we expand the random variables $\boldsymbol{u}(\omega)^T = [u_1(\omega), \ldots, u_N(\omega)]$ in a PC-series:

$$\forall k : \quad u_k(\omega) = \sum_\alpha u_k^{(\alpha)} H_\alpha(\omega) = \mathsf{u}_k \mathsf{H}(\omega), \tag{22}$$

where $\mathsf{H}(\omega)^T = [\ldots, H_\alpha(\omega), \ldots]$, and $\mathsf{u}_k = [\ldots, u_k^\alpha, \ldots]$.

For the purpose of actual computation, truncate the PCE Eq. (22) after finitely many terms $\alpha \in \mathcal{J}_{M,p}$, thus introducing a finite dimensional approximation $\mathcal{S}_{M,p} = \mathrm{span}\{H_\alpha | \alpha \in \mathcal{J}_{M,p}\} \subset \mathcal{S}$. The set $\mathcal{J}_{M,p} \subset \mathcal{J}$ is here defined for $M, p \in \mathbb{N}$ as (see also [73]):

$$\mathcal{J}_{M,p} = \{\alpha \in \mathcal{J} | \forall \jmath > M : \ \alpha_\jmath = 0, \ |\alpha| \le p\}, \qquad \text{where} \quad |\alpha| := \sum_{\jmath=1}^{\infty} \alpha_\jmath. \tag{23}$$

Just as we require for $\bigcup_N \mathcal{V}_N$ to be dense in $\mathcal{V}$—see [100, 18]—here we rely on the fact that the closure of $\bigcup_{M,p} \mathcal{S}_{M,p}$ is all of $\mathcal{S}$, see [73].

The ansatz for the Galerkin method may be written concisely as

$$\boldsymbol{u}(\omega) = \sum_{\alpha} \boldsymbol{u}^{(\alpha)} H_\alpha(\omega), \tag{24}$$

and the task is to compute the $\boldsymbol{u}^{(\alpha)}$.

3.2.3. *Linear Systems*

The "input information", the matrix $\boldsymbol{K}(\omega)$ is also expanded in a polynomial chaos:

$$\boldsymbol{K}(\omega) = \sum_{\gamma} \boldsymbol{K}^{(\gamma)} H_\gamma(\omega) \tag{25}$$

where of course $\boldsymbol{K}^{(0)} = \mathsf{E}(\boldsymbol{K}(\cdot)) = \overline{\boldsymbol{K}}$, and each $\boldsymbol{K}^{(\gamma)}$ is a stiffness matrix computed with a "conductivity" $\kappa^{(\gamma)}(x)$. The right hand side $\boldsymbol{f}(\omega)$ is expanded analogously.

From this one has

$$\left(\sum_{\gamma} \boldsymbol{K}^{(\gamma)} H_\gamma(\omega)\right) \sum_{\alpha} \boldsymbol{u}^{(\alpha)} H_\alpha(\omega) = \sum_{\beta} \boldsymbol{f}^{(\beta)} H_\beta(\omega). \tag{26}$$

To obtain a finite dimensional approximation, truncate the series for $\boldsymbol{u}(\omega)$ by letting only $\alpha \in \mathcal{J}_{M,p}$—this subset was defined in relation Eq. (23). Having done that, of course the equation can not be satisfied any more for all ω, and one may use Galerkin projection conditions to obtain conditions for all $\beta \in \mathcal{J}_{M,p}$:

$$\sum_{\alpha \in \mathcal{J}_{M,p}} \left(\sum_{\gamma \in \mathcal{J}} \mathsf{E}\left(H_\beta(\cdot) H_\gamma(\cdot) H_\alpha(\cdot)\right) \boldsymbol{K}^{(\gamma)} \right) \boldsymbol{u}^{(\alpha)} = \mathsf{E}\left(H_\beta(\cdot) \boldsymbol{f}\right) = \boldsymbol{f}^{(\beta)}. \tag{27}$$

Defining new matrices $\boldsymbol{\Delta}^{(\gamma)}$ with elements $(\Delta^{(\gamma)})_{\alpha,\beta} = \mathsf{E}\left(H_\beta(\cdot) H_\gamma(\cdot) H_\alpha(\cdot)\right)$, and defining block vectors $\mathbf{u} = [\ldots, \boldsymbol{u}^{(\alpha)}, \ldots]$ and similarly $\mathbf{f}$, one can write this equation with the Kronecker product as

$$\mathbf{K}\mathbf{u} := \left[\sum_{\gamma} \boldsymbol{\Delta}^{(\gamma)} \otimes \boldsymbol{K}^{(\gamma)} \right] \mathbf{u} = \mathbf{f}. \tag{28}$$

The question comes whether, and if so when, one should truncate the PCE of Eq.(25), as has to be done in an actual computation. It is good to know that this problem is resolved when looking at the projected equation Eq. (28) (see [73] for a proof): *The series in γ in Eqs. (27, 28) is a finite sum, even when the PCE Eq. (25) is not.*

The sum and Kronecker product structure allow savings in memory usage and coarse grain parallelisation in the numerical solution process, and we refer to the next section for references and more details on this topic.

Unfortunately, often many terms have to be used in the PCE, and we would like to reduce this. One possibility is to use a direct KLE of the matrix, obtained via a KLE of the coefficient of conductivity κ. The appeal of this direct KLE is that a similar Kronecker product structure as Eq. (28) with very few terms in the sum may be achieved, and the computation

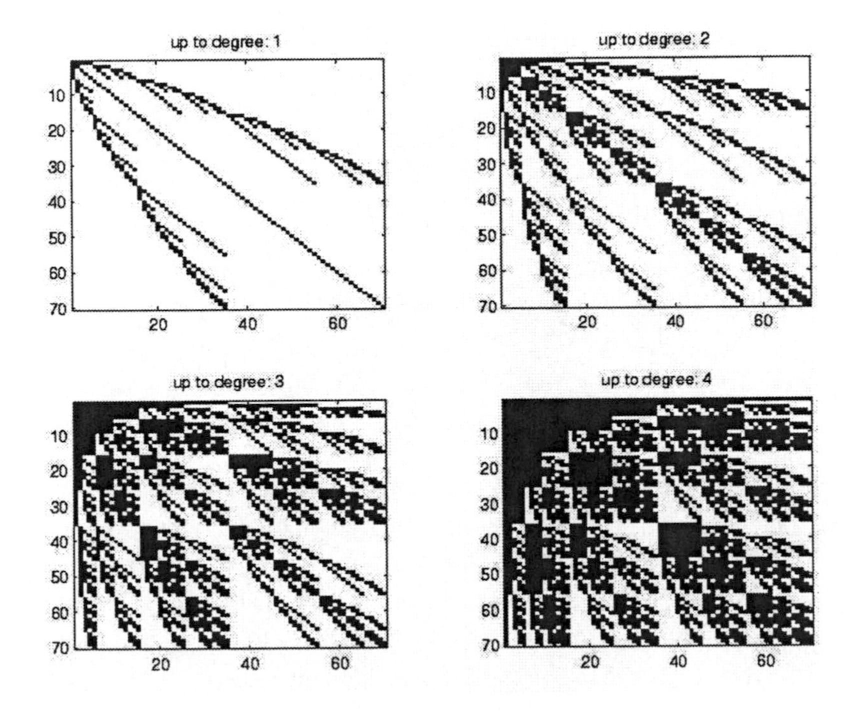

Figure 3. Sparsity Pattern of Kronecker Factor $\mathbf{\Delta}$.

of these terms is completely analogous to the normal computation of a stiffness matrix. It can be shown [73] that, similarly to the result on the PCE, the Galerkin projection enables the direct use of the KLE without any further hypotheses on the covariance function $C_\kappa(x, y)$ of $\kappa(x, \omega) = \sum_j \varsigma_j \xi_j(\omega) \kappa_j(x)$.

When the space discretisation is performed for such an expansion of the conductivity as in Eq. (12), one may see right away that

$$\boldsymbol{K}(\omega) = \overline{\boldsymbol{K}} + \sum_{\jmath=1}^{\infty} \varsigma_\jmath \, \xi_\jmath(\omega) \boldsymbol{K}_\jmath, \tag{29}$$

where $\boldsymbol{K}_\jmath$ is computed by using the KL-eigenfunction $\kappa_\jmath(x)$ as conductivity instead of $\kappa(x)$, and $\varsigma_\jmath$ and $\xi_\jmath(\omega)$ are the singular values and eigenfunctions of the KLE of $\kappa(x,\omega)$. Note that this may be usually computed with existing software, all one has to do to supply another "material", namely $\kappa_\jmath(x)$.

For the discrete $\boldsymbol{u}(\omega)$ use again the PCE as before, and impose the Galerkin conditions to obtain the coefficients:

$$\forall \alpha \in \mathcal{J}_{M,p} : \qquad \sum_{\beta \in \mathcal{J}_{M,p}} \mathsf{E}\left(H_\alpha(\cdot)\boldsymbol{K}(\cdot)H_\beta(\cdot)\right) \boldsymbol{u}^{(\beta)} = \mathsf{E}\left(H_\alpha(\cdot)\boldsymbol{f}\right) = \boldsymbol{f}^{(\alpha)}, \tag{30}$$

Expanding Eq. (30) with the series in Eq. (29) gives for all $\alpha \in \mathcal{J}_{M,p}$:

$$\sum_{\beta \in \mathcal{J}_{M,p}} \mathsf{E}\left(H_\alpha(\cdot) \left[\overline{\boldsymbol{K}} + \sum_{\jmath=1}^{\infty} \varsigma_\jmath \xi_\jmath(\cdot) \boldsymbol{K}_\jmath \right] H_\beta(\cdot) \right) = \sum_{\beta \in \mathcal{J}_{M,p}} \left[\overline{\boldsymbol{K}} + \sum_{\jmath=1}^{\infty} \Delta_{\alpha,\beta}^{(\jmath)} \boldsymbol{K}_\jmath \right] \tag{31}$$

where

$$\Delta_{\alpha,\beta}^{(\jmath)} = \varsigma_\jmath \mathsf{E}\left(H_\alpha(\cdot)\xi_\jmath(\cdot)H_\beta(\cdot)\right). \tag{32}$$

To compute such an expectation, one may again use the PCE of $\xi_\jmath(\omega) = \sum_\gamma c_\jmath^{(\gamma)} H_\gamma(\omega)$ as in Eq. (14). From the previous discussion we know that the PCE series are in this case only finite sums: $\mathsf{E}\left(H_\alpha(\cdot)\xi_\jmath(\cdot)H_\beta(\cdot)\right) = \sum_{\gamma \in \mathcal{J}_{M,2p}} c_\jmath^{(\gamma)} \Delta_{\alpha,\beta}^{(\gamma)}$. Now define the matrices $\boldsymbol{\Delta}^{(\jmath)}$ with elements $\Delta_{\alpha,\beta}^{(\jmath)}$ from Eq. (32), and set $\boldsymbol{\Delta}^{(0)} = \boldsymbol{I}$, $\boldsymbol{K}_0 = \overline{\boldsymbol{K}}$. Using again the block vectors $\mathbf{u} = [\ldots, \boldsymbol{u}^{(\alpha)}, \ldots]$ and $\mathbf{f}$, one may write this equation as

$$\mathbf{K}\mathbf{u} := \left[\sum_{\jmath=0}^{\infty} \boldsymbol{\Delta}^{(\jmath)} \otimes \boldsymbol{K}_\jmath \right] \mathbf{u} = \mathbf{f}. \tag{33}$$

One may further expand for $\jmath > 0$: $\boldsymbol{\Delta}^{(\jmath)} = \sum_{\gamma \in \mathcal{J}_{M,2p}} \varsigma_\jmath c_\jmath^{(\gamma)} \boldsymbol{\Delta}^{(\gamma)}$, such that with Eq. (33)

$$\mathbf{K}\mathbf{u} = \left[\sum_{\jmath=0}^{\infty} \sum_{\gamma \in \mathcal{J}_{M,2p}} \varsigma_\jmath c_\jmath^{(\gamma)} \boldsymbol{\Delta}^{(\gamma)} \otimes \boldsymbol{K}_\jmath \right] \mathbf{u} = \mathbf{f}. \tag{34}$$

In Fig. 3 we see the sparsity pattern of $\boldsymbol{\Delta}^{(\jmath)}$, depending on how many terms were used in the PCE, produced with the *MATLAB* spy function.

White space corresponds to zeros, whereas each dot represents one full spatial matrix with the size and sparsity pattern of $\boldsymbol{K}$. A remark similar as before applies here as well: There is at least one PC-term for each KL-term, hence we can expect that in an actual computation we have many less terms in Eq. (33) than in Eq. (28)—although formally there is still an infinite series in Eq. (33).

As noted before, the KLE converges in $L_2(\mathcal{R})$, whereas there is only stability against perturbations in $L_\infty(\mathcal{R})$. We need uniform convergence to be able to truncate the series and still guarantee the conditions Eq. (18). But the Galerkin projection helps again, as may be shown (see [73] for a proof): *The series in Eq.* (33) *resp. Eq.* (34) *converges uniformly. Hence a finite number of terms suffices to keep the discrete operators* $\mathbf{K}$ *uniformly—in the discretisation of* $\mathcal{V}\otimes\mathcal{S}$*—positive definite, and therefore their inverses uniformly bounded, assuring the stability of the approximation process.*

The Eq. (33) is again in tensor- or Kronecker product form, and for the computation it is definitely kept in this way [34, 85, 71, 49, 51, 73, 55]. Solution methods used are usually of the Krylov subspace type [38], where only multiplication with the system matrix $\mathbf{K}$ is necessary. In our example all the matrices $\boldsymbol{K}_\jmath$ and $\boldsymbol{\Delta}^{(\jmath)}$ are symmetric, hence so is $\mathbf{K}$. The matrices are also positive definite, therefore preconditioned conjugate gradients may be used. There are again plenty of opportunities for coarse level parallelism, obvious in the sum and in the Kronecker product, this is described in more detail in [49, 51, 73, 55].

The PCE also gives a natural multi-level structure to the equations, which can be used in the solution process [70, 73, 55]. An additional possibility is to select the approximating subspaces adaptively according to the functional which one wants to compute [52, 55]. An example for a realisation for a coefficient $\kappa(x,\omega)$ and a solution $u(x,\omega)$ is shown in Fig. 4.

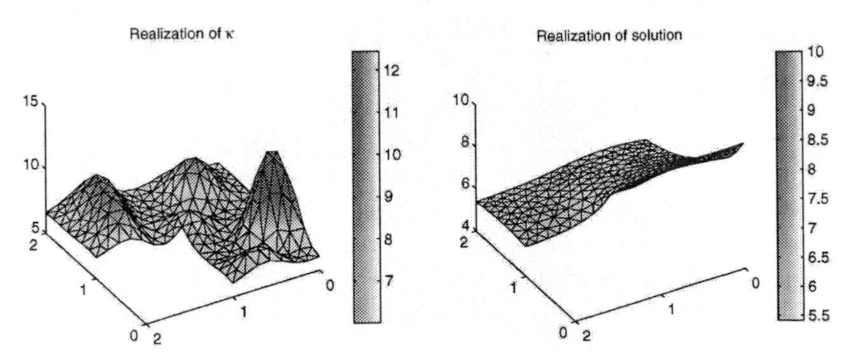

Figure 4. Realisations of Material and Solution on an L-shaped Domain.

Since these are random variables at each point, it might be more instructive to consider the pointwise mean $\bar{u}(x)$ and the pointwise variance $C_u(x,x)$, shown in Fig. 5

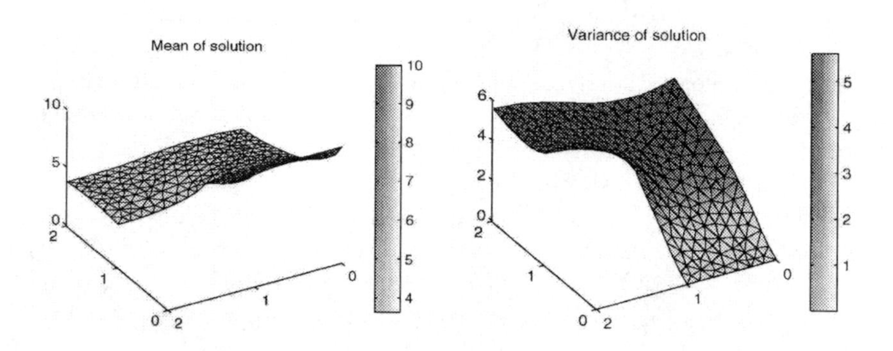

Figure 5. Mean and Variance of Solution on an L-shaped Domain.

3.2.4. *Non-Linear Systems*

As often, the ability to solve the linear or at least linearised problem is the "workhorse" also for non-linear problems. Look back and consider again our model problem Eq. (3). Assume now that the hydraulic conductivity $\kappa(x,u)$ depends also on the hydraulic head and on soil properties which are described by two other fields, $\hat{\kappa}(x)$ and $\check{\kappa}(x)$. We use the model $\kappa(x,u) = \hat{\kappa}(x) + \check{\kappa}(x)u(x)^2$, and we assume that both fields satisfy the boundedness and positivity conditions Eq. (18). This should be seen as a first approximation to more accurately modelled non-linear behaviour.

As before, a standard spatial discretisation—the same finite element spaces may be used [18, 100, 113] —leads to a problem of the form

$$\boldsymbol{A}(\omega)[\boldsymbol{u}(\omega)] = \boldsymbol{f}(\omega), \tag{35}$$

where both the non-linear operator $\boldsymbol{A}(\omega)[\cdot]$ and the RHS $\boldsymbol{f}(\omega)$ are random, and hence so is $\boldsymbol{u}(\omega)$. Once we write the solution $\boldsymbol{u}(\omega)$ as a linear combination of Hermite polynomials in $\boldsymbol{u}(\omega) = \sum_\alpha \boldsymbol{u}^{(\alpha)} H_\alpha(\omega)$, and now one has to compute the coefficients $\boldsymbol{u}^{(\alpha)}$.

The Galerkin method is obtained by inserting the stochastic ansatz for the PCE of $\boldsymbol{u}(\omega)$ into Eq. (35). In general there will be a residuum

$$\boldsymbol{R}(\omega)[\boldsymbol{u}(\omega)] = \boldsymbol{f}(\omega) - \boldsymbol{A}(\omega)[\boldsymbol{u}(\omega)], \tag{36}$$

which is then projected in a standard Galerkin manner onto the finite dimensional stochastic subspace $\text{span}\{H_\alpha | \alpha \in \mathcal{J}_{M,p}\}$, and one requires the projection to vanish. This results in

$$\mathbf{r}(\mathbf{u}) = [\ldots, \mathsf{E}\left(H_\alpha(\cdot)\boldsymbol{R}(\cdot)[\mathsf{uH}(\cdot)]\right), \ldots] = \mathbf{0}, \tag{37}$$

where the same block vectors—$\mathbf{u} = [\ldots, \boldsymbol{u}^{(\alpha)}, \ldots]$ —as before are used.

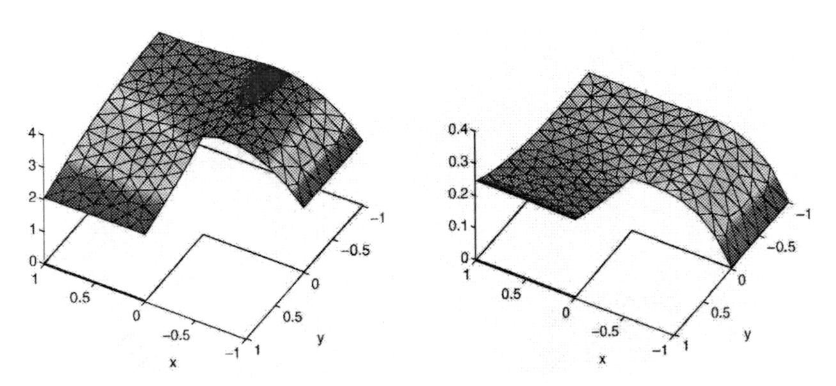

Figure 6. Mean and Standard Deviation of Solution to Non-linear Model.

Now Eq. (37) is a huge non-linear system, and one way to approach it is through the use of Newton's method, which involves linearisation and subsequent solution of the linearised system, employing the methods of the preceeding sections 3.2.2 and 3.2.3.

Another possibility, avoiding the costly linearisation and solution of a new linear system at each iteration, is the use of Quasi-Newton methods [67, 22]. This was done in [53], and the Quasi-Newton method used—as we have a symmetric positive definite or potential minimisation problem this was the *BFGS*-update—performed very well. The Quasi-Newton methods produce updates to the inverse of a matrix, and these low-rank changes [22] are also best kept in tensor product form [67]; so that we have tensor products here on two levels, which makes for a very economical representation.

But in any case, in each iteration the residual Eq. (37) has to be evaluated at least once, which means that for all $\alpha \in \mathcal{J}_{M,p}$ the integral

$$\mathsf{E}\left(H_\alpha(\cdot)\boldsymbol{R}(\cdot)[\mathsf{uH}(\cdot)]\right) = \int_\Omega H_\alpha(\omega)\boldsymbol{R}(\omega)[\mathsf{uH}(\omega)]\,dP(\omega)$$

has to be computed. In general this can not be done analytically as before in the case of linear equations, and one has to resort to numerical quadrature rules:

$$\int_\Omega H_\alpha(\omega)\boldsymbol{R}(\omega)[\mathsf{u}\mathsf{H}(\omega)]\,dP(\omega) \approx \sum_{z=1}^{Z} w_z H_\alpha(\boldsymbol{\theta}_z)\boldsymbol{R}(\boldsymbol{\theta}_z)[\boldsymbol{u}(\boldsymbol{\theta}_z)].$$

What this means is that for each evaluation of the residual Eq. (37) the spatial residual Eq. (36) has to be evaluated Z times—once for each $\boldsymbol{\theta}_z$ where one has to compute $\boldsymbol{R}(\boldsymbol{\theta}_z)[\boldsymbol{u}(\boldsymbol{\theta}_z)]$. Certainly this can be done independently and in parallel without any communication. But we would like to point out that instead of solving the system every time for each $\boldsymbol{\theta}_z$ as in the section 3.1.1 on MC and direct integration methods, here we only have to compute the residual.

Error $\cdot 10^4$ in Mean for Galerkin $\qquad$ $Prob\{u > 3.25\}$

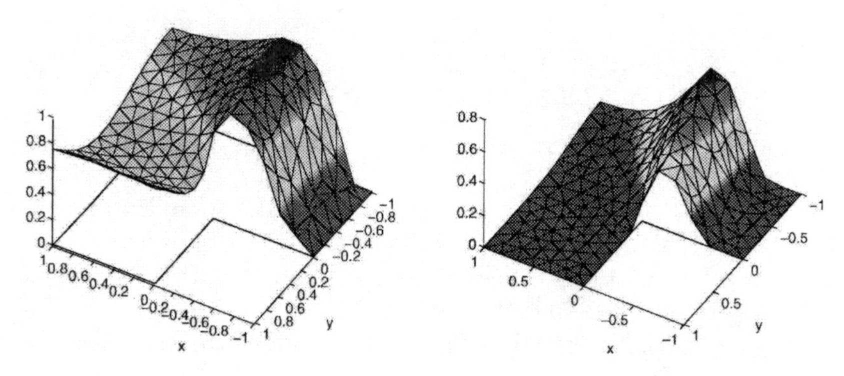

Figure 7. Error in Mean for PC-Galerkin and Example Statistic.

We compute the solution for the non-linear groundwater flow model. The soil parameter $\kappa(x,\omega)$ is chosen beta-distributed as indicated before. As a reference, the mean and standard deviation were computed by Smolyak quadrature S_6^6 —see [72, 73]—in altogether $Z = 6,188$ integration points. They are show in Fig. 6.

Next we compute the PC-expansion via the Galerkin method explained in this section, the error of which for the mean is shown in Fig. 7. We choose a polynomial chaos of degree 2 in 6 independent Gaussian variables as ansatz (28 stochastic functions). A spatial discretisation in 170 degrees of freedom was performed, totalling 4,760 non-linear equations. The BFGS solver required 19 iterations, and as the first iterations required line-searches, the residual had to be evaluated 24 times. The residual was

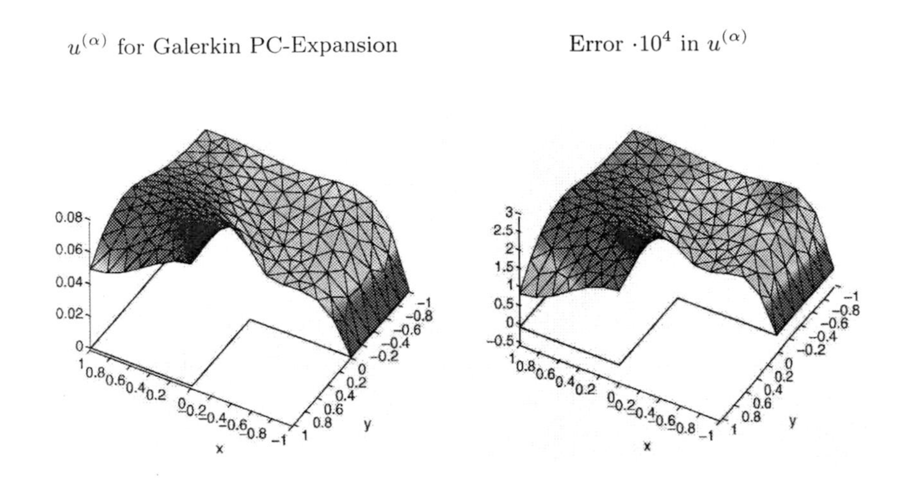

Figure 8. PC-Galerkin Vector and its Error.

integrated by the 5-stage Smolyak quadrature S_5^6 in $Z = 1,820$ integration points. As the evaluation in each integration point requires one integration in the spatial dimension, 43,680 spatial integrations were performed.

As we now have a "response surface", we show also in Fig. 7 as an example the computation of $p_{u_0}(x) = \Pr\{u(x) > 3.25\}$ for all points in the domain R. Next we show one of the PC-vectors in Fig. 8, and its error. It is small, at least in the "eye-ball" norm.

3.3. SPARSE REPRESENTATION OF THE MODEL OUTPUT

Just as for the model input, as described in section 2.3, a sparse representation in tensor products is desirable for the output. The procedure presented so far already allows the solution $\boldsymbol{u}(\omega)$ to be represented as a sum in products of vectors $\boldsymbol{u}^{(\alpha)}$ due to the spatial discretisation, and polynomials $H_\alpha(\omega)$ in random variables. This is already a tensor product structure, but it could be made even sparser without sacrificing too much accuracy.

3.3.1. *Error Estimators*

In order to be able to drop some of the terms in such a tensor sum, error estimators are needed. They allow to have a controlled error with a minimum amount of numerical effort, just as for spatial discretisations. Similarily to this case, error estimators can be constructed for different error measures. These can be different norms of the solution, or more generally seminorms.

These arise often as absolute values of linear functionals, and are already very close to goal-oriented error estimators.

These latter estimators arise from the idea that often one is just interested in a functional of the solution Eq. (17). Assume for simplicity that this functional is linear $\Psi(\mathsf{u}) = \langle\langle \psi, \mathsf{u} \rangle\rangle$; in case of a non-linear functional we have to look at its derivative.

With this defining vector ψ one solves the adjoint equation to the state equation $\mathsf{Au} = \mathsf{f}$:

$$\mathsf{A}^* \upsilon = \psi, \tag{38}$$

where A^* is the adjoint operator defined by $\forall \mathsf{u}, \upsilon : \langle\langle \mathsf{Au}, \upsilon \rangle\rangle = \langle\langle \mathsf{u}, \mathsf{A}^* \upsilon \rangle\rangle$. The solution υ of Eq. (38) may be regarded as sensitivity of the functional. Assume that $\hat{\mathsf{u}}$ is an approximate solution of Eq. (1). With this one has

$$\Psi(\mathsf{u}) - \Psi(\hat{\mathsf{u}}) = \langle\langle \psi, \mathsf{u} \rangle\rangle - \langle\langle \psi, \hat{\mathsf{u}} \rangle\rangle = \langle\langle \psi, \mathsf{u} - \hat{\mathsf{u}} \rangle\rangle = \\ \langle\langle \mathsf{A}^* \upsilon, \mathsf{u} - \hat{\mathsf{u}} \rangle\rangle = \langle\langle \upsilon, \mathsf{Au} - \mathsf{A}\hat{\mathsf{u}} \rangle\rangle = \langle\langle \upsilon, \mathsf{f} - \mathsf{A}\hat{\mathsf{u}} \rangle\rangle = \langle\langle \upsilon, \mathsf{r} \rangle\rangle, \tag{39}$$

where r is the residuum. One sees that it is possible to compute the error in the functional via the dual solution abd the residual. This method is termed therefore the dual weighted residual method.

3.3.2. *Adaptive Approximation*

The basic idea embodied in Eq. (39) may now be used to advantage in an adaptive solution process. By projecting the sensitivity υ onto each basis vector of the approximating subspace, we see whether this is an important contribution or not. In this way an adaptive refinement and coarsening may be carried out.

4. Conclusion and Outlook

This article has tried to illuminate some aspects of current computational methods for uncertainty quantification. Here the idea of random variables as functions in an infinite dimensional space which have to be approximated by elements of finite dimensional spaces has brought a new view to the field. It allows the Galerkin methods, which have been so useful in the approximation of partial differential and integral equations, to be applied to this case as well. With this comes the whole range of techniques developed in that area for fast and efficient solution, error estimation, adaptivity, etc. On a more philosophical level, a stochastic problem is in that way converted to a large deterministic one.

It has been attempted to contrast these new methods with the more traditional Monte Carlo approaches. For the new Galerkin methods, the polynomial chaos expansion—or something equivalent—seems to be fundamental. The Karhunen-Loève expansion can in that light rather be seen as a model reduction strategy. Its full use as a *data reduction* device is yet to be explored. The possibility of computing the coefficients of the stochastic expansion both as the solution of a large coupled system, and as evaluations of certain integrals of the response brings the circle back to the well-known Monte Carlo methods when seen as an integration rule. It remains yet to be seen which of these approaches is more effective in which situation.

References

1 R. J. Adler: *The Geometry of Random Fields.* John Wiley & Sons, Chichester, 1981.

2 K. E. Atkinson: *The Numerical Solution of Integral Equations of the Second Kind.* Cambridge University Press, Cambridge, 1997.

3 G. Augusti, A. Baratta and F. Casciati: *Probabilistic Methods in Structural Engineering.* Chapman and Hall, London, 1984.

4 I. Babuška: On randomized solutions of Laplace's equation. *Časopis pro Pěstováni Matematiky* **86** (1961) 269–275.

5 I. Babuška and P. Chatzipantelidis: On solving elliptic stochastic partial differential equations. *Comp. Meth. Appl. Mech. Engrg.* **191** (2002) 4093–4122.

6 I. Babuška, R. Tempone, and G. E. Zouraris: Galerkin finite element approximations of stochastic elliptic partial differential equations. *SIAM J Num. Anal.* **XX** (2004) xxx–yyy. Technical report, TICAM report 02-38, Texas Institute for Computational and Applied Mathematics, University of Texas, Austin, TX, 2002. `http://www.ticam.utexas.edu/reports/2002/0238.pdf`

7 I. Babuška, K.-M. Liu, and R. Tempone: Solving stochstic partial differential equations based on the experimental data. *Math. Models & Meth. in Appl. Sciences* **13** (3) (2003) 415–444

8 I. Babuška, F. Nobile, J. T. Oden and R. Tempone: Reliability, Uncertainty, Estimates: Validation and Verification. Technical report, TICAM report 04-05, Texas Institute for Computational Engineering and Sciences, University of Texas, Austin, TX, 2004. `http://www.ticam.utexas.edu/reports/2004/0405.pdf`

9 I. Babuška, R. Tempone and G. E. Zouraris: Solving elliptic boundary value problems with uncertain coefficients by the finite element method: the stochastic formulation. *Comp. Meth. Appl. Mech. Engrg.* **194** (2005) XXXX–YYYY.

10 Y. Ben-Haim and I. Elishakoff: *Convex models of uncertainty in applied mechanics.* Elsevier, Amsterdam, 1990.

11 J. S. Bendat and A. G. Piersol: *Engineering Applications of Correlation and Spectral Analysis.* John Wiley & Sons, Chichester, 1980.

12 F. E. Benth and J. Gjerde: Convergence rates for finite element approximations for stochastic partial differential equations. *Stochastics and Stochochastics Reports* **63** (1998) 313–326.

13 A. V. Balakrishnan: *Stochastic Differential Systems.* Lecture Notes in Economics and Mathematical Systems **84**. Springer-Verlag, Berlin, 1973.

14 P. Besold: *Solutions to Stochastic Partial Differential Equations as Elements of Tensor Product Spaces.* Doctoral thesis, Georg-August-Universität, Göttingen, 2000.

15 R. N. Bracewell: *The Fourier Transform and Its Applications.* McGraw-Hill, New York, NY, 1978.

16 R. E. Caflisch: Monte Carlo and quasi-Monte Carlo methods. *Acta Numerica* **7** (1998) 1–49.

17 G. Christakos: *Random Field Models in Earth Sciences.* Academic Press, San Diego, CA, 1992.

18 P. G. Ciarlet: *The Finite Element Method for Elliptic Problems.* North-Holland, Amsterdam, 1978.

19 R. W. Clough and J. Penzien: *Dynamics of Structures.* McGraw-Hill, New York, NY, 1975.

20 R. Courant and D. Hilbert: *Methods of Mathematical Physics.* John Wiley & Sons, Chichester, 1989.

21 M. K. Deb, I. Babuška, and J. T. Oden: Solution of stochastic partial differential equations using Galerkin finite element techniques. *Comp. Meth. Appl. Mech. Engrg.* **190** (2001) 6359–6372.

22 J. E. Dennis, Jr. and R. B. Schnabel: *Numerical Methods for Unconstrained Optimization and Nonlinear Equations.* SIAM, Philadelphia, PA, 1996.

23 A. Der Kuireghian and J.-B. Ke: The stochastic finite element method in structural reliability. *J. Engrg. Mech.* **117** (1988) 83–91.

24 I. Doltsinis: Inelastic deformation processes with random parameters—methods of analysis and design. *Comp. Meth. Appl. Mech. Engrg.* **192** (2003) 2405–2423.

25 J. L. Doob: *Stochastic Processes.* John Wiley & Sons, Chichester, 1953.

26 I. Elishakoff: Are probabilistic and antioptimization methods interrelated? In [27].

27 I. Elishakoff (ed.): *Whys and Hows in Uncertainty Modelling—Probability, Fuzziness, and Anti-Optimization.* Springer-Verlag, Berlin, 1999.

28 Ph. Frauenfelder, Chr. Schwab and R. A. Todor: Finite elements for elliptic problems with stochastic coefficients. *Comp. Meth. Appl. Mech. Engrg.* **194** (2005) 205–228.

29 C. W. Gardiner: *Handbook of Stochastic Methods.* Springer-Verlag, Berlin, 1985.

30 T. Gerstner and M. Griebel: Numerical integration using sparse grids. *Numer. Algorithms* **18** (1998) 209–232.

31 R. Ghanem: Ingredients for a general purpose stochastic finite elements implementation. *Comp. Meth. Appl. Mech. Engrg.* **168** (1999) 19–34.

32 R. Ghanem: Stochastic finite elements for heterogeneous media with multiple random non-Gaussian properties. *ASCE J. Engrg. Mech.* **125** (1999) 24–40.

33 R. Ghanem and P. D. Spanos: *Stochastic Finite Elements—A Spectral Approach.* Springer-Verlag, Berlin, 1991.

34 R. Ghanem and R. Kruger: Numerical solution of spectral stochastic finite element systems. *Comp. Meth. Appl. Mech. Engrg.* **129** (1999) 289–303.

35 I. I. Gihman and A. V. Skorokhod: *Stochastic Differential Equations*. Springer-Verlag, Berlin, 1972.

36 I. I. Gihman and A. V. Skorokhod: *Introduction to the Theory of Random Processes*. Dover Publications, New York, NY, 1996.

37 J. Glimm and D. H. Sharp: Predicition and the quantification of uncertainty. *Physica D* **133** (1999) 152–170.

38 G. Golub and C. F. Van Loan: *Matrix Computations*. Johns Hopkins University Press, Baltimore, MD, 1996.

39 M. Grigoriu: *Applied non-Gaussian Processes*. Prentice Hall, Englewood Cliffs, NJ, 1995.

40 M. Grigoriu: *Stochastic Calculus—Applications in Science and Engineering*. Birkhäuser Verlag, Basel, 2002.

41 A. Haldar and S. Mahadevan: *Reliability assessment using stochastic finite element analysis*. John Wiley & Sons, Chichester, 2000.

42 H. Holden, B. Øksendal, J. Ubøe, and T.-S. Zhang: *Stochastic Partial Differential Equations*. Birkhäuser Verlag, Basel, 1996.

43 S. Janson: *Gaussian Hilbert Spaces*. Cambridge University Press, Cambridge, 1997.

44 H. Jeggle: *Nichtlineare Funktionalanalysis*. BG Teubner, Stuttgart, 1979.

45 M. Jardak, C.-H. Su, and G. E. Karniadakis: Spectral polynomial chaos solutions of the stochastic advection equation. *SIAM J. Sci. Comput.* **17** (2002) 319–338.

46 G. Kallianpur: *Stochastic Filtering Theory*. Springer-Verlag, Berlin, 1980.

47 D. Kavetski, S. W. Franks and G. Kuczera: Confronting input uncertainty in environmental modelling. pp. 49–68 in: Q. Duan, H. V. Gupta, S. Sorooshian, A. N. Rousseau and R. Turcotte (eds.), *Calibration of Watershed Models*, AGU Water Science and Application Series, Vol. 6.

48 A. Keese: A review of recent developments in the numerical solution of stochastic PDEs (stochastic finite elements). Technical report, Informatikbericht 2003-6, Institute of Scientific Computing, Department of Mathematics and Computer Science, Technische Universität Braunschweig, Brunswick, 2003.
`http://opus.tu-bs.de/opus/volltexte/2003/504/`

49 A. Keese and H. G. Matthies: Parallel solution of stochastic PDEs. *Proc. Appl. Math. Mech.* **2** (2003) 485–486.

50 A. Keese and H. G. Matthies: Efficient solvers for nonlinear stochastic problems. In *Proceedings of WCCM V—5th World Congress Comput. Mech.*, Vienna, (7–12 July 2002).
`http://wccm.tuwien.ac.at/`

51 A. Keese and H. G. Matthies: Parallel solution methods for stochastic systems. pp. 2023–2025 in K.-J. Bathe (ed.), *Computational Fluid and Solid Mechanics—Proceedings of the 2nd MIT conference*. Elsevier, Amsterdam, 2003.

52 A. Keese and H. G. Matthies: Adaptivity and sensitivity for stochastic problems. pp. 311–316 in P. D. Spanos and G. Deodatis (eds.), *Computational Stochastic Mechanics 4*. Millpress, Rotterdam, 2003.

53 A. Keese and H. G. Matthies: Numerical methods and Smolyak quadrature for nonlinear stochastic partial differential equations. Technical report, Informatikbericht 2003-5, Institute of Scientific Computing, Department of Mathematics and Computer Science, Technische Universität Braunschweig, Brunswick, 2003.
`http://opus.tu-bs.de/opus/volltexte/2003/471/`

54 A. Keese and H. G. Matthies: Parallel Computation of Stochastic Groundwater Flow. Technical report, Informatikbericht 2003-9, Institute of Scientific Computing, Department of Mathematics and Computer Science, Technische Universität Braunschweig, Brunswick, 2003.
`http://opus.tu-bs.de/opus/volltexte/2003/505/`

55 A. Keese: *Numerical Solution of Systems with Stochastic Uncertainties—A General Purpose Framework for Stochastic Finite Elements.* Doctoral thesis, Technische Universität Braunschweig, Brunswick, 2003.

56 M. Kleiber and T. D. Hien: *The Stochastic Finite Element Method. Basic Perturbation Technique and Computer Implementation.* John Wiley & Sons, Chichester, 1992.

57 P. E. Kloeden and E. Platen: *Numerical Solution of Stochastic Differential Equations.* Springer-Verlag, Berlin, 1995.

58 A. Khuri and J. Cornell: *Response Surfaces: Designs and Analyses.* Marcel Dekker, New York, NY, 1987.

59 I. Kramosil: *Probabilistic Analysis of Belief Functions.* Kluwer, Rotterdam?, 2001.

60 P. Krée and C. Soize: *Mathematics of Random Phenomena.* D. Reidel, Dordrecht, 1986.

61 W.-K. Liu, T. Belytschko, and A. Mani: Probabilistic finite elements for nonlinear structural dynamics. *Comp. Meth. Appl. Mech. Engrg.* **56** (1986) 61–86.

62 M. Loève: *Probability Theory.* Springer-Verlag, Berlin, 1977.

63 L. D Lutes: A perspective on state-space stochastic analysis. *Proc. 8th ASCE Speciality Conf. Prob. Mech. Struct. Reliability,* 2000.

64 H. O. Madsen, S. Krenk and N.C. Lind: *Methods of Structural Safety.* Prentice Hall, Englewood Cliffs, NJ, 1986.

65 G. Maglaras, E. Nikolaidis, R. T. Hafka and H. H. Cudney: Analytical-experimental comparison of probabilistic and fuzzy set methods for designing under uncertainty. In [75].

66 P. Malliavin: *Stochastic Analysis.* Springer-Verlag, Berlin, 1997.

67 H. Matthies and G. Strang: The solution of nonlinear finite element equations. *Int. J. Numer. Meth. Engrg.* **14** (1979) 1613–1626.

68 H. G. Matthies, C. E. Brenner, C. G. Bucher, and C. Guedes Soares: Uncertainties in probabilistic numerical analysis of structures and solids—stochastic finite elements. *Struct. Safety* **19** (1997) 283–336.

69 H. G. Matthies and C. G. Bucher: Finite elements for stochastic media problems. *Comp. Meth. Appl. Mech. Engrg.* **168** (1999) 3–17.

70 H. G. Matthies and A. Keese: Multilevel solvers for the analysis of stochastic systems. pp. 1620–1622 in K.-J. Bathe (ed.), *Computational Fluid and Solid Mechanics—Proceedings of the 1st MIT conference.* Elsevier, Amsterdam, 2001.

71 H. G. Matthies and A. Keese: Fast solvers for the white noise analysis of stochastic systems. *Proc. Appl. Math. Mech.* **1** (2002) 456–457.

72 H. G. Matthies: Computational aspects of probability in non-linear mechanics. In: A. Ibrahimbegović and B. Brank (eds.), *Multi-physics and multi-scale computer models in non-linear analysis and optimal design of engineering structures under extreme conditions*, NATO-ARW Bled 2004. Narodna i univerzitetna knjižica, Ljubljana, 2004. `http://arw-bled2004.scix.net/Files/acceptedpapers/stoch-mech.pdf`

73 H. G. Matthies and A. Keese: Galerkin methods for linear and nonlinear elliptic stochastic partial differential equations. *Comp. Meth. Appl. Mech. Engrg.* **194** (2005) 1295–1331. `http://opus.tu-bs.de/opus/volltexte/2003/489/`

74 R. E. Melchers: *Structural Reliability Analysis and Prediction.* John Wiley & Sons, Chichester, 1999.

75 H. G. Natke and Y. Ben-Haim (eds.): *Uncertainty: Models and Measures.* Akademie-Verlag, Berlin, 1997.

76 A. Naess and S. Krenk (eds.): *IUTAM Symposium on Advances in Nonlinear Stochastic Mechanics.* Series SOLID MECHANICS AND ITS APPLICATIONS : Vol. 47, Kluwer, Dordrecht, 1996.

77 H. Niederreiter: *Random Number Generation and Quasi-Monte Carlo Methods.* SIAM, Philadelphia, PA, 1992.

78 E. Novak and K. Ritter: The curse of dimension and a universal method for numerical integration. pp. 177–188 in G. Nürnberger, J. W. Schmidt, and G. Walz (eds.), *Multivariate Approximation and Splines, ISNM.* Birkhäuser Verlag, Basel, 1997.

79 E. Novak and K. Ritter: Simple cubature formulas with high polynomial exactness. *Constr. Approx.* **15** (1999) 499–522.

80 J. T. Oden: *Qualitative Methods in Nonlinear Mechanics.* Prentice-Hall, Englewood Cliffs, NJ, 1986.

81 J. T. Oden, I. Babuška, F. Nobile, Y. Feng and R. Tempone: Theory and methodology for estimation and control of errors due to modeling, approximation, and uncertainty. *Comp. Meth. Appl. Mech. Engrg.* **194** (2005) 195–204.

82 B. Øksendal: *Stochastic Differential Equations.* Springer-Verlag, Berlin, 1998.

83 M. Papadrakakis and V. Papadopoulos: Robust and efficient methods for stochastic finite element analysis using Monte Carlo simulation. *Comp. Meth. Appl. Mech. Engrg.* **134** (1996) 325–340.

84 A. Papoulis: *Probability, Random Variables, and Stochastic Processes.* McGraw-Hill, New York, NY, 1984.

85 M. Pellissetti and R. Ghanem: Iterative solution of systems of linear equations arising in the context of stochastic finite elements. *Adv. Engrg. Softw.* **31** (2000) 607–616.

86 K. Petras: Smolpack—a software for Smolyak quadrature with delayed Clenshaw-Curtis basis-sequence. `http://www-public.tu-bs.de:8080/~petras/software.html`.

87 K. Petras: Fast calculation of coefficients in the Smolyak algorithm. *Numer. Algorithms* **26** (2001) 93–109.

88 F. Riesz and B. Sz.-Nagy: *Functional Analysis.* Dover Publications, New York, NY, 1990.

89 K. K. Sabelfeld: *Monte Carlo Methods in Boundary Value Problems.* Springer-Verlag, Berlin, 1991.

90 S. Sakamoto and R. Ghanem: Simulation of multi-dimensional non-Gaussian non-stationary random fields. *Prob. Engrg. Mech.* **17** (2002) 167–176.

91 G. I. Schuëller: A state-of-the-art report on computational stochastic mechanics. *Prob. Engrg. Mech.* **14** (1997) 197–321.

92 G. I. Schuëller: Recent developments in structural computational stochastic mechanics. pp. 281–310 in B.H.V. Topping (ed.), *Computational Mechanics for the Twenty-First Century.* Saxe-Coburg Publications, Edinburgh, 2000.

93 G. I. Schuëller and P. D. Spanos (eds.): *Monte Carlo Simulation.* Balkema, Rotterdam, 2001.

94 C. Schwab and R.-A. Todor: Sparse finite elements for stochastic elliptic problems—higher order moments. *Computing* **71** (2003) 43–63.

95 G. Shafer: *A Mathematical Theory of Evidence.* Princeton University Press, Princeton, 1976.

96 M. Shinozuka and G. Deodatis: Simulation of stochastic processes and fields. *Prob. Engrg. Mech.* **14** (1997) 203–207.

97 A. V. Skorokhod: *Studies in the Theory of Random Processes.* Dover, New York, NY, 1985.

98 S. A. Smolyak: Quadrature and interpolation formulas for tensor products of certain classes of functions. *Sov. Math. Dokl.* **4** (1963) 240–243.

99 G. Strang: *Introduction to Applied Mathematics.* Wellesley-Cambridge Press, Wellesley, MA, 1986.

100 G. Strang and G. J. Fix: *An Analysis of the Finite Element Method.* Wellesley-Cambridge Press, Wellesley, MA, 1988.

101 B. Sudret and A. Der Kiureghian: Stochastic finite element methods and reliability. A state-of-the-art report. Technical Report UCB/SEMM-2000/08, Department of Civil & Environmental Engineering, University of California, Berkeley, CA, 2000.

102 T. G. Theting: Solving Wick-stochastic boundary value problems using a finite element method. *Stochastics and Stochastics Reports* **70** (2000) 241–270.

103 G. Våge: Variational methods for PDEs applied to stochastic partial differential equations. *Math. Scand.* **82** (1998) 113–137

104 E. Vanmarcke, M. Shinozuka, S. Nakagiri and G. I. Schuëller: Random fields and stochastic finite elements. *Struct. Safety* **3** (1986) 143–166.

105 E. Vanmarcke: *Random Fields: Analysis and Synthesis.* The MIT Press, Cambridge, MA, 1988.

106 P. Walley: *Statistical Reasoning with Imprecise Probabilities.* Chapman and Hall, Boca Raton, 1991.

107 N. Wiener: The homogeneous chaos. *Amer. J. Math.* **60** (1938) 897–936.

108 D. Xiu and G. E. Karniadakis: The Wiener-Askey polynomial chaos for stochastic differential equations. *SIAM J. Sci. Comput.* **24** (2002) 619–644.

109 D. Xiu, D. Lucor, C.-H. Su, and G. E. Karniadakis: Stochastic modeling of flow-structure interactions using generalized polynomial chaos. *ASME J. Fluid Engrg.* **124** (2002) 51–69.

110 D. Xiu and G. E. Karniadakis: Modeling uncertainty in steady state diffusion problems via generalized polynomial chaos. *Comp. Meth. Appl. Mech. Engrg.* **191** (2002) 4927–4948.

111 X. F. Xu and L. Graham-Brady: Nonlinear analysis of heterogeneous materials with GMC technique. *Proc. 16th ASCE Engrg. Mech. Conf.*, Seattle, WA, 2003.

112 L. A. Zadeh: Fuzzy sets. *Information and Control* **8** (1965) 338–353.

113 O. C. Zienkiewicz and R. L. Taylor: *The Finite Element Method.* Butterwort-Heinemann, Oxford, 5th ed., 2000.

OPTIMUM PERFORMANCE-BASED RELIABILITY DESIGN OF STRUCTURES

M. Papadrakakis *(e-mail:mpapadra@central.ntua.gr)*

M. Fragiadakis, N.D. Lagaros
Institute of Structural Analysis & Seismic Research
National Technical University of Athens
Zografou Campus, Athens 15780, Greece

Abstract. The objective of this paper is to present a performance based design procedure for steel structures in the framework of structural optimization. The structural performance is evaluated by means of the reliability demand and resistance methodology of FEMA-350 (Federal Emergency Management Agency) guidelines where the uncertainties and randomness in capacity and seismic demand are taken into account in a consistent manner. The structure has to be able to respond for different hazard levels with a desired confidence. Both Nonlinear Static and Nonlinear Dynamic analysis procedures are used in order to obtain the response for two hazard levels. The design procedure is performed in a structural optimization environment, where the Evolution Strategies algorithm is implemented for the solution of the optimization problem. In order to handle the excessive computational cost the inelastic time history analyses are performed in a parallel computing environment. The objective of the study is to obtain the design with the least material weight, and thus with less cost, that is capable to respond with the desired confidence for each performance level following the specifications of FEMA-350.

Key words: Performance-based design, structural optimization, evolutionary algorithms, inelastic time history analysis.

1. Introduction

Recent advances in the field of computational mechanics and more specifically in structural optimization and parallel computing have made possible the move from the traditional trial and error design procedures towards fully automated design procedures where a structural optimization search

A. Ibrahimbegovic and I. Kozar (eds.),Extreme Man-Made and Natural Hazards in Dynamics of Structures, 137–159.

engine is used. This is mostly attributed to the rapid development of evolutionary based optimizers, which are capable of handling complicated structural problems at the expense of more optimization cycles. These algorithms imitate biological evolution in nature and combine the concept of the survival of the fittest with evolutionary operators to form a robust global search mechanism. For the complex and realistic non-linear structural optimization problems that are investigated in this study these optimizers are the only reliable approach, since most mathematical programming optimizers will converge to a local optima or may not converge at all. Furthermore, Evolutionary Algorithms (EA) do not require the calculation of gradients of the constraints, as opposed to mathematical programming algorithms, and thus structural design code checks can be implemented in an optimization environment as constraints in a straightforward manner.

Since the early seventies structural optimization has been the subject of intensive research and several different approaches have been advocated for the optimal design of structures in terms of optimization methods or problem formulation. Most of the attention of the engineering community has been directed towards the optimum design of structures under static loading conditions with the assumption of linear elastic structural behaviour. For a large number of real-life structural problems assuming linear response and ignoring the dynamic characteristics of the seismic action during the design phase may lead to structural configurations highly vulnerable to future earthquakes. Furthermore, seismic design codes suggest that under severe earthquake events the structures should be designed to deform inelastically due to the large intensity inertia loads imposed.

Performance-Based Design (PBD) has been introduced recently to increase the structural safety against natural hazards, and in the case of earthquakes, to make them having a predictable and reliable performance. In other words, the structures should be able to resist earthquakes in a quantifiable manner and to present levels of desired possible damage. Therefore, the modern conceptual approach of seismic structural design is that the structure should meet performance-based objectives for a number of different hazard levels ranging form earthquakes with a small intensity but also with a small return period to more destructive events with large return periods. The current state of practice in performance-based engineering for steel structures is defined by US guidelines FEMA-350 (Federal Emergency Management Agency) for the design of new buildings and by FEMA-356 for the seismic rehabilitation of existing buildings. These guidelines introduce procedures that can be considered as the first significant diversification from prescriptive building design codes such as EC8 and IBC. Prescriptive codes require that the engineer has to ensure that the structure satisfies a number of checks, and in so doing, the structure is supposed to be safe for occupancy. According to a prescriptive code the strength of the structure is evaluated at one loading level corresponding to a single limit state. Serviceability is usually checked as well to ensure that the structure will not deflect or vibrate excessively during its service life.

The structural performance during an earthquake depends highly on a number of parameters which in essence are uncertain. Such parameters are material properties and workmanship as well as the hysteretic behaviour of the members and the joints. Large randomness also exists in the intensity and the characteristics of earthquake ground motion. Furthermore, uncertainty exists in the analytical procedure that would be adopted and on the mathematical model of the structure. In order to account for as many as possible of the above uncertainties a reliability-based, performance-oriented approach is proposed by FEMA-350 guidelines. This approach explicitly accounts for uncertainties and randomness is seismic demand and capacities in a consistent manner and satisfies the defined reliability performance objectives that correspond to various damage states and seismic hazards. In other words during the optimization process each candidate design must be able to respond with a predefined confidence level.

In the present study it is demonstrated how structural optimization can be applied within the reliability based framework of FEMA-350 for the design of steel frames, where the demand and resistance factor concept is adopted. The Evolution Strategies algorithm (Lagaros *et al.* 2004), specifically tailored to meet the characteristics of the problem at hand, is adopted for the solution of the optimization problem. The proposed methodology is applied using two alternative analytical procedures specified by FEMA-350 as the Nonlinear Static Procedure (NSP) and Nonlinear Dynamic Procure (NDP). In order to handle the increased computational cost of the large number of inelastic time history analyses required by the NDP procedure, parallel processing in a cluster of PCs is performed. In the numerical study presented it is demonstrated how structural optimization can be used to provide more economic designs having better control on structural performance.

The rest of the paper is organized as follows. In the second section, the performance-based design procedure employed in this study is discussed. The acceptance criteria of FEMA-350 and the constraints of the optimization problem are also outlined in this section. The formulation of the optimization problem is subsequently presented where the objective function and the ES optimization algorithm considered in the current study are briefly presented. Finally, a case study demonstrating the efficiency of the proposed design procedure is presented in Section 4 followed by the concluding remarks.

2. Progress on Seismic Design Using Structural Optimization Procedures

Structures built according to the provisions of contemporary seismic design codes are designed to respond inelastically under strong earthquakes. However, this behaviour is taken into account during the design phase only implicitly. This practice can be mainly attributed to the fact that non-linear timehistory analysis results in increased computational complexity and requires excessive computational resources. In general, a comparatively limited number of studies have been performed for the solution of structural optimization problems under dynamic loading conditions considering inelastic behaviour. One of the earliest studies on the subject is the work of Polak *et al.* (1976) who minimized the cost of multi-storey frame structures using inelastic timehistory analysis of simplified structural models. Bhatti and Pister (1981) and Balling *et al.* (1981) introduced optimization procedures based on nonlinear timehistory analysis. Several limit-states were considered, while different performance objectives are adopted for each limit-state. Pezeshk (1998) presented an integrated non-linear analysis and optimal minimum weight design methodology, for a simple 3D frame under an equivalent pseudo-static loading scheme.

Following recent developments in structural design procedures (e.g. Charney 2000, Fragiadakis *et al.* 2006a,b), a number of researchers have integrated structural optimization in the framework of Performance-Based Earthquake Engineering (PBEE). Ganzerli *et al.* (2000) implemented a performance-based optimization procedure of RC frames using convex optimum design models and a standard mathematical optimizer. Esteva *et al.* (2002) presented a performance and reliability-based optimization under life-cycle considerations using special type damage functions in conjunction with pushover analysis. Gong (2003) combined pushover analysis and a dual optimization algorithm for the optimal design of steel building frames for various performance levels. Chan and Zou (2004) proposed a two-stage, optimality criteria-based, optimization procedure for 2D concrete frame structures.

The advancements achieved by Evolutionary Algorithms over the past years made possible the solution of real-scale structural optimization problems incorporating design code-based constraints. Two of the earliest studies, where EA were employed for the optimum seismic design of structures, are those of Kocer and Arora (1999) and (2002) for the optimal design of H-frame transition poles and latticed towers conducting nonlinear timehistory analysis. They proposed the use of Genetic Algorithms (GA) and Simulated Annealing (SA) for the solution of discrete variable problems although the computational time required was excessive. Beck *et al.* (1999) proposed a multi-criteria GA-based structural optimisation approach under uncertainties in both structural properties and seismic hazard.

Cheng *et al.* (2000) used a multi-objective GA-based formulation incorporating game and fuzzy set theory for the optimum design of seismically loaded 2D frames. Game theory is used in order to achieve a compromise solution that satisfies all competing objectives in the multi-objective optimization problem, while with the implementation of fuzzy set theory they transformed the constrained optimization problem into an unconstrained one which in turn was solved with a Pareto GA-based optimizer. Liu *et al.* (2003) proposed a GA-based multi-objective structural optimization procedure for steel frames using pushover analysis considering minimum weight, life-cycle cost and design complexity criteria as the objective functions of the optimization problem. The same authors recently extended their methodology in the framework of PBEE (Liu *et al.* 2004 and 2005).

As opposed to the aforementioned studies on the optimum seismic design of structures and the different approaches proposed in the past, this work is focused on using structural optimization as a tool for comparing alternative design procedures. The set up of the optimization problem differs according to the design procedure examined, while several issues regarding nonlinear timehistory analysis and structural optimization, in the framework of earthquake engineering practice, are discussed. European seismic design regulations for steel moment resisting frames are used as the testbed of the optimum design procedures, while the provisions of U.S. guidelines (FEMA-356) are also taken into consideration. In total three alternative design procedures for seismic design are evaluated in the framework of structural optimization: modal analysis, linear timehistory analysis and nonlinear timehistory analysis. The dynamic analysis based procedures are considered using both natural and artificial ground motion records.

3. Performance-Based Design Optimization

Prescriptive building codes cannot provide acceptable levels of building life-cycle performance because they include a great number of provisions designed to ensure adequate strength of structural members and, indirectly or implicitly, of the overall structural strength. The basic philosophy of a PBD procedure is to allow engineers to determine explicitly the seismic demands at predefined performance levels by introducing higher level design checks.

3.1. PERFORMANCE LEVELS AND OBJECTIVES

Building performance is a combination of the performance of both structural and nonstructural components. FEMA-350 describes the overall levels of structural and nonstructural damage that buildings are expected to meet using two performance levels, the *Collapse Prevention (CP)* level and *Immediate Occupancy (IO)* level. The CP performance level corresponds to a hazard level with a 2% probability of exceedance in 50 years (2/50) and is defined as the post earthquake damage state in which the structure is on the verge of experiencing partial or total collapse. Substantial damage to the structure has occurred, potentially including significant degradation in the stiffness and strength of the lateral-force-resisting system and large permanent lateral deformation of the structure. The IO performance level corresponds to a hazard level with a 50% probability of exceedance in 50 years (50/50) and is defined as the post earthquake damage state in which only limited structural damage has occurred. Damage is anticipated to be so slight that no repair is required after the earthquake. Buildings that meet this performance level should be safe for immediate post earthquake occupancy.

FEMA-350 guidelines provide the framework for the estimation of the level of confidence that a structure will be able to achieve a desired performance objective. The main difference with previous FEMA-356 guidelines is that the performance objectives are not expressed in a deterministic manner. Each performance objective set by FEMA-350 guidelines consist of a hazard level, for which the corresponding performance is to be achieved. For example, a design may be determined to provide at least a 90% level of confidence of the performance objective for the CP, for earthquake hazards with a 2% probability of exceedance in 50 years, and with a 50% or higher level of confidence for IO, for earthquake hazards with a 50% probability of exceedance in 50 years. The confidence levels are obtained by means of the factored demand to capacity ratio where a number of parameters are considered that introduce explicitly the uncertainty in prediction of seismic demand and in the analytical procedure adopted as well as the uncertainty and randomness in the prediction of structural capacity.

3.2. ACCEPTANCE CRITERIA – OPTIMIZATION CONSTRAINTS

A number of constraints must be expressed for the structural optimization problem in order the optimum design to be acceptable in practice. FEMA-350 and EC3 provisions are adopted in order to formulate these constraints. Each candidate design is checked against these constraints, while care should be taken during the optimization process in order to perform first the checks that require less computational effort. Therefore, prior to any analysis the column to beam strength ratio is calculated and also a check is performed on whether the sections chosen are of class 1, as EC3 suggests. The later check is necessary in order to ensure that the members have the capacity to develop their full plastic moment and rotational ductility, while the former is necessary in order to have a design consistent with the 'strong column-weak beam' design philosophy. Subsequently, all EC3 checks must be satisfied for the gravity loads using the following load combination:

$$S_d = 1.35\sum_j G_{kj} \text{"+"} 1.50\sum_i Q_{ki} \tag{1}$$

where "+" implies "to be combined with", the summation symbol "Σ" implies "the combined effect of", G_{kj} denotes the characteristic value "k" of the permanent action j and Q_{ki} refers to the characteristic value "k" of the variable action i.

If the above constraints are satisfied, then nonlinear static or dynamic analysis is performed for both hazard levels for the selected ground motion records. When the dynamic analysis procedure is adopted, the maximum interstorey drift value or the median drift value is obtained depending on the number of records used. According to FEMA standards, if less than seven records are used the maximum drift value must be considered, otherwise the median value must be used instead. For the analysis step, for both analytical procedures earthquake loading is applied using the following load combination:

$$S_d = \sum_j G_{kj} \text{"+"} E_d \text{"+"} \sum_i \psi_{2i} Q_{ki} \tag{2}$$

where E_d is the design value of the seismic action for the two hazard levels respectively and ψ_{2i} is the combination coefficient for quasi permanent value of the variable action i, here taken equal to 0.30.

Table1. Parameters for the Evaluation of Confidence for mid-rise Special Moment frames according to FEMA-350.

	IO Performance Level		CP Performance Level	
	NSP	NDP	NSP	NDP
β_{UT}	0.20	0.20	0.40	0.40
C	0.02	0.02	0.10	0.10
ϕ	1.00	1.00	0.85	0.85
γ	1.40	1.40	1.20	1.20
γ_a	1.45	1.02	0.99	1.06

Acceptability of the structural performance is evaluated using the FEMA-350 procedure where the level of confidence that the building has the ability to meet the desired performance objectives is determined. The confidence level is determined through evaluation of the factored-demand-to-capacity ratio which is given by the expression:

$$\lambda = \frac{\gamma \cdot \gamma_\alpha \cdot D}{\phi \cdot C} \tag{3}$$

where:

C is the median estimate of the capacity of the structure, expressed as the maximum interstorey drift demand.

D is the calculated demand for the structure, can be obtained either from nonlinear static or nonlinear dynamic analysis.

γ is the demand variability factor that accounts for the variability inherent in the prediction of demand related to assumptions made in structural modelling and prediction of the character of ground motion.

γ_α is the analysis uncertainty factor that accounts for bias and uncertainty, associated with the specific analytical procedure used to estimate demand as a function of ground shaking intensity.

φ is the resistance factor that accounts for the uncertainty and variability inherent in the prediction of structural capacity as a function of ground shaking intensity.

Detailed description and the theoretical background of these parameters are provided by Yun *et al.* (2002) and by FEMA-350 (Appendix A). In the present study the above parameters take values from FEMA-350 tables. Table 1 lists the values adopted in this study for mid-rise moment frames with rigid connections for both nonlinear static and nonlinear dynamic analysis. Once λ of Eq. (3) is known, the following expression is solved for K_X :

$$\lambda = \exp\left\{-b\beta_{UT}(K_X - k\beta_{UT}/2)\right\} \tag{4}$$

where:

K_X is the standard Gaussian variant associated with probability x of not being exceeded.

b is a coefficient relating the incremental change in demand to an incremental change in ground shaking intensity, at the hazard level of interest, typically taken as having a value of 1.0.

β_{UT} is an uncertainty measure coefficient dependent on a number of sources of uncertainty in the estimation of structural demands and capacities. Sources of uncertainty include, for example, the actual material properties and the effective damping. If the uncertainty associated with each source i is denoted as β_{Ui}, then β_{UT} is obtained as:

$$\beta_{UT} = \sqrt{\sum_i \beta_{Ui}^2} \tag{5}$$

k is the slope of the hazard curve when plotted on a log-log scale, where the hazard curve is of the form:

$$H_{Si}(S_i) = k_0 S_i^{-k} \tag{6}$$

where: $H_{Si}(S_i)$ is the probability of ground shaking having a spectral response acceleration greater than S_i and k_0 is a constant that depends on the seismicity of the site.

The confidence level for each performance objective is obtained once the factored demand to capacity ratio (Eq. (3)) is computed and Eq. (4) is solved for K_X. The confidence level for each performance level can be easily obtained from conventional probability tables. For example, if K_X is 1.28 the confidence level is 90%. According to FEMA-350 when the global behaviour is limited by interstorey drift, the minimum confidence levels are 90% for the CP and 50% for the IO level. In order to implement the above procedure in a structural optimization environment the factored demand to capacity ratio is obtained for each performance level and a constraint in the maximum interstorey drift is formulated by rearranging Eq. (3):

$$\frac{\lambda \cdot \phi \cdot C}{\gamma \cdot \gamma_a} - D \leq 0 \tag{7}$$

3.3. FINITE ELEMENT MODELING

In order to perform dynamic analysis considering inelastic behaviour there is a need for a detailed and accurate simulation of the structure in the areas where plastic deformations are expected to occur. Given that the plastic hinge approach has limitations in terms of accuracy, especially under dynamic loading, the fiber approach was adopted in this study. A fiber beam element based on the natural mode method has been employed proposed by Argyris *et al.* (1998). This finite element formulation is based on the natural mode method where the description of the displacement field along the beam is performed with quantities having a clear physical meaning.

Each structural element is discretized into a number of sections, and each section is further divided into a number of fibers (Figure 1), which are restrained to beam kinematics. The sections are located either at the centre of the element or at its Gaussian integration points. The main advantage of the fiber approach is that each fiber has a simple uniaxial material model allowing an easy and efficient implementation of the inelastic behaviour. This approach is considered suitable for inelastic beam-column elements under dynamic loading and provides reliable solution compared to other formulations for inelastic analysis of frame structures at the expense of increased computational demands.

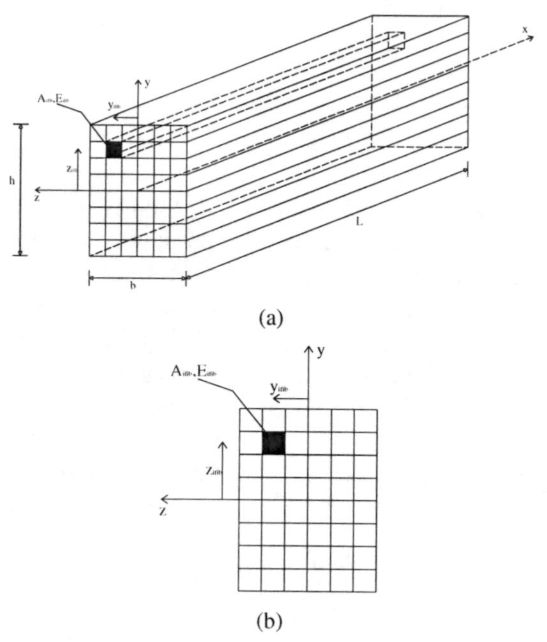

Figure 1. Modelling of inelastic behaviour-The fiber approach.

An adaptive discretization is used where a detailed simulation is restricted to the regions of the joints. Thus, the vicinity (usually 2.5% to 10% of element's length) of beam and column joints is discretized with a denser mesh of beam elements, while the remaining part of the member is discretized with linear elastic elements.

A simple bilinear stress-strain relationship with kinematic hardening is adopted. Studies have shown that this law is adequate and gives accurate results for many practical applications. The stress-strain relationship that allows for strain hardening is:

$$\begin{aligned} \sigma &= E \cdot \varepsilon &&, \text{ for } \varepsilon \le \varepsilon_y \\ \sigma &= E \cdot \varepsilon_y + E_{st} \cdot \left(\varepsilon - \varepsilon_y\right), &&\text{ for } \varepsilon > \varepsilon_y \end{aligned} \tag{8}$$

where σ_y and ε_y are the yield strain and the corresponding yield stress, E is the elastic Young's modulus and E_{st} is the strain-hardening modulus.

4. Structural Optimization

The objective of structural optimization is to find a proper selection of the design variables in order to achieve the minimum weight of the structure under some behavioural constraints (acceptance criteria) which are imposed by structural design codes. Three are the main classes of structural optimization problems: sizing, shape and topology. In sizing optimization problems the aim is usually to minimize the weight of the structure under certain behavioural constraints on stresses and displacements. The design variables are most frequently chosen to be dimensions of the cross-sectional areas of the members of the structure. Due to engineering practice demands the members are divided into groups corresponding to the same design variable. This linking of elements results in a trade-off between the use of more material and the need of symmetry and uniformity of structures due to practical considerations. Furthermore, the fact that due to fabrication limitations the design variables are not continuous but discrete, since cross-sections belong to a certain set, has to be to taken into account. A discrete structural optimization problem can be formulated in the following form:

$$\begin{aligned} &\min && F(s) \\ &\text{subject to} && g_j(s) \leq 0 \quad j=1,...,m \\ & && s_i \in R^d, \quad i=1,...,n \end{aligned} \tag{9}$$

where F(s) and g(s) denote the objective and constraints functions respectively. R^d is a given set of discrete values, the design variables s_i (i=1,...,n) can take values only from this set.

4.1. EVOLUTIONARY COMPUTATION

Evolutionary Computation (EC) encompasses methods of simulating evolution processes on computing systems. Evolutionary algorithms belong to EC and represent the probabilistic category of optimization methods. Evolutionary algorithms have been found capable to produce very powerful and robust search algorithms although the similarity between these algorithms and the natural evolution is based on a crude imitation of biological reality. The resulting evolutionary algorithms are based on a population of individuals, which are subjected to processes of mutation, recombination/crossover and selection.

GA and ES are the most popular evolutionary algorithms. Both algorithms imitate biological evolution in nature and have three characteristics that differ from mathematical programming optimization methods: (i) In place of the usual deterministic operators, they use randomized operators: mutation, selection and recombination/crossover, (ii) Instead of a single design point, they work simultaneously with a population of design points in the space of design variables, (iii) They can handle continuous, discrete or mixed optimization problems, (iv) They do not need gradient information. In this work the ES method is used, the selection of this optimization algorithm by the authors was based on their, as well as other researchers', experience regarding the relative superiority of ES over the rest of the methods in the problems considered (Lagaros *et al.* 2002, Bäck and Schwefel 1993, Papadrakakis *et al.* 2003).

In engineering practice the design variables are not continuous because the structural parts are, usually, constructed with certain variation of their dimensions. Thus design variables can only take values from a predefined discrete set. The multi-membered ES adopted in the current study (Papadrakakis *et al.* 1998) uses the three basic genetic operators: *recombination*, *mutation* and *selection*.

4.2. DISCRETE EVOLUTION STRATEGIES (DES)

The ES optimization procedure starts with a set of parent vectors. If any of these parent vectors gives an infeasible design then this parent vector is modified until it becomes feasible. Subsequently, the offsprings are generated and checked if they are in the feasible region. According to $(\mu+\lambda)$ selection scheme in every generation the values of the objective function of the parent and the offspring vectors are compared and the worst vectors are rejected, while the remaining ones are considered to be the parent vectors of the new generation. On the other hand, according to (μ,λ) selection scheme only the offspring vectors of each generation are used to produce the new generation. This procedure is repeated until the chosen termination criterion is satisfied. The number of parents and offsprings involved affects the computational efficiency of the multi-membered ES discussed in this work. It has been observed that values of μ and λ equal to the number of the design variables produce better results (Papadrakakis *et al.* 1998). The ES algorithm for structural optimization applications can be stated as follows:

1. *Initial generation:*
   - 1a. *Generate* s_k $(k = 1,\ldots,\mu)$ vectors
   - 1b. *Structural analysis step (NSP or NDP)*

1c. *Acceptance criteria check:* if satisfied k=k+1 else k=k. Go to *step 1a*

2. *New generation:*
   2a. *Generate* s_ℓ ($\ell = 1,...,\lambda$) vectors
   2b. *Structural analysis step (NSP or NDP)*
   2c. *Acceptance criteria check:* if satisfied $\ell=\ell+1$ else $\ell=\ell$. Go to *step 2a*
3. *Selection step:* selection of the next generation parents according to $(\mu+\lambda)$ or (μ,λ) scheme
4. *Convergence check:* If satisfied stop, else go to *step 2*

4.3. PERFORMANCE-BASED STRUCTURAL OPTIMIZATION USING NONLINEAR STATIC ANALYSIS (NSP)

Pushover analysis is based on the assumption that the response of the structure is related to the response of an equivalent single degree of freedom system with properties proportional to the first mode of the structure, The Pushover analysis step is initiated as soon as the structure has satisfied the initial static analysis step, while a lateral load distribution that follows the fundamental mode is adopted. The analysis is terminated as soon as the 150% of the target displacement that corresponds to the 2% in 50 (2/50) years earthquake is reached or earlier if the algorithm fails to converge because a collapse mechanism has been formed. The target displacement is obtained from the FEMA-356 formula:

$$\delta_t = C_0 C_1 C_2 C_3 Sa \frac{T_e^2}{4\pi^2} g \tag{10}$$

where C_0, C_1, C_2, C_3, are modification factors. C_0 relates the spectral displacement to the likely building roof displacement. C_1 relates the expected maximum inelastic displacements to the displacements calculated for linear elastic response. C_2 represents the effect of the hysteresis shape on the maximum displacement response and C_3 accounts for P-Δ effects. For the test case considered in this study, C_1 and C_3 are both equal to 1.0 while C_2 takes the values of 1.0, 1.1 and 1.2 depending on the three structural performance levels considered. T_e is the effective fundamental period of the building in the direction under consideration. Sa is the response spectrum acceleration corresponding to the T_e period. The pushover curve is converted to a bilinear curve with a horizontal post-yield branch that balances the area above and below the curve.

4.4. PERFORMANCE-BASED STRUCTURAL OPTIMIZATION USING NONLINEAR DYNAMIC ANALYSIS (NDP)

In the performance-based design procedure proposed, two performance levels are considered, while ten ground motion records are selected for each level. Totally twenty nonlinear dynamic analyses have to be performed for each candidate design. Therefore, in order to apply the proposed procedure in real world structures, parallel processing is implemented in a PC cluster environment. Depending on the number of available processors there are two stages that parallelization can be performed: the *analysis* and the *optimizer* stage. If the number of available processors permits, the advantages of the natural parallelization of the ES optimizer (defined as optimizer parallelization stage) can also be exploited (Papadrakakis *et al.* 2003).

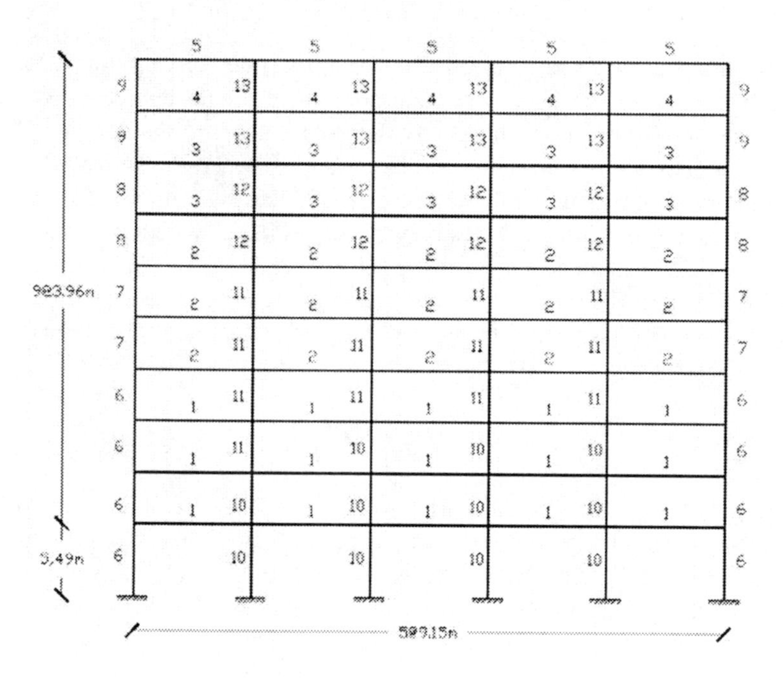

Figure 2. Geometry and member grouping.

In this work parallel processing is used in order to accelerate the performance-based optimization procedure. Parallel processing is performed in the analysis level using a cluster of twenty nodes. In each node a single nonlinear dynamic analysis is performed and the maximum interstorey drift values are obtained. Median interstorey drift and the corresponding maximum median values are obtained for both performance levels. The hardware platform that was used for the parallel computing implementation consists of a PC cluster with 20 nodes Pentium III in 500 Mhz interconnected through Fast Ethernet, with every node in a separate 100Mbit/sec switch port. Message passing is performed with the programming platforms PVM working over FastEthernet. Due to the number of nodes of the PC cluster only the analysis parallelization stage was implemented in this work.

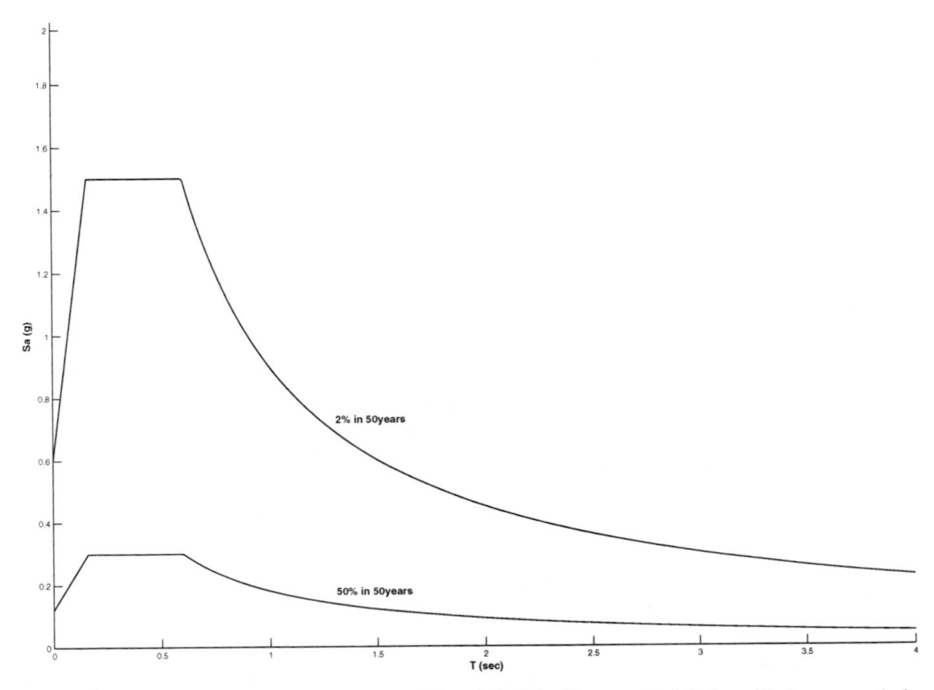

Figure 3. Uniform EC8 hazard spectra with 2% and 50% in 50 yrs probabilities of being exceeded

5. Numerical Results

A test example has been considered in order to illustrate the efficiency of the proposed design procedure. The structure considered is the five-bay, ten-storey moment resisting plane frame shown in Figure 2. Its geometric characteristics and the group members are also shown in Figure 2. The modulus of elasticity is equal to 210GPa and the yield stress is σ_y=235MPa. The constitutive material law is bilinear with a strain hardening ratio of 0.01, while the frame is assumed to have rigid connections and fixed supports. The permanent load is equal to 5kN/m^2 and the live load is taken as Q = 2kN/m^2. The gravity loads are contributed from an effective area of 5m.

The problem consists of 13 design variables which represent the cross sections of the members as shown in Figure 1. The cross-sections are W-shape available from manuals of the American Institute of Steel Construction (AISC). More specifically all beams are chosen from a database of 23 W21 sections and all columns are chosen from a database of 37 W14 cross sections.

Seismic hazard is represented by the EC8 spectrum where two uniform hazard spectra for the 2% and the 50% earthquakes are obtained using peak ground acceleration values of 0.60g and 0.12g, respectively (Figure 3). Moreover, the importance factor γ_I was taken equal to 1 and the characteristic periods T_A and T_B of the spectrum were considered equal to 0.15 and 0.60sec (soil type B), respectively. All records are scaled to match the 5% damped EC8 elastic response spectrum at the first mode $Sa(T_1,5\%)$. The slope k of the hazard curve of Eq. (6) is obtained from the expression (FEMA-350, Appendix A):

$$k = \frac{\ln\left(H_{Si}^{50/50} / H_{Si}^{2/50}\right)}{\ln\left(Sa(T_1)^{50/50} / Sa(T_1)^{2/50}\right)} = \frac{3.53}{\ln\left(Sa(T_1)^{50/50} / Sa(T_1)^{2/50}\right)} \tag{11}$$

where $H_{Si}^{50/50}$ and $H_{Si}^{2/50}$ are the probabilities of exceedance for the two earthquakes and $Sa(T_1)$ is the 5% damped spectral acceleration at the first mode of each candidate design obtained from the EC8 spectrum.

Table 2. Records for the 2/50 hazard level.

Earthquake	Mw	Station	Distance	Soil
Valparaiso,	8.0	Llolleo	34	rock
Chile 1985/5/3		Pichilemu	24	rock
		Valparaiso – UTFSM	28	rock
		Valparaiso - Ventanas	28	soil
		Vina del Mar	30	soil
		Zapallar	30	rock
Michoacan,	8.0	Caletade Campos	12	rock
Mexico 985/9/19		La Union	22	rock
		La Villita	18	rock
		Zihuatenejo	21	rock

For the nonlinear dynamic procedure, two sets of ten strong ground motion records (Tables 2 and 3) that were generated for the Los Angeles area as parts of the FEMA/SAC Steel Project (Somerville *et al.* 1997) were used. The two sets correspond to 2% and 50% probabilities of exceedance in 50 years, respectively, while from the available dataset the fault parallel component was selected. These records are used as input for the analysis in order to compute the median response quantities for the performance levels associated with record probabilities. One may argue that in order to reduce the dispersion in the results of the time-history analysis more earthquake records should have been used. However, in order to reduce the dispersion by a factor of two we would need four times more records (Benjamin and Cornell 199=70). In our case forty records for each performance level would reduce the dispersion only by half. Thus, for the nonlinear dynamic procedure, a reasonable balance between the number of records used and the computational cost is achieved.

Table 3. Records for the 50/50 hazard level.

Earthquake	Mw	Station	Distance	Soil
Honeydew	6.1	Cape Mendocino	20	rock
1991/8/17		Petrolia	17	soil
Cape Mendocino	6.8	Bunker Hill	8.8	rock
1992/4/25		Butler Valley	37	rock
Cape Mendocino	6.6	Bunker Hill	27	rock
aftershock,		Centerville	27	soil
4/26/92 07:41		Eureka College	46	soil
Cape Mendocino	6.6	Bunker Hill	27	rock
aftershock,		Ferndale	34	soil
4/26/92 11:18		Fortuna	43	soil

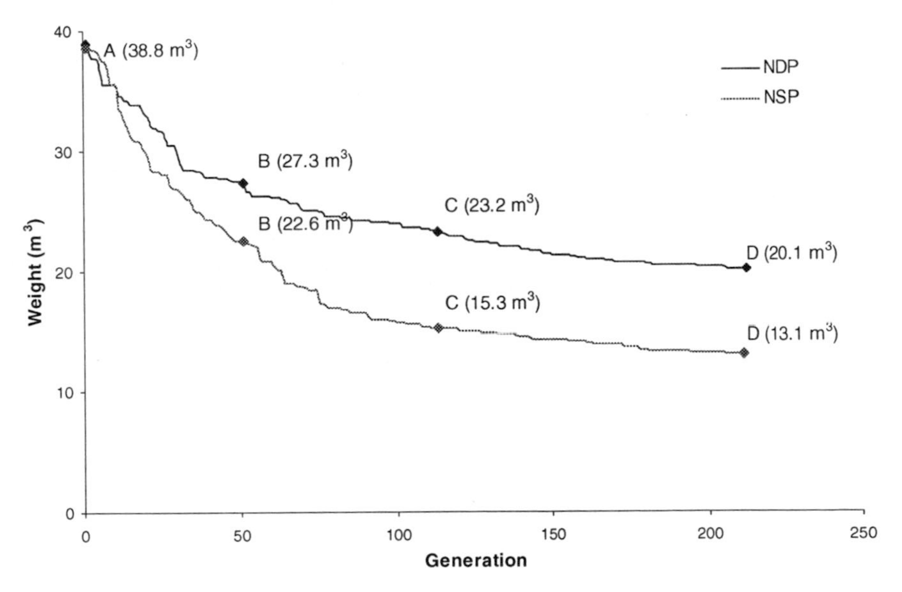

Figure 4. Generation history.

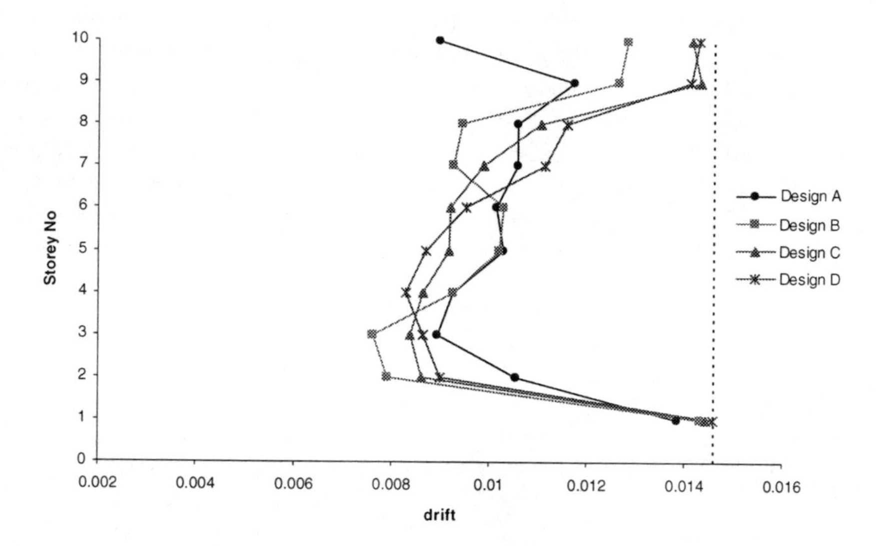

Figure 5. Drift profiles for the 50/50 performance level – Designs obtained with Nonlinear Dynamic Procedure.

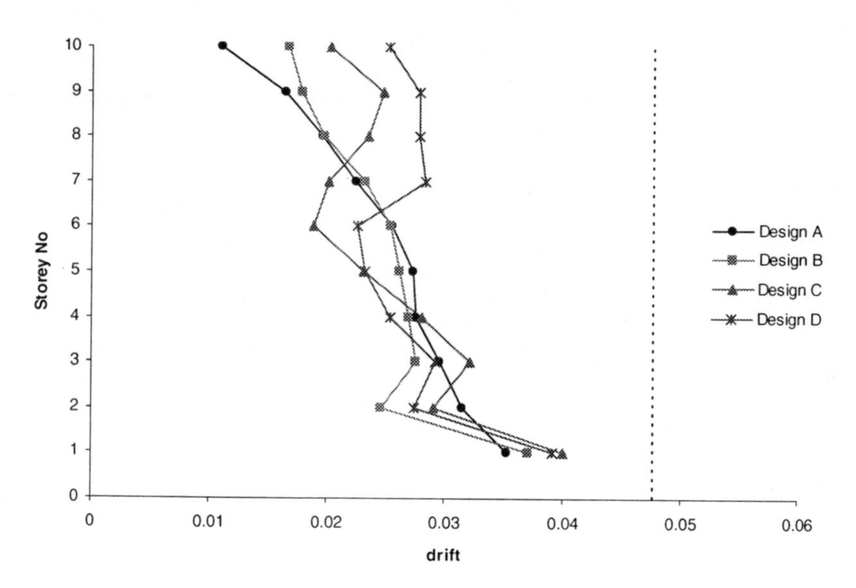

Figure 6. Drift profiles for the 2/50 performance level– Designs obtained with Nonlinear Dynamic Procedure.

All time history analyses were performed using a force based beam-column fiber element while geometric nonlinearities were also considered in the analysis. Rayleigh damping is used to obtain a damping ratio of 2% for the first and the fourth mode. For the calculation of maximum interstorey drift limits the values shown in Table 1 were adopted.

Following the optimization procedure described in the previous sections the ES optimization history is shown in Figure 4 for both analysis procedures where the reduction in material weight with respect to the number of ES generations is shown. The four designs A, B, C and D identified in Figure 4 for each procedure are listed in Table 4. It can be seen that compared to the static analysis procedure the dynamic procedure produces more conservative results. Although, the variability in the results due to the analysis procedure adopted is accounted for by means of the γ_a parameter of Eq. (3), still the design obtained with static analysis is less expensive.

Table 4. Characteristic Optimal Solutions.

	Nonlinear Static Analysis				*Nonlinear Dynamic Analysis*			
Section	A	B	C	D	A	B	C	D
1	W21x57	W21x101	W21x101	W21x101	W21x128	W21x147	W21x122	W21x132
2	W21x68	W21x83	W21x68	W21x73	W21x146	W21x83	W21x101	W21x101
3	W21x57	W21x44	W21x44	W21x44	W21x217	W21x93	W21x68	W21x62
4	W21x57	W21x68	W21x83	W21x44	W21x161	W21x50	W21x68	W21x44
5	W21x57	W21x44	W21x44	W21x44	W21x307	W21x73	W21x57	W21x44
6	W14x655	W14x257	W14x99	W14x99	W14x455	W14x257	W14x193	W14x109
7	W14x730	W14x311	W14x233	W14x99	W14x730	W14x550	W14x342	W14x193
8	W14x550	W14x455	W14x426	W14x257	W14x500	W14x455	W14x426	W14x193
9	W14x730	W14x370	W14x145	W14x109	W14x500	W14x311	W14x283	W14x120
10	W14x808	W14x342	W14x211	W14x211	W14x808	W14x550	W14x550	W14x550
11	W14x605	W14x257	W14x211	W14x176	W14x605	W14x311	W14x311	W14x311
12	W14x808	W14x426	W14x211	W14x99	W14x808	W14x455	W14x257	W14x145
13	W14x605	W14x398	W14x132	W14x99	W14x505	W14x283	W14x109	W14x109
Vol.(m^3)	38.49	22.44	15.28	13.06	38.85	27.31	23.22	20.07
50/50drift	0.0194	0.0211	0.0284	0.0296	0.0138	0.0143	0.0145	0.0146
2/50drift	0.0504	0.0596	0.0745	0.0718	0.0353	0.0371	0.0400	0.0392

Figures 5 and 6 show the distribution of drift demand along the height of the frame for the two performance levels for the designs given in Figure 4 (or Table 4). The drift demand that corresponds to the 90% and the 50% confidence level for the 2/50 and the 50/50 performance levels for k=2.5 are also shown with a dotted line. A high concentration of drift demands in the first steps of the optimization process is observed at the first story level due to the fact that this story is 1.53m higher than the other stories. However, more optimized solutions with less material weight tend to have a more uniform distribution of drift demand compared to the demand on heavier designs. Furthermore, it can be seen that the constraint for the 50/50 design level was more critical during the optimization process.

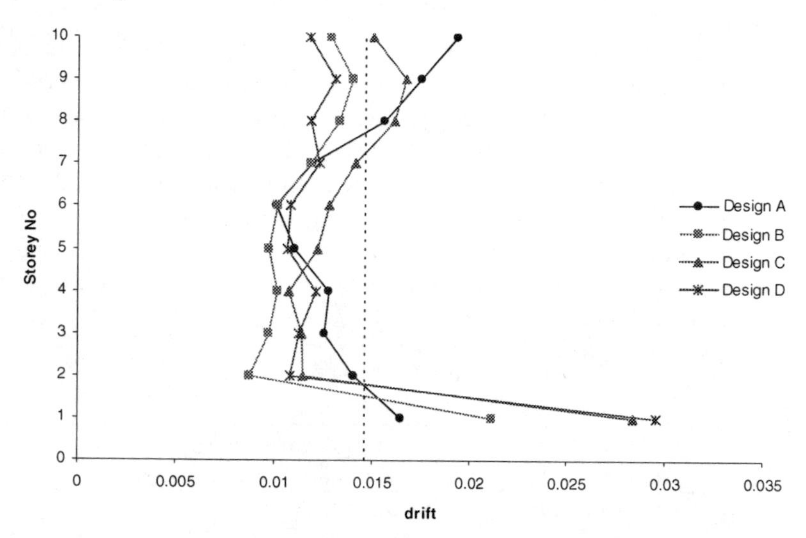

Figure 7. Drift profiles for the 50/50 performance level– Designs obtained with Nonlinear Static. Procedure

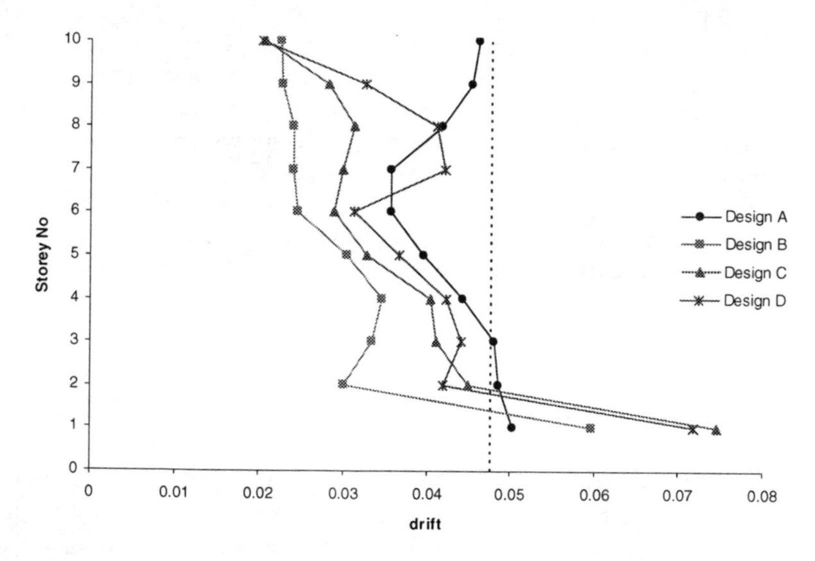

Figure 8. Drift profiles for the 2/50 performance level– Designs obtained with Nonlinear Static Procedure.

Nonlinear dynamic analysis with the records of Tables 2 and 3 has also been performed for the designs obtained with the nonlinear static analysis procedure (Figure 4). The maximum interstorey drift profiles obtained are shown in Figures 7 and 8 for the 50/50 and the 2/50 performance levels, respectively. As can be seen, these designs when subjected to nonlinear dynamic analysis fail to meet the performance objectives of FEMA-350, implying that the analysis procedure uncertainty assumed by FEMA-350 for nonlinear static analysis was underestimated. Thus, the optimum designs obtained with the nonlinear static analysis procedure is not feasible in the framework of the guidelines.

6. Conclusions

A framework for the performance-based optimal design of steel structures has been proposed in this study. Two different analytical procedures, nonlinear static analysis and nonlinear dynamic analysis, were implemented in order to determine the level of damage for two performance levels. The procedure has been applied for the design of a ten-story moment resisting frame. In order to account for a number of uncertainties which are critical for the determination of the structural response for each performance level, the FEMA-350 methodology has been adopted. It is shown that structural code design checks can be implemented in a straightforward manner when nonlinear analysis procedures are adopted and designs that meet the specified performance objectives with the desired confidence can be easily obtained.

A modified evolution strategies algorithm was implemented for the solution of the optimization problem. The trial and error design procedure that is performed on a heuristic fashion by structural engineers can be replaced by an automatic procedure based on a structural optimization search engine while increased control of structural performance can be obtained since more elaborate analysis procedures are adopted for the design. For the nonlinear dynamic analysis case, in order to handle the increased computational cost of the large number analyses required, the procedure has been implemented in a parallel processing environment.

From the numerical study considered it is shown that considerable variation in structural response is obtained when static or dynamic analysis procedures are adopted. Although, the FEMA-350 framework accounts for the variability raised from different analysis procedures still the outcome of analysis varies considerably.

References

Argyris J, Tenek L, Mattssonn A. BEC: A 2-node fast converging shear-deformable isotropic and composite beam element based on 6 rigid-body and 6 straining modes. *Computer Methods in Applied Mechanics in Engineering* 1998; **152**: 281-336.

Bäck, T., and Schwefel, H.-P. "An Overview of Evolutionary Algorithms for Parameter Optimization", Journal of Evolutionary Computation, 1(1): 1-23, (1993).

Balling RJ, Ciampi V, Pister KS, Polak E. Optimal design of seismic-resistant planar steel frames. Earthquake Engineering Research Center, University of California, Berkeley. Report No. UCB/EERC-81/20, 1981.

Beck JL, Chan E, Irfanoglu A, Papadimitriou C. Multi-criteria optimal structural design under uncertainty. *Earthquake Engineering and Structural Dynamics* 1999; **28**: 741-761.

Benjamin J.R., and Cornell C.A. "Probability, Statistics and Decision for Civil Engineers", McGraw-Hill, New York, 1970.

Bhatti MA, Pister KS. A dual criteria approach for optimal design of earthquake resistant structural systems. *Earthquake Engineering and Structural Dynamics* 1981; **9**: 557-572.

Chan C-M, Zou X-K. Elastic and inelastic drift performance optimization for reinforced concrete buildings under earthquake loads, *Earthquake Engineering and Structural Dynamics* 2004; **33**: 929–950.

Charney FA. Needs in the Development of a comprehensive performance based optimization process. In Elgaaly M. (Ed.) ASCE Structures 2000 Conference Proceedings. May 8–10; 2000, Philadelphia, Pennsylvania, USA; Paper No. 28.

Cheng FY, Li D, Ger J. Multiobjective optimization of seismic structures. In Elgaaly M. (Ed.) ASCE Structures 2000 Conference Proceedings. May 8–10; 2000, Philadelphia, Pennsylvania, USA; Paper No. 24.

Esteva L, Dıaz-Lopez O, Garcıa-Perez J, Sierra G, Ismael E. Life-cycle optimization in the establishment of performance-acceptance parameters for seismic design. *Structural Safety* 2002; **24**: 187–204.

Eurocode 3, Design of steel structures. Part1.1: General rules for buildings. CEN- ENV, 1993.

Eurocode 8, Design of structures for earthquake resistance, Part 1, European standard CEN 1998-1, Draft No. 6, European Committee for Standardization, Brussels, 2003.

FEMA 350: Recommended Seismic Design Criteria for New Steel Moment-Frame Buildings. Federal Emergency Management Agency, Washington DC, 2000.

FEMA 356: Prestandard and commentary for the seismic rehabilitation of buildings. Federal Emergency Management Agency, Washington DC, SAC Joint Venture, 2000.

Fragiadakis, M., Lagaros, N.D. Papadrakakis, M. Performance-based optimum design of steel structures considering life cycle cost, Str. Mult. Opt., (to appear), 2006a.

Fragiadakis, M., Lagaros, N.D. Papadrakakis, M. Performance based earthquake engineering using structural optimization tools, International Journal of Reliability and Safety, (to appear), 2006b.

Ganzerli S, Pantelides CP, Reaveley LD. Performance-based design using structural optimization. *Earthquake Engineering and Structural Dynamics* 2000; **29**: 1677-1690.

Gong Y. *Performance-based design of steel building frameworks under seismic loading*. PhD Thesis, Dep. of Civil Eng. University of Waterloo, Canada, 2003.

International Code Council (ICC). International Building Code (IBC), Falls Church, VA, 2000.

Kocer FY, Arora JS. Optimal design of H-frame transmission poles for earthquake loading. *ASCE Journal of Structural Engineering* 1999; **125**(11): 1299-1308.

Kocer FY, Arora JS. Optimal design of latticed towers subjected to earthquake loading, ASCE Journal of Structural Engineering 2002; 128(2):197-204.

Lagaros N.D., Fragiadakis M., Papadrakakis M. "Optimum design of shell structures with stiffening beams", AIAA Journal, 42(1): 175-184, (2004).

Lagaros, N.D., Papadrakakis, M., Kokossalakis, G. "Advances in Structural Optimization with Evolutionary Algorithms", Computer & Structures, 80(7-8): 571-587, (2002).

Liu M, Burns SA, Wen YK. Multiobjective optimization for performance-based seismic design of steel moment frame structures. *Earthquake Engineering and Structural Dynamics* 2005; **34**: 289-306.

Liu M, Burns SA, Wen YK. Optimal seismic design of steel frame buildings based on life cycle cost considerations. *Earthquake Engineering and Structural Dynamics* 2003; **32**: 1313-1332.

Liu M, Wen YK, Burns SA. Life cycle cost oriented seismic design optimization of steel moment frame structures with risk-taking preference. *Engineering Structures* 2004; **26**: 1407-1421.

Papadrakakis M, Lagaros ND, Thierauf G, Cai J., "Advanced solution methods in structural optimization based on Evolution Strategies", Engineering Computations Journal, 15: 12-34, (1998).

Papadrakakis M., Lagaros N.D., Fragakis Y. "Parallel computational strategies for structural optimization", International Journal for Numerical Methods in Engineering, 58(9): 1347-1380, (2003).

Pezeshk S. Design of framed structures: an integrated non-linear analysis and optimal minimum weight design. *International Journal for Numerical Methods in Engineering* 1998; **41**: 459-471.

Polak E, Pister KS, Ray D. Optimal design of framed structures subjected to earthquakes. *Engineering Optimization* 1976; **2**: 65-71.

Somerville P, Smith N, Punyamurthula S, Sun J. Development of Ground Motion Time Histories for Phase 2 of the FEMA/SAC Steel Project. Report No. SAC/BD-97/04, SAC Joint Venture, Richmond CA, 1997.

Yun S.-Y., Hamburger R.O., Cornell C.A., Foutch D.A. "Seismic Performance Evaluation for Steel Moment Frames", J. Struct. Engrg. 128(4): 534-545, (2002).

Part III: FIRE AND EXPLOSION INDUCED EXTREME LOADING CONDITIONS

COMPUTATIONAL MODELLING OF SAFETY CRITICAL CONCRETE STRUCTURES AT ELEVATED TEMPERATURES

N. Bicanic* *(bicanic@civil.gla.ac.uk)*
Department of Civil Engineering, University of Glasgow, Glasgow, UK

C. Pearce
Department of Civil Engineering, University of Glasgow, Glasgow, UK

C. Davie
School of Civil Engineering and Geosciences, University of Newcastle upon Tyne, UK

Abstract. Multiphysics computational modelling framework for safety critical concrete structures is considered through the development of a comprehensive thermo-hygro-mechanical coupling (T-H-M), in order to account for the development of spallation pressures at exposures to high temperatures. Simulation of the ageing process in prestressed concrete pressure vessels is also considered.

Key words: thermo-hygro-mechanical modelling of concrete at high temperature.

1. Introduction

Computational modelling of structural concrete (plain, reinforced and/or prestressed) and concrete structures with nonlinear finite elements has been a subject of research, development and of practical industrial relevance for many years. Despite a long 'track record' it has often been argued that the impact of extensive research in constitutive modelling of concrete, as well as in nonlinear solution algorithms on engineering practice has been relatively moderate – the difficulties lay mainly in translating the continuous field quantities from nonlinear finite element analyses into the design code related data, like crack widths, crack spacing, residual section capacities, damage levels etc. Nonlinear analyses are almost exclusively applied in a safety assessment of non standard and/or safety critical structures, as well as in fornesic studies of structural failures.

These issues are no means new[1] and it is perhaps humbling to note that the current issues remain surprisingly similar, despite the fact that many advances have indeed happened in the meantime. The early, often intuitive developments to account for the most significant nonlinear effects – associated with concrete cracking – have gradually been replaced by complex concepts which have brought together various modelling

A. Ibrahimbegovic and I. Kozar (eds.),Extreme Man-Made and Natural Hazards in Dynamics of Structures, 163–176.

formulations and theoretical frameworks, and where strict boundaries between the traditional concepts like plasticity, damage mechanics or fracture mechanics no longer apply. Recent developments comprise various novel constitutive models, which are cast in a more rigorous thermodynamic setting and these developments and are associated with more robust and stable algorithms, have increased confidence and helped to narrow the gap between the research and practice, as demands for complex and sophisticated nonlinear analyses of concrete structures grow.

An increased need for reliable, robust and - above all - industry relevant analyses has identified marked differences in numerical predictions emanating from different constitutive models and have highlighted an apparent lack of experimental data to fully support capabilities of sophisticated analyses. This has in turn led to a whole series of well documented comparative studies[2,3] and more complex benchmark model problems - on the material point level (Fig 1), on the component level, as well on the full scale structures (Fig 2) [4,5].

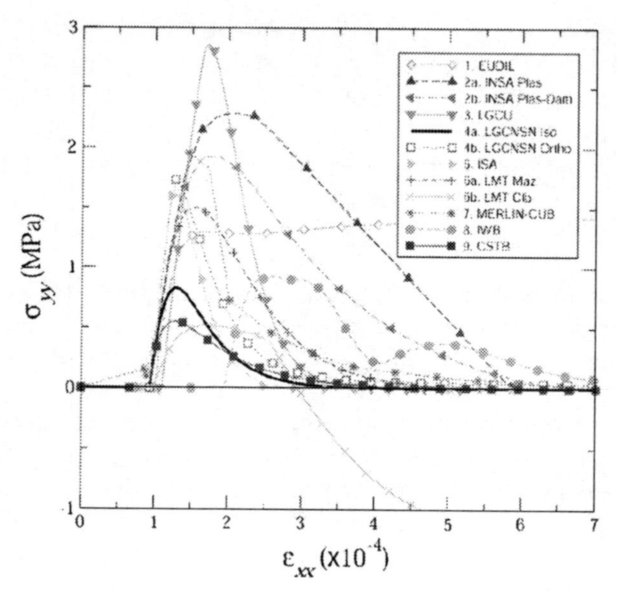

Figure 1. Wide range and marked differences in predictions, Willam's Material Point Test (from Ghavaminan et al 2004).

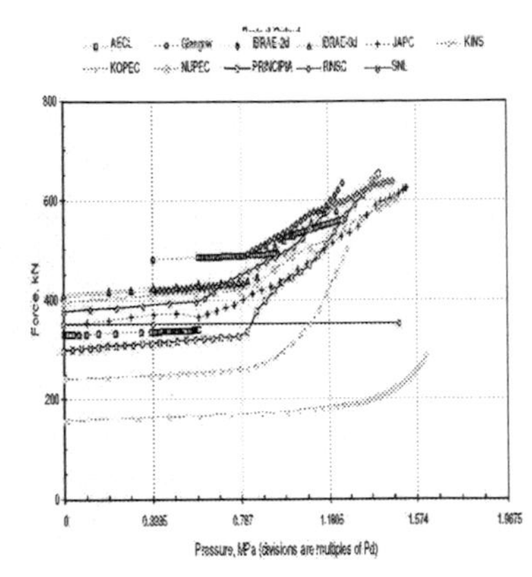

Figure 2. Sandia National Laboratories, Robin Analysis of a 1:4 Scaled Pre-stressed Concrete Containment Vessel Model, 2000.

In the overall context of continuing advancement in computational modelling of safety critical concrete structures, there is a notable increase in the importance and the potential of *multiphysics (multifield)* and *multiscale* modelling frameworks, as it transpires that some crucial macroscopic manifestations often cannot be adequately described by considering only the macro level - especially in the consideration of exceptional and/or extreme conditions or degradation processes.

2. Multi-Physics Formulations for Concrete at High Temperatures

Consideration of multiphysics is clearly needed when the behaviour of concrete at high temperature needs to be considered. Despite it's low tech image, concrete is a complex, multi-phase material, consisting of aggregates bound by a highly porous, hygroscopic solid cement paste skeleton, which can be up to 68% pore space by volume (~28% gel pores ≤ 2.6nm in diameter and up to 40% capillary pores in the range of 1μm in diameter). Under normal environmental conditions the pore space is filled with fluids, typically including dry air and water. The water is characteristically present in the form of vapour, liquid and adsorbed water, which is physically bound onto the surface of the solid skeleton.

Concrete can experience high temperatures through a number of accidental or deliberate scenarios, e.g. fire or high temperature industrial applications. In all these scenarios the concrete and the fluids within it will, to some degree, undergo the processes described above and damage may occur. The severity of the damage will depend on numerous factors including the intensity of heating, the material and physical properties of the concrete and the environmental conditions to which the concrete is exposed over its lifetime.

The behaviour of concrete under exposure to high temperatures is greatly dependent on its composite structure and in particular, on the physical and chemical composition of the cement paste, which is a highly porous, hygroscopic material. At room temperature, the pores in the paste may be fully or partially filled with fluids typically including free (evaporable) water, water vapour and dry air. On the surfaces of the pores the free water exists as adsorbed water, physically bound to the solid, and as such does not behave as a liquid. The solid skeleton of the paste itself is composed of various chemical compounds and chemically bound water.

When exposed to high temperature, heat is transported through the material, resulting in changes in the chemical composition, physical structure and fluid content of the cement paste which in turn result in changes to the overall mechanical properties (strength, stiffness, fracture energy, etc.) and physical properties (thermal conductivity, permeability, porosity, etc.) of the concrete. When exposed to high temperatures, heat is conducted and convected through the material, leading to various changes in the fluids, including phase changes (both evaporation and condensation) and transportation, through several mechanisms including pressure driven flow and diffusion. Furthermore, the fluid content may be affected by temperature dependent changes in the structure of the concrete. These may be chemical changes such as (de)hydration, or physical changes, for example to the porosity and permeability of the concrete, all of which have a direct effect on the fluid transport behaviour.

As these processes act on the pore fluids, they result in changes to the pore pressures. These pressures have a direct mechanical affect on the concrete in that they modify the effective stress state. Through this mechanism, pore pressures are generally thought to be a major contributing factor to the development of damage, and spalling.

2.1. MATHEMATICAL FORMULATION – GOVERNING EQUATIONS

A computational model for concrete subject to thermal loading needs to consider as many of the above (mostly non-linear) phenomena and their coupled interactions, as possible[6-7]. The governing mass conservation equations to describe heat and moisture transport in concrete containing free (evaporable) water, water vapour and dry air can be defined by (1-3). Furthermore, the energy conservation for the system can be defined by (4).

$$\frac{\partial(\varepsilon_L \rho_L)}{\partial t} = -\nabla \cdot \mathbf{J}_L - \dot{E}_L + \frac{\partial(\varepsilon_D \rho_L)}{\partial t} \tag{1}$$

$$\frac{\partial(\varepsilon_G \tilde{\rho}_V)}{\partial t} = -\nabla \cdot \mathbf{J}_V + \dot{E}_L \tag{2}$$

$$\frac{\partial(\varepsilon_G \tilde{\rho}_A)}{\partial t} = -\nabla \cdot \mathbf{J}_A \tag{3}$$

$$\underline{(\rho C)} \frac{\partial T}{\partial t} = -\nabla \cdot (-k\nabla T) - \underline{(\rho C\mathbf{v})} \cdot \nabla T - \lambda_E \dot{E}_L - \lambda_D \frac{\partial(\varepsilon_D \rho_L)}{\partial t} \tag{4}$$

From equations (1-4), a system of coupled differential equations can be derived in terms of the chosen primary variables; e.g. temperature, T, gas pressure, P_G, and the vapour content, $\tilde{\rho}_V$. It may be noted that the convection term, $\underline{(\rho C\mathrm{v})} \cdot \nabla T$ in equation (4), is ignored on the assumption that the transfer of energy by convection is accounted for within the empirical relationship for the thermal conductivity, $k(T)$.

The mass fluxes of dry air, water vapour and free water can be expressed in terms of pressure and concentration gradients (re Darcy's Law and Fick's Law) (5-8). The water flux equation is affected by the capillary pressure and the model adopted for the diffusion of adsorbed water

$$\mathbf{J}_V = \varepsilon_G \tilde{\rho}_V \mathbf{v}_G - \varepsilon_G \tilde{\rho}_G D_{VA} \nabla \left(\frac{\tilde{\rho}_V}{\tilde{\rho}_G} \right) \tag{5}$$

$$\mathbf{J}_A = \varepsilon_G \tilde{\rho}_A \mathbf{v}_G - \varepsilon_G \tilde{\rho}_G D_{AV} \nabla \left(\frac{\tilde{\rho}_A}{\tilde{\rho}_G} \right) \tag{6}$$

$$\mathbf{J}_L = \left(1 - \frac{S_B}{S}\right) \varepsilon_L \rho_L \mathbf{v}_L + \left(\frac{S_B}{S}\right) \varepsilon_L \rho_L \mathbf{v}_B \tag{7}$$

where

$$\mathbf{v}_G = -\frac{KK_G}{\mu_G} \nabla P_G \qquad \mathbf{v}_L = -\frac{KK_L}{\mu_L} \nabla P_L \tag{8}$$

$$\mathbf{v}_B = -D_B \nabla S_B \qquad \phi = \varepsilon_L + \varepsilon_G \qquad S = \frac{\varepsilon_L}{\phi}$$

$$S_B = \begin{cases} S & \textbf{for } S \le S_{SSP} \\ S_{SSP} & \textbf{for } S > S_{SSP} \end{cases} \tag{9}$$

The free water content in the concrete is determined from sorption isotherms, as defined by Bažant and Kaplan[8], which are a function of the concrete cement content and the relative humidity (P_V / P_{Sat}) and temperature in the pores (10).

$$\varepsilon_L = \frac{\varepsilon_{Cem}\rho_{Cem}}{\rho_L} \cdot f\left(\frac{P_V}{P_{Sat}}, T\right) \tag{10}$$

The gas volume fraction can then be determined from equation (9):The dry air and water vapour are assumed to behave as ideal gases and their pressures and partial densities are considered to be additive (re Dalton's law) (11-12):

$$\tilde{\rho}_G = \tilde{\rho}_A + \tilde{\rho}_V \qquad P_G = P_A + P_V \tag{11}$$

$$P_V = R_V \tilde{\rho}_V T \qquad P_A = R_A \tilde{\rho}_A T \tag{12}$$

The liquid pressure is the difference between the gas pressure and the capillary pressure (13). The two remaining variables, fundamental to this modified formulation, are the coefficient of bound water diffusion, D_B, given[9] by the empirical relationship (14), and the capillary pressure, P_C, which is calculated via the Kelvin-Laplace Equation (15).

$$P_L = P_G - P_C \tag{13}$$

$$D_B = D_B^0 \exp\left(-2.08 \frac{S}{S_{SSP}} \frac{T}{T_{Ref}}\right) \quad \textbf{for } S \le S_{SSP} \tag{14}$$

$$P_C = -R_V T \rho_L \ln\left(\frac{P_V}{P_{Sat}}\right) \tag{15}$$

From the governing equations (1-4) a system of coupled differential equations can be developed, with reference to an appropriate set of primary variables, such that, in matrix-vector form (16)

$$\mathbf{C}\dot{\mathbf{u}} - \nabla \cdot (\mathbf{K}\nabla \mathbf{u}) = 0 \tag{16}$$

where $\mathbf{u} = (T \quad P_G \quad \tilde{\rho}_V)^{\mathrm{T}}$.

2.2. SIMULATION OF AGEING IN CONCRETE PRESSURE VESSELS

Ageing of concrete nuclear reactor pressure vessels (Fig 3) is a subject of renewed interest. During their normal working life the internal chambers of these vessels are exposed to temperatures up to ~80°C, with an ambient temperature on the outside. Some of these vessels have currently been in operation for more than 30 years Another consideration for these vessels is the possibility of accidental and unplanned excursions of exposure to higher temperatures (up to ~400°C), as may be experienced due to a loss of cooling water.

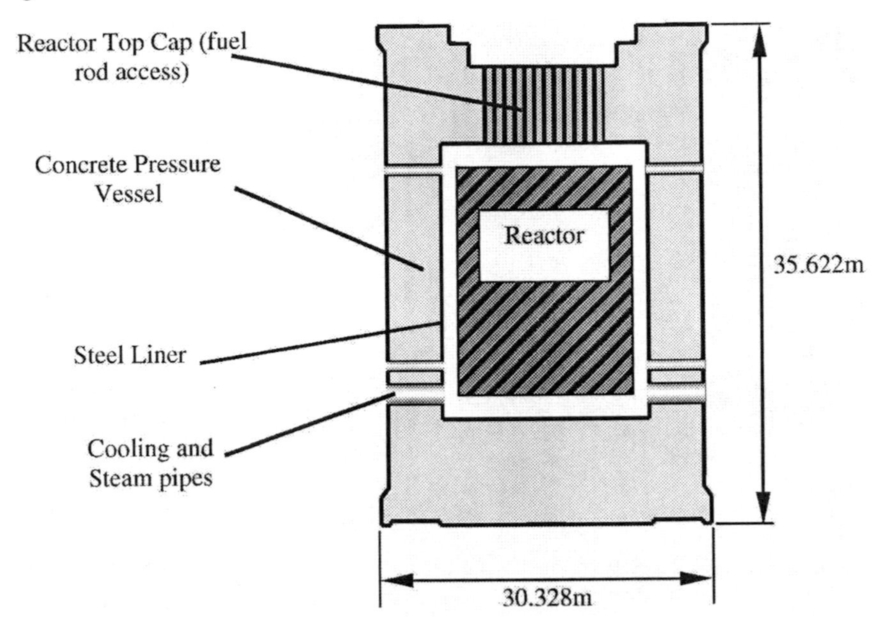

Figure 3. Section through Cylindrical Concrete Nuclear Reactor Pressure Vessel.

Simulations to follow focus on the heat and moisture transport in concrete nuclear pressure vessels and the direct effect that this could have on their structural integrity, both under normal operating conditions and under temperature excursions. The investigation was carried out using the fully coupled hygro-thermal-mechanical numerical model developed during the MAECENAS[10] project and described in detail in Davie et al[11]. Although reference is made to the mechanical consequences of the transport behaviour, mechanical behaviour is not fully considered here.

Detailed 3D and Axisymmetric finite element meshes, representative of a specific type of nuclear pressure vessel were developed for use during the MAECENAS project (Figs 4a, 4b). However, an initial coupled hygro-thermal-mechanical model presented here was conducted on a smaller mesh, equivalent to a slice through the full axi-symmetric mesh (Fig 4c).

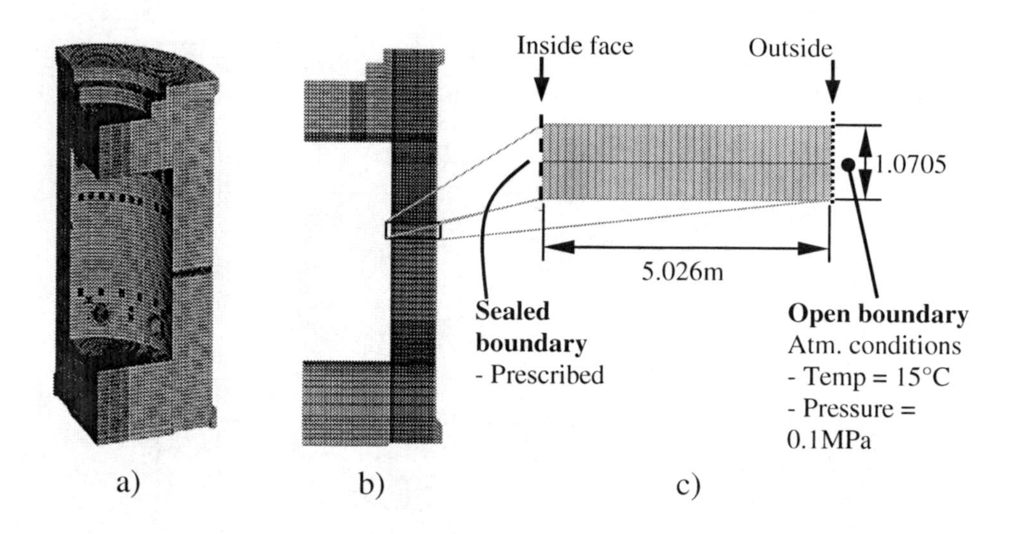

Figure 4. MAECENAS Pressure Vessel FE meshes a) One Quarter 3D Mesh (20 noded elements), b) Axi-symmetric Mesh (8 noded elements), c) Axi-symmetric 'Slice' Mesh (8 noded elements).

In order to reproduce the external conditions experienced by a nuclear pressure vessel, Cauchy type boundary conditions to simulate the free exchange of heat and fluid with the atmosphere were applied at the outside face of the mesh, while for the inside face, only a prescribed (Dirichlet type) temperature boundary was defined. Boundary conditions were not defined for the fluid phases in order to represent the sealed conditions imposed by the steel liner on the inside of the vessel. The top and bottom sides of the mesh were considered as symmetric boundaries to represent the large, continuous structure above and below the 'slice'.

The first of the two parametric series comprised 5 analyses in which a 33 year heating cycle (from data recorded for an existing pressure vessel) (Fig 5a) was applied to the inside face and the resulting fluid transport behaviour monitored.

The parameters investigated were the initial permeability K_0, and the initial porosity ϕ_0, of the concrete. It should also be noted that both permeability K, and porosity, ϕ_0, increase irreversibly with increasing temperature. The second study involved two further analyses in which heating cycles representative of typical temperature excursions were applied (Figures 5b, 5c) and again the resulting fluid transport behaviour was monitored.

All other parameters were kept constant and a summary of all the analyses can be found in Tables 1 and 2.

Table 1. - Summary of Analyses for Parametric Series for the Standard Temperature Profile (Fig 5a).

Series	Analysis	Initial Permeability, K_0	Initial Porosity,• ϕ_0	Temperature Profile
1	1	5.0×10^{-17} m^2	0.099	Standard
1	2	1.0×10^{-21} m^2	0.099	Standard
1	3	2.0×10^{-18} m^2	0.099	Standard
1	4	2.0×10^{-18} m^2	0.090	Standard
1	5	2.0×10^{-18} m^2	0.120	Standard

Table 2. - Summary of Analyses for Parametric Series for Exceptional Temperature Scenarios (Figs 5b and c).

Series	Analysis	Initial Permeability, K_0	Initial Porosity,• ϕ_0	Temperature Profile
2	1	2.0×10^{-18} m^2	0.099	Excursion Scenario 1
2	2	2.0×10^{-18} m^2	0.099	Excursion Scenario 2

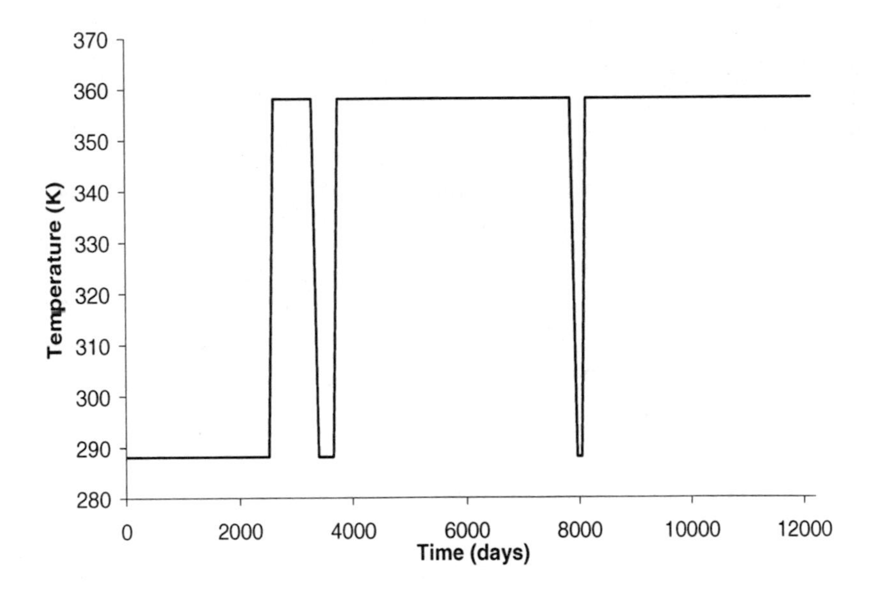

Figure 5. (a) Temperature History Standard Heating Cycle.

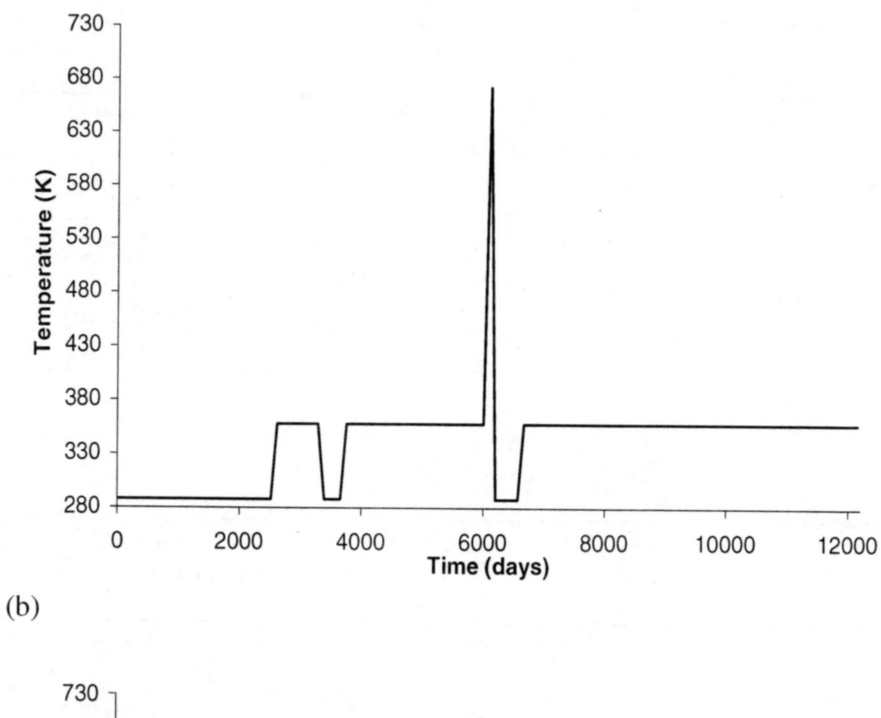

(b)

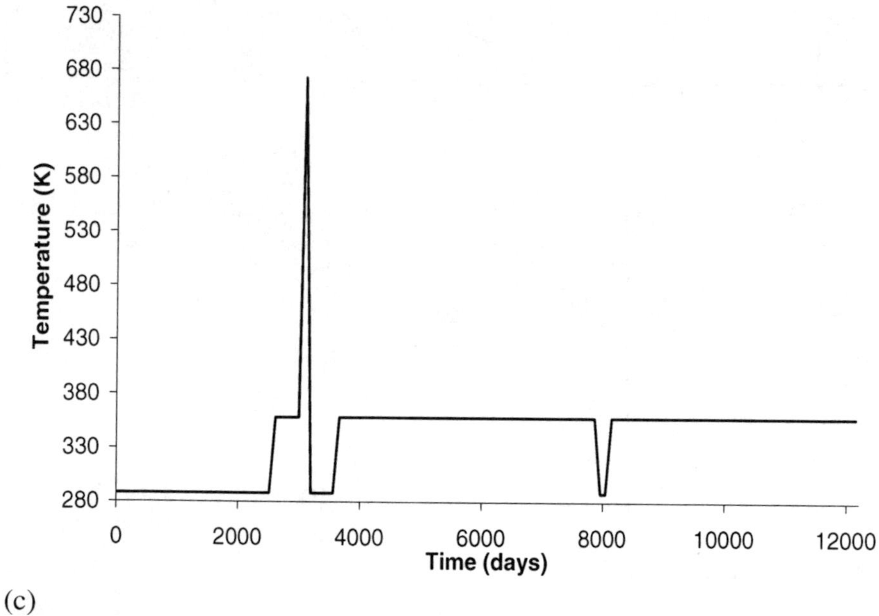

(c)

Figure 5. Temperature Histories (b) Excursion 1 Heating Cycle, (c) Excursion 2 Heating Cycle.

As discussed previously, the development and the level of pore pressures in the concrete were of particular interest. However, for simplicity only gas pressures, which can be shown to be largely analogous with the overall pore pressure behaviour, are reported here.

2.2.1. *Parametric Series 1*

As can be seen from the results shown in Figures 6 and 7 the gas pressures predicted at the inside face of the pressure vessel over its 33 year life, vary considerably depending on the initial values of both permeability and porosity. As it would be expected, both higher permeabilities and higher porosities lead to lower maximum predicted pressures as fluids can move more easily away from the hot face, towards the atmosphere.

Despite the variations, the magnitudes of the gas pressures (~0.2-0.8MPa) are generally not high enough to exceed the tensile strength of the concrete and cause structural damage on their own. However, in conjunction with mechanical stresses produced during operation of the reactor, these pressures may become significant and should be taken into account.

Of more immediate concern is the gas pressure predicted with the lowest permeability ($1.0 \times 10^{-21} m^2$). After 33 years this has reached ~1.4MPa and continues to rise. While not enough to cause fracture of the concrete on its own, this pressure may threaten the structural integrity of the steel liner in the vessel. If this liner is ruptured, radioactive gases will escape into the concrete and eventually into the atmosphere.

While both parameters can be seen to have a significant effect, the porosity can often be related to the design mix of the concrete and is therefore readily known and accounted for. However, values for permeability are less easily estimated and vary considerably (over about 4 orders of magnitude) in the literature [2, 3]. This uncertainty is clearly a concern when its potential effects are considered.

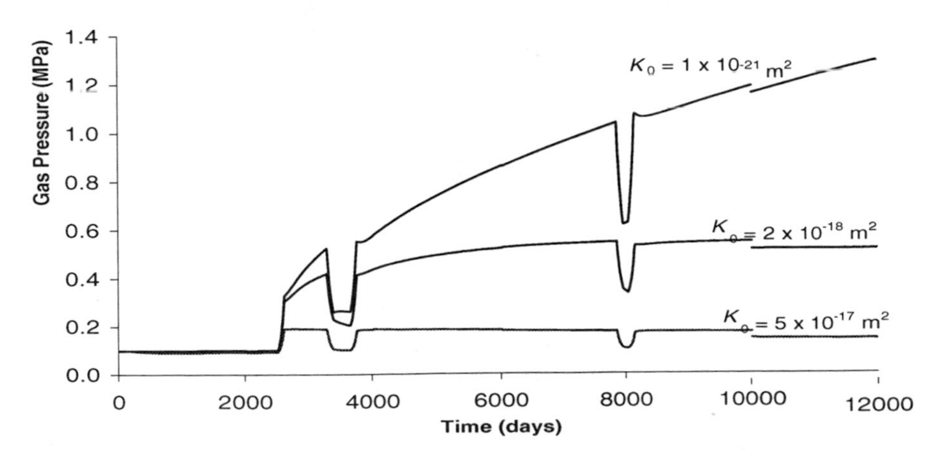

Figure 6. Gas Pressure Histories for the 33 years cycles. Standard temperature profile. Influence of changing permeability for constant initial porosity.

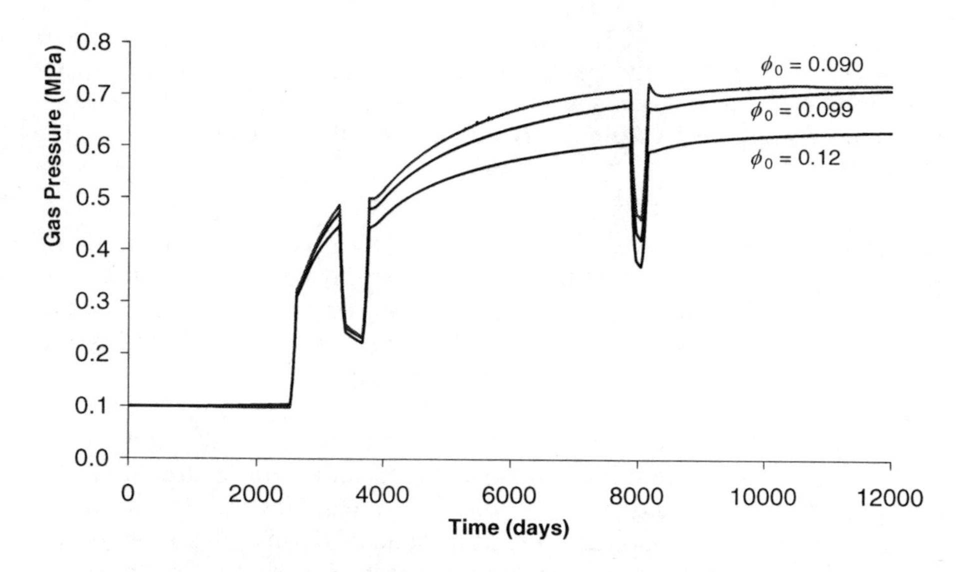

Figure 7. Gas Pressure Histories for the 33 years cycles. Standard temperature profile. Influence of changing initial porosity for constant permeability.

2.2.2. *Parametric Series 2*

As can be seen from Figure 8, the two transient temperature excursions have a significant effect on the predicted gas pressures. During both excursions, pressures of ~10MPa were predicted.

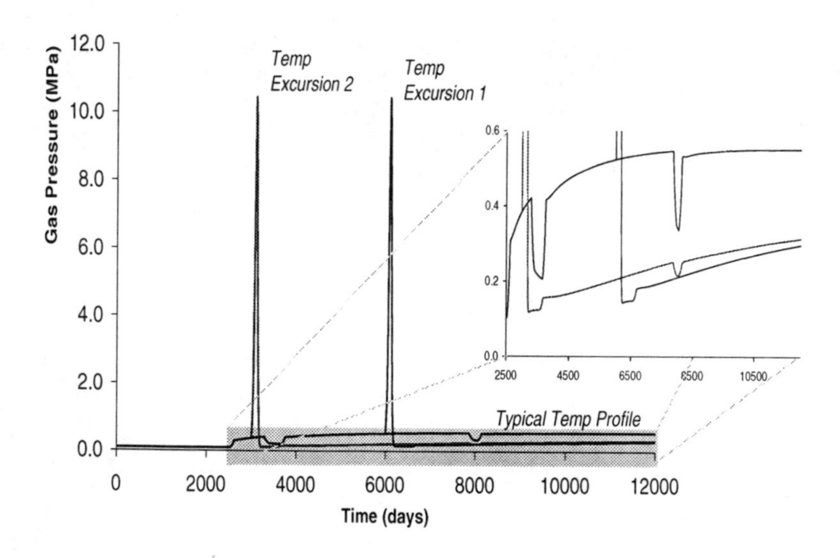

Figure 8. Gas Pressure Histories for the 33 years cycles. Two representative exceptional transient temperature profiles.

If these values were realistic they would potentially compromise the structural integrity of the pressure vessel both by causing fracturing, since the pressure will exceed the tensile strength of most concretes, and by rupturing the steel liner.

A further point of note is that in both cases the gas pressures predicted after the temperature excursion are considerably lower than those predicted under the normal operating heat cycle (Fig 8 inset). This is because the permeability and porosity of the concrete have been increased significantly by the irreversible damage caused by the temperature excursion and the fluids can more easily flow away from the inside face of the pressure vessel.

It should also be noted, however that, although the temperature excursions reached the same temperatures and therefore caused the same permeability and porosity increases, the gas pressures predicted after the two incidents were different.

This highlights the importance of considering the full heat and fluid transport history of a concrete structure when predicting the potential effects on its structural integrity.

3. Conclusions

The multiphysics modelling framework was argued through the development of a comprehensive thermo-hygro-mechanical coupling (T-H-M), in order to account for the development of spallation pressures at exposures to high temperatures. Capillary pressure and adsorbed water diffusion are incorporated into a computational model in order to investigate their influence on the finite element analysis of heat and moisture transfer in concrete exposed to high temperatures. Comparative studies were carried out by a simulation of ageing process in prestressed concrete pressure vessels.

References

J. H. Argyris, G. Faust, J. Szimmat, E. P. Warnke and K. J. Willam, Recent developments in the finite element analysis of prestressed concrete reactor vessels, Nuclear Engineering and Design, Volume 28, Issue 1 , July 1974, Pages 42-75.

S. Ghavamian et al, Civil engineering structures of nuclear power plants - Modelling issues, NAFEMS Awareness Seminar, Modelling Concrete Structures with Finite Elements, London, June 2003.

Sandia National Laboratories, Pre-test Round Robin Analysis of a Pre-stressed Concrete Containment Vessel Model, NUREG/CR-6678, SAND 00-1535, Aug 2000.

R.A. Dameron et al, Posttest Analysis of the NUPEC/NRC 1:4 Scale Prestressed Concrete Containment Vessel Model, (NUREG/CR-6809), June 2003. http://www.nrc.gov/reading-rm/doc-collections/nuregs/contract/cr6809/

N. K. Prinja, D. Shepherd, Simulating Structural Collapse of a PWR Containment, Paper # H04-6, Transactions of the 17th International Conference on Structural Mechanics in Reactor Technology (SMiRT 17), Prague, Czech Republic, August 17 –22, 2003.

Tenchev, R. T., Li, L. Y. & Purkiss, J. A., Finite Element Analysis of Coupled Heat and Moisture Transfer in Concrete Subjected to Fire, Numerical Heat Transfer, Part A, 2001; 39: pp 685 - 710.

Baroghel-Bouny, V., Mainguy, M. Lassabatere, T. & Coussy, O., Characterization and Identification of Equilibrium and Transfer Moisture Properties for Ordinary and High-Performance Cementitious Materials, Cement and Concrete Research, 1999; 29: pp 1225 – 1238.

Bažant, Z. P. & Kaplan, M. F., Concrete at High Temperatures: Material Properties and Mathematical Models; Longman: Harlow, 1996, ISBN 0 582 08626 4.

Gawin, D., Majorana, C. E. & Schrefler, B. A., Numerical Analysis of Hygro-Thermal Behaviour and Damage of Concrete at High Temperature, Mechanics of Cohesive-Frictional Materials, 1999; 4: pp 37 – 74.

"Modelling of ageing in concrete nuclear power plants" (MAECENAS Consortium), European Community EURATOM programme (Contract no. FIKS-CT-2001-00186)

Davie, C. T., Pearce, C. J. and Bi ani , N., Influence of Capillary Pressure and Adsorbed Water Diffusion on Heat and Moisture Transport in Concrete at High Temperatures, Proc. of 12th Ann. Conf. of the Assoc. for Comp. Mech. Eng. - UK, University of Cardiff, Cardiff, 2004.

THREE-DIMENSIONAL FE ANALYSIS OF HEADED STUD ANCHORS EXPOSED TO FIRE

J. Ozbolt* *(ozbolt@icm.univ-stuttgart.de)*
Institut of Construction Materials, University of Suttgart, Germany

I. Kozar
Department of Civil Engineering, University of Rijeka, Croatia

G. Periškić
Institut of Construction Materials, University of Suttgart, Germany

Abstract- In the present paper a transient three-dimensional thermo-mechanical model for concrete is presented. For given boundary conditions, temperature distribution is calculated by employing a three-dimensional transient thermal finite element analysis. Thermal properties of concrete are assumed to be constant and independent of the stress-strain distribution. In the thermo-mechanical model for concrete the total strain tensor is decomposed into pure mechanical strain, free thermal strain and load induced thermal strain. The mechanical strain is calculated by using temperature dependent microplane model for concrete (Ožbolt et al., 2001). The dependency of the macroscopic concrete properties (Young's modulus, tensile and compressive strengths and fracture energy) on temperature is based on the available experimental database. The stress independent free thermal strain is calculated according to the proposal of Nielsen et al. (2001). The load induced thermal strain is obtained by employing the bi-parabolic model, which was recently proposed by Nielsen et al. (2004). It is assumed that the total load induced thermal strain is irrecoverable, i.e. creep component is neglected. The model is implemented into a three-dimensional FE code. The performance of headed stud anchors exposed to fire was studied. Three-dimensional transient thermal FE analysis was carried out for three embedment depths and for four thermal loading histories. The results of the analysis show that the resistance of anchors can be significantly reduced if they are exposed to fire. The largest reduction of the load capacity was obtained for anchors with relatively small embedment depths. The numerical results agree well with the available experimental evidence.

Key words: Concrete; high temperature; 3D finite element analysis; microplane model; thermo-mechanical model; headed studs.

1. Introduction

Concrete does not burn, however, when its temperature increases for a couple of hundred of degrees Celsius its behavior changes significantly. The concrete mechanical properties, such as strength, elasticity modulus and fracture energy, are at high temperatures rather different than for the concrete at normal temperature. At high temperature large temperature gradients lead in concrete structures to temperature-induced stresses, which cause damage. Furthermore, creep and relaxation of concrete that is due to

A. Ibrahimbegovic and I. Kozar (eds.), Extreme Man-Made and Natural Hazards in Dynamics of Structures, 177–198.

high temperature play also an important role. The main reason for the complexity of the behavior of concrete at high temperature is due to the fact that concrete contains water, which at high temperature changes its aggregate state. Moreover, at high temperature the aggregate can change its structure or it can loose its weight through the emission of CO_2, such as calcium based stones. Although the behaviour of concrete at high temperature is in the literature well documented (Bažant and Kaplan, 1996; Khoury et al., 1985a; Schneider, 1986, 1988; Thelandersson, 1983) further tests are needed to clarify the tensile post-peak behaviour of concrete, which has significant influence on the response of concrete structures. The main problem in the experimental investigations is due to the fact that such experiments are rather demanding, i.e. one has to perform loading and measurement at extremely high temperatures. Furthermore, such experiments can be carried out only on relatively small structures. To better understand behavior of concrete structures, as an alternative to the experiments one can employ numerical analysis. However, one needs models which can realistically predict behavior of concrete at high temperature.

There are principally two groups of models: (i) Thermo-mechanical models and (ii) Thermo-hydro-mechanical models (Gawin et al., 1999; Pearce et al., 2003; Stabler, 2000; Terro, 1998). The first group of the models are phenomenological. In these models the mechanical properties of concrete are temperature (humidity) dependent whereas the temperature (humidity) distribution is independent of the mechanical properties of concrete. The second group of the models are from the physical point of view more realistic. Namely, in these models the physical processes that take place at concrete micro structural level are coupled, i.e. the interaction between mechanical properties, temperature, humidity, pore pressure and hydration is accounted for. These models are interesting from the theoretical point of view. They are rather complex and therefore for practical engineering applications one has to employ the first group of the models.

In the present paper a three-dimensional (3D) model that is based on the thermo-mechanical coupling between mechanical properties of concrete and temperature is discussed. The isothermal microplane model is used as a constitutive law for concrete with model parameters being made temperature dependent. The model is implemented into a three-dimensional finite element code and its performance is first compared with the experimental results known from the literature. Subsequently, the influence of high temperature on the pullout concrete cone resistance of a headed stud anchors is investigated. The finite element analysis is performed in two steps. For given temperature boundary conditions (air temperature and/or concrete surface temperature) it is first calculated distribution of temperature. In the second step the required load history is applied with taking into account the influence of temperature on the concrete mechanical properties.

2. Transient Thermal Analysis

As the first step of coupling between mechanical properties of concrete and temperature, for given thermal boundary conditions at time t it has to be calculated temperature distribution over a solid structure of volume Ω. In each point of continuum, which is defined by the Cartesian coordinates (x,y,z), the conservation of energy has to be fulfilled. This can be expressed by the following differential equation:

$$\lambda \Delta T(x,y,z,t) + W(x,y,z,t,T) - c\rho \frac{\partial T}{\partial t}(x,y,z,t) = 0 \tag{1}$$

where T = temperature, λ = conductivity, c = heat capacity, ρ = density, W = internal source of heating and Δ = Laplace-Operator. The surface boundary condition that has to be satisfied reads:

$$\lambda \frac{\partial T}{\partial \mathbf{n}} = \alpha (T_M - T) \tag{2}$$

where $\mathbf{n}$ = normal to the boundary surface Γ, α = transfer or radiation coefficient and T_M = temperature of the media in which surface Γ of the solid Ω is exposed to (for instance temperature of air). To solve the problem by the finite element method the above differential equations (1) and (2) has to be written in the weak (integral) form that reads (Belytschko et al. 2001):

$$\int_{\Omega} \lambda \left(\frac{\partial v}{\partial x}\frac{\partial T}{\partial x} + \frac{\partial v}{\partial y}\frac{\partial T}{\partial y} + \frac{\partial v}{\partial z}\frac{\partial T}{\partial z} \right) d\Omega + \int_{\Omega} v \left(c\rho \frac{\partial T}{\partial t} \right) d\Omega + \int_{\Gamma} v\alpha \left(T - T_M \right) d\Gamma = 0 \tag{3}$$

where v is trial function. After introducing the condition that the functional is stationary one obtains the following system of linear equations (Voigt notation):

$$[\mathrm{C}]\{\dot{\mathrm{T}}\} + ([\mathrm{K}] + [\mathrm{H}])\{\mathrm{T}\} = \{\mathrm{R}\} \tag{4a}$$

with

$$[\mathrm{C}] = \int_{\Omega} c\rho [\mathrm{N}]^T [\mathrm{N}] d\Omega; \; [\mathrm{K}] = \int_{\Omega} [\mathrm{B}]^T [\lambda][\mathrm{B}] d\Omega;$$

$$[\mathrm{H}] = \int_{\Gamma} \alpha [\mathrm{N}]^T [\mathrm{N}] d\Gamma; \; [\mathrm{R}] = \int_{\Gamma} [\mathrm{N}]^T \alpha T_M d\Gamma \tag{4b}$$

where [N] is the column matrix of shape functions that relates temperature field and nodal temperatures and [B] relates the field of temperature gradients and nodal temperatures. Eq. (4a) is solved using direct method based on the following assumption for the solution in the $(n+1)^{\text{th}}$ time step

$$\{\mathbf{T}\}_{n+1} = \{\mathbf{T}\}_n + \Delta t\left((1-\beta)\dot{\mathbf{T}}_n + \beta\dot{\mathbf{T}}_{n+1}\right) \tag{5}$$

Parameter β has been set to $\beta = 0.5$ what yields to the unconditionally stable Crank-Nicolson method that reads (Cook et al., 2002):

$$\begin{aligned}&\left(\frac{1}{\Delta t}[\mathrm{C}] + \beta([\mathrm{K}]+[\mathrm{H}])\right)\{\mathrm{T}\}_{n+1} = \\ &\left(\frac{1}{\Delta t}[\mathrm{C}] - (1-\beta)([\mathrm{K}]+[\mathrm{H}])\right)\{\mathrm{T}\}_n + (1-\beta)\{\mathrm{R}\}_n + \beta\{\mathrm{R}\}_{n+1}\end{aligned} \tag{6}$$

The above equation has been programmed for 3D solid finite elements.

3. Decomposition of Strain

In the present model the total strain tensor ε_{ij} (indicial notation) for stressed concrete exposed to high temperature can be decomposed as (Khoury et al., 1985a; Schneider, 1986; Thelandersson, 1987):

$$\varepsilon_{ij} = \varepsilon_{ij}^{m}(T, \sigma_{kl}) + \varepsilon_{ij}^{ft}(T) + \varepsilon_{ij}^{tm}(T, \sigma_{kl}) + \varepsilon_{ij}^{c}(T, \sigma_{kl}) \tag{7}$$

where ε_{ij}^{m} = mechanical strain tensor, ε_{ij}^{ft} = free thermal train tensor, ε_{ij}^{tm} = thermo-mechanical strain tensor and ε_{ij}^{c} are strains that are due to the temperature dependent creep of concrete.

In general, the mechanical strain component can be decomposed into elastic, plastic and damage parts. In the present model these strain components are obtained from the constitutive law. Free thermal strain is stress independent and it is experimentally obtained by measurements on a load-free specimen. In such experiments it is not possible to isolate shrinkage of concrete. Therefore, the temperature dependent shrinkage is contained in the free thermal strain. The thermo-mechanical strain is stress and temperature dependent. It appears only during the first heating and not during the subsequent cooling and heating cycles (Khoury et al., 1985a). This strain is irrecoverable and lead in concrete structures to severe tensile stresses during cooling. Temperature dependent creep stain is of the same nature as the thermo-mechanical strain except that it is partly recoverable. In an experiment it is not possible to isolate this. For low temperature rates, which is normal case in the experiments, this strain component compared to the thermo-mechanical strain is small. Therefore, temperature dependent creep stain is in the present model neglected.

4. Mechanical Strain

The mechanical strain components are obtained from the constitutive law of concrete. In the present model for the temperature independent (isothermal) constitutive law the microplane model is used (Ožbolt et al., 2001). The temperature dependency is adopted such that the macroscopic properties of concrete (Young's modulus, compressive and tensile strengths and fracture energy) are made time dependent.

4.1. ISOTHERMAL CONSTITUTIVE LAW FOR CONCRETE – MICROPLANE MODEL

In the microplane model the material is characterized by a relation between the stress and strain components on planes of various orientations. These planes may be imagined to represent the damaged planes or weak planes in the microstructure, such as those that exist at the contact between aggregate and the cement matrix. In the model the tensorial invariance restrictions need not be directly enforced. Superimposing in a suitable manner the responses from all the microplanes automatically satisfies them. The basic concept behind the microplane model was advanced in 1938 by Taylor (Taylor, 1938) and developed in detail for plasticity by Batdorf and Budianski in 1949 under the name "slip theory of plasticity" (Batdorf and Budianski, 1949). The model was later extended by Bažant and co-workers for modelling of quasi-brittle materials that exhibit softening (Bažant et al. 1988, 1990; Carol et al. 2001).

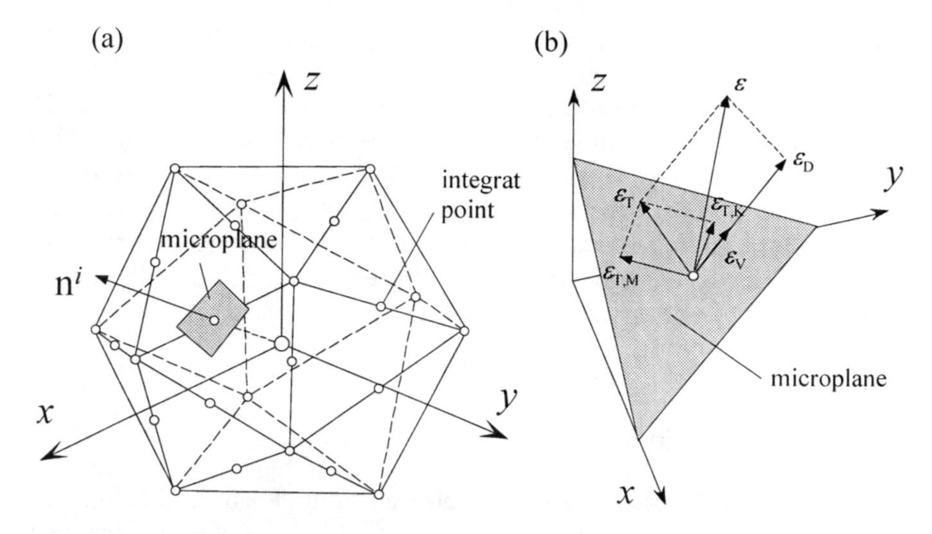

Figure 1. The concept of the microplane model: (a) Discretization of the unit volume sphere for each finite element integration point (21 microplane directions – integration points) (b) Microplane strain components.

The microplane model used in the present study, was recently proposed by Ožbolt et al. (2001). The model is based on the so-called relaxed kinematic constraint concept and it is a modification of the M2 microplane model proposed by Bažant and Prat (1988). Let ignore the effect of temperature and assume that the total strain tensor is identical to the mechanical strain tensor, i.e. $\varepsilon_{ij}^{m} = \varepsilon_{ij}$. In the model the microplane (see Fig. 1) is defined by its unit normal vector of components n_i. Microplane strains are assumed to be the projections of ε_{ij} (kinematic constraint). Normal and shear stress and strain components (σ_N, σ_{Tr}, ε_N, ε_{Tr}) are considered on each plane. Based on the virtual work approach (weak form of equilibrium), the macroscopic stress tensor is obtained as an integral over all possible, in advance defined, microplane orientations (S denotes the surface of the unit sphere):

$$\sigma_{ij} = \frac{3}{2\pi}\int_S \sigma_N n_i n_j dS + \frac{3}{2\pi}\int_S \frac{\sigma_{Tr}}{2}(n_i\delta_{rj} + n_j\delta_{ri})\ dS \qquad (8)$$

To realistically model concrete, the normal microplane stress and strain components have to be decomposed into volumetric and deviatoric parts ($\sigma_N = \sigma_V + \sigma_D$, $\varepsilon_N = \varepsilon_V + \varepsilon_D$; see Fig. 1), which leads to the following expression for the macroscopic stress tensor:

$$\sigma_{ij} = \sigma_V \delta_{ij} + \frac{3}{2\pi} \int_S \sigma_D (n_i n_j - \frac{\delta_{ij}}{3}) dS + \frac{3}{2\pi} \int_S \frac{\sigma_{Tr}}{2} (n_i \delta_{rj} + n_j \delta_{ri}) \, dS \tag{9}$$

For each microplane component, the uniaxial stress-strain relations read:

$$\sigma_V = F_V(\varepsilon_V) \quad ; \quad \sigma_D = F_D(\varepsilon_{D,eff}) \quad ; \quad \sigma_{Tr} = F_{Tr}(\varepsilon_{Tr,eff}, \varepsilon_V) \tag{10}$$

where F_V, F_D and F_{Tr} are the uniaxial stress-strain relationships for volumetric, deviatoric and shear components, respectively. For the deviatoric and shear microplane strain components in Eq. (10) only effective parts of microplane strains ($\varepsilon_{D,eff}$ and $\varepsilon_{Tr,eff}$), defined below, are used to calculate microplane stresses. Finally, the macroscopic stress tensor is obtained from Eq. (9). The integration over all microplane directions (21 directions, symmetric part of the sphere) is performed numerically.

To model concrete cracking for any load history realistically, the effective microplane strains are introduced in Eq. (10). They are calculated as:

$$\varepsilon_{M,eff} = \varepsilon_M \psi_M (\varepsilon_M, \sigma_I) \tag{11}$$

where subscript M denotes the corresponding microplane components (V, D, Tr), ε_M is the microplane strain obtained from the projection of the total strain tensor (kinematic constraint) and ψ is the so-called discontinuity function which depends on the microplane strain components and maximum principal stress σ_I. On the individual microplanes this function accounts for discontinuity of the macroscopic strain field (cracking). It is calculated such that for dominant tensile load the ratio between the volumetric and deviatoric stiffness remains constant for the entire load history. The function "relaxes" the kinematic constraint, which is in the case of strong localization of strains physically unrealistic. Consequently, in the smeared fracture type of the analysis the discontinuity function ψ enables localization of strains not only for tensile fracture, but also for dominant compressive type of failure. Detailed discussion of the features, development and problems related to various versions of the microplane models are beyond the scope of the present paper. For more detail refer to the above cited literature.

4.2. THERMO-MECHANICAL COUPLING

To account for the effect of temperature the macroscopic mechanical properties of concrete need to be temperature dependent. The nonlinear finite element analysis is incremental and the load increment is defined by the time step Δt in which the load, the boundary conditions, the temperature, etc. change. In the present model it is assumed that during load increment the temperature is constant. Consequently, the material parameters that are temperature dependent are during the load step constant as well.

4.2.1. *Young's Modulus*

The experiments show that with the increase of temperature Young's modulus E decreases (Thelandersson, 1983). It is assumed that at relatively low temperatures decrease of E is caused by the loss of capillary water (vaporisation). However, at higher temperatures decrease of E is due to the decomposition of individual concrete components (cement paste and aggregate). In the present model, temperature dependent Young's modulus follows the proposal of Stabler (2000), i.e. E is assumed to be a scalar function of temperature that reads:

$$
\begin{aligned}
&E(T) = \left[1 - \max(\omega_{t,E})\right] \cdot E_0 \\
&\mathit{for}\ 0 \le \theta \le 10 \qquad \omega_{t,E} = 0.2\theta - 0.01\theta^2 \\
&\mathit{for}\ \quad \theta > 10 \qquad \omega_{t,E} = 1
\end{aligned}
\tag{12}
$$

where E_0 = Young's modulus at temperature T_0 = 20°C and $\theta = (T-T_0)/100^\circ C$ is the relative temperature. Eq. (12) is plotted in Fig. 2. As can be seen, it shows good agreement with the experimental evidence. Note that $\max(\omega_{t,E})$ corresponds to the maximal temperature ever reached, i.e. by cooling Young's modulus does not increases.

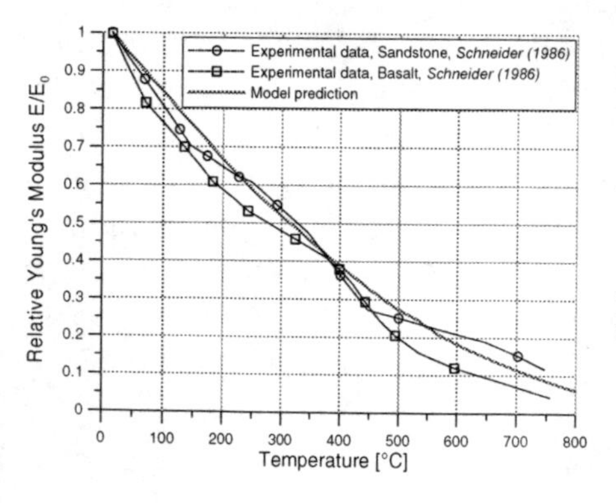

Figure 2. Relationship between Young's modulus and temper-ature.

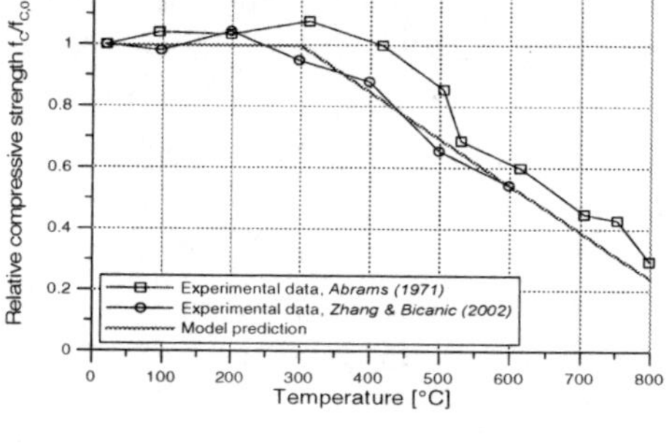

Figure 3. Relationship between the concrete compressive strength and temperature.

4.2.2. *Compressive Strength of Concrete*

According to the experimental evidence (Abrams, 1971; Schneider, 1986; Zhang and Bićanić, 2002), for temperatures up to 300°C the concrete compressive strength slightly increases with increase of temperature. However, with further increase of temperature, the concrete strength decreases almost linearly. Namely, at lower temperatures hydration of cement paste is more advanced. Moreover, due to the thermal strains the frictional and aggregate locking phenomena are even stronger than for the concrete at normal temperature. Due to these effects the compressive strength does not decrease. At extremely high temperature, microcracks, vaporisation and decomposition of cement paste and aggregate cause decrease of the concrete compressive strength. In the present model it is assumed that up to $T = 300°C$ the cylinder compressive strength f_c is temperature independent and for higher temperature it decreases as a linear function of temperature:

$$\begin{aligned} &f_c(T) = \max(\omega_{t,f_c}) f_{c,0} \\ &for \quad 0 \le \theta \le 2.80 \qquad \omega_{t,f_c} = 1.0 \\ &for \quad \theta > 2.80 \qquad \omega_{t,f_c} = 1.43 - 0.153\theta \end{aligned} \tag{13}$$

where $f_{c,0}$ = uniaxial compressive strength at $T = 20°C$. The adopted dependency is plotted in Fig. 3 and compared with experimental results. As can be seen the comparison shows good agreement.

4.2.3. *Tensile Strength of Concrete*

The experimental evidence indicates that the tensile strength of concrete decreases almost linearly with increase of temperature (Schneider, 1986; Zhang and Bićanić, 2002). At lower temperatures thermal strains lead to micro cracking and damage of the aggregate-cement paste interface which, reduces tensile strength of concrete. With increase of temperature, micro-cracks, vaporisation and decomposition of cement paste and aggregate also lead to further decrease of the concrete tensile strength. In the present model the following dependency of tensile strength on temperature is adopted:

$$f_t(T) = \max(\omega_{t,f_t}) f_{t,0} \qquad \omega_{t,f_t} = 1 - 0.131\theta \tag{14}$$

where $f_{t,0}$ = uniaxial compressive strength at $T = 20°C$. The plot of Eq. (14) is shown and compared with the test data in Fig. 4.

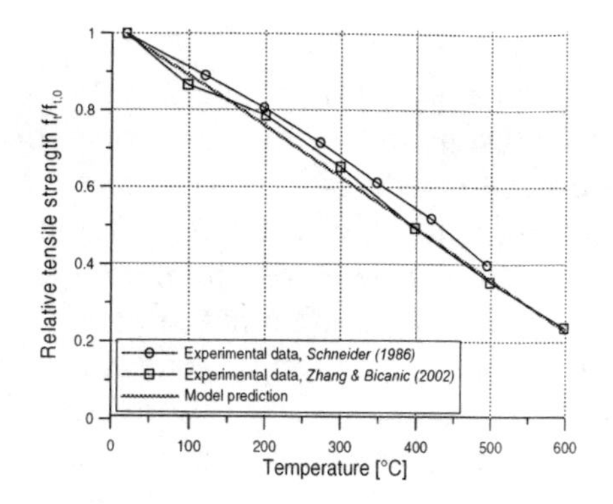

Figure 4. The relative tensile strength as a function of temperature.

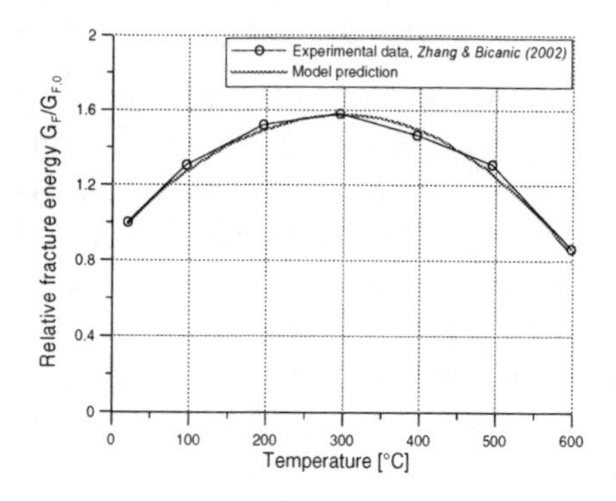

Figure 5. The relative concrete fracture energy as a function of temperature.

4.2.4. *Fracture Energy of Concrete*

A recent experimental investigation (Zhang and Bićanić, 2002) show that with the increase of temperature up to approximately 300°C the concrete fracture energy increases for approximately 60%. However, with further increase of temperature it starts to decrease and at approximately 600°C reaches about 90% of its initial value. This can be explained by the fact that at temperatures between approximately 100 and 300°C, the degree of hydration is higher than at normal room temperature. Moreover, at this stage temperature strains contribute to the frictional effects and aggregate interlock, which increases ductility. At higher temperatures the microcracks, vaporisation and decomposition of cement paste and aggregate cause decrease of concrete ductility. In the present model, the dependency of the concrete fracture energy G_F on the temperature is obtained by fitting of the test data by Zhang and Bićanić (2002). The adopted dependency reads:

$$G_F(T) = \max(\omega_{t,G_F})G_{F,0}$$
$$\begin{aligned} &for \quad 0 \leq \theta \leq 2.80 \quad && \omega_{t,G_F} = 1 + 0.407\theta - 0.0727\theta^2 \\ &for \quad \theta > 2.80 \quad && \omega_{t,G_F} = 0.917 + 0.467\theta - 0.0833\theta^2 \end{aligned} \tag{15}$$

where $G_{F,0}$ = concrete fracture energy at $T = 20°C$. Eq. (15) is plotted in Fig. 5 and compared with experimental results. As can be seen, the comparison shows good agreement.

5. Thermal Strain

As mentioned before, the total thermal strain generated as a consequence of heating of concrete can be decomposed into strains that are stress independent (free thermal strains) and strains, which are stress dependant (stress induced thermal strains).

5.1. FREE THERMAL STRAIN

The experimental evidence (Schneider, 1986) indicates that free thermal strains in concrete specimen mainly depend on the type and amount of the aggregate. As can be seen from Fig. 6, the relationship between the free thermal strain and temperature is highly non-linear and dependent on the thermal stability of the aggregate. Although the experiments indicate that the free thermal strain depends on the rate of temperature, in the present model is assumed that this strain depends only on the temperature. Moreover, it is assumed that in the case of a stress free specimen, thermal strains in all three mutually perpendicular directions are the same (isotropic thermal strains). In the present model the rate of the free thermal strain is calculated as (indicial notation) (Nielsen et al., 2001):

$$
\begin{aligned}
&\dot{\varepsilon}_{ij}^{ft} = \alpha \dot{T} \delta_{ij} \\
&for \quad 0 \le \theta \le 6 \quad \alpha = \frac{6.0 \times 10^{-5}}{7.0 - \theta} \\
&for \quad \theta > 6 \qquad \alpha = 0
\end{aligned}
\tag{16}
$$

where δ_{ij} is Kronecker delta. The above equation is plotted in Fig. 6. In the same figure are also shown the experimental results for concrete made of three different aggregate types (Schneider, 1986). It can be seen that the free thermal strain very much depends on the aggregate tape. Up to approximately 600°C the free thermal strain increases with increase of temperature, however, further increase of temperature causes no further increase of thermal strain. The reason for this is that beyond 600°C the change of the crystalline structure of the aggregate takes place (Bažant and Kaplan, 1996). The relationship (16) is chosen such that only the main trend is covered but not the exact development of free thermal strains for particular concrete type.

5.2. STRESS INDUCED THERMAL STRAIN – CREEP

When a concrete specimen is first loaded and than exposed to high temperature, the resulting thermal strain is different from the one when the specimen is not loaded (Bažant and Chern, 1987; Khoury et al., 1985b; Thelandersson, 1987; Thienel, 1993; Thienel and Rostassy 1996). The difference can be obtained if the free thermal strain is subtracted from the resulting thermal strain. This difference is in the literature known as stress induced thermal strain. As already mentioned, the stress induced thermal strain consists of two parts – irrecoverable part and partly recoverable part (temperature dependent creep). Since the partly recoverable part has only theoretical meaning and is much smaller than the irrecoverable part, in the present model it is neglected, i.e. the total stress dependent thermal strain is assumed to be irrecoverable.

Based on the experimental evidence, in the present model the bi-parabolic thermo-mechanical strain model is used (Nielsen et al., 2004). The uniaxial (scalar) rate form of this model reads:

$$\dot{\varepsilon}^{tm}(T,\sigma) = \frac{\sigma}{f_c^{T_0}}\beta\dot{T}$$

$$\beta = 0.01 \cdot \begin{cases} 2\cdot A\cdot\theta + B & for \quad 0 \le \theta \le \theta^* = 4.5 \\ 2\cdot C\cdot(\theta-\theta^*) + 2\cdot A\cdot\theta^* + B & for \quad \theta > \theta^* \end{cases} \tag{17}$$

where θ^* is a dimensionless transition temperature between the two expressions which correspond to 470°C. The above two expressions are introduced to account for abrupt change in behavior detected in the experiments. A, B and C are experimentally obtained constants that are in the present model set as: $A = 0.0005$, $B = 0.00125$ and $C = 0.0085$.

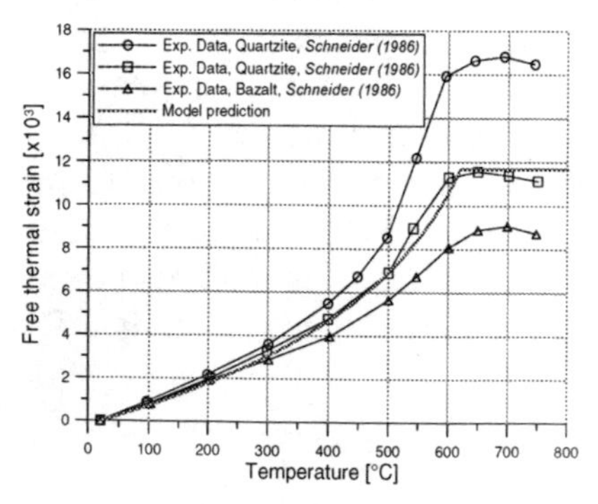

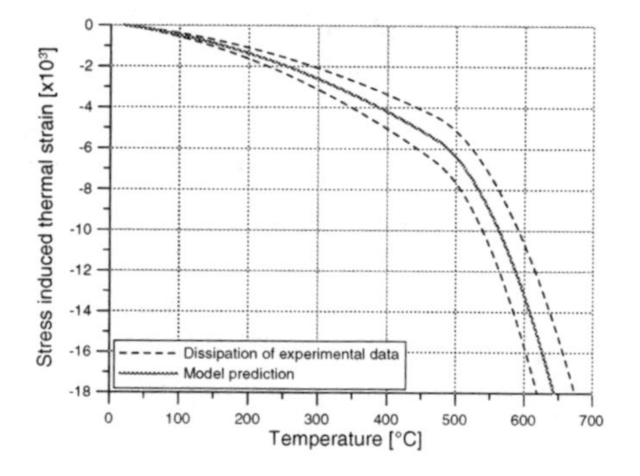

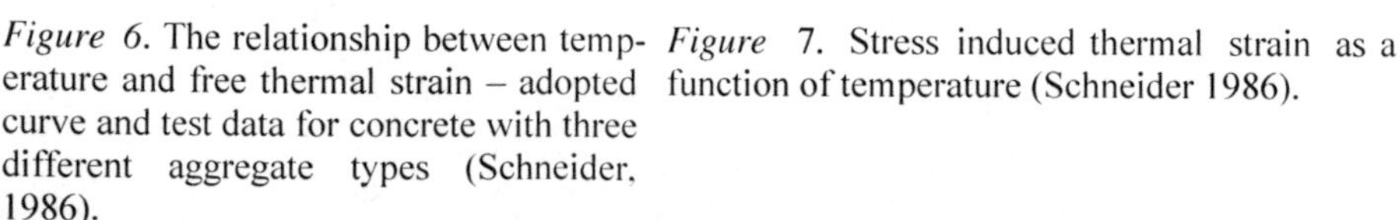

Figure 6. The relationship between temperature and free thermal strain – adopted curve and test data for concrete with three different aggregate types (Schneider, 1986).

Figure 7. Stress induced thermal strain as a function of temperature (Schneider 1986).

All experimental investigations for stress induced thermal strain are performed for sustained compressive load and there is no test available for sustained tensile load. Because of this and the fact that the tensile stress is limited by the relative small tensile strength of concrete, it is assumed that Eq. (17) applies only on compressive stress. Furthermore, it is assumed that the Poisson's ratio, which relates axial and lateral stress induced thermal strains, is a material constant equal to the Poisson's ratio of undamaged concrete. Based on these assumptions the three-dimensional form of Eq. (17) reads:

$$\dot{\varepsilon}_{ij}^{tm}(T,\sigma_{ij}) = \frac{\beta}{f_c^{T_0}}\left((1+\nu)\sigma_{ij}^{-} - \nu\sigma_{kk}^{-}\delta_{ij}\right)\dot{T}(T_{\max})$$
$$\dot{T}(T_{\max}) = \dot{T} \;\; for \;\; T \geq T_{\max}; \;\; \dot{T}(T_{\max}) = 0 \;\; for \;\; T < T_{\max} \tag{18}$$

where '–' indicates compressive stress, i.e. tensile stress components of the stress tensor are set to zero. T_{max} is the maximal temperature reached so far and it is introduced in Eq. (18) to recognize the irreversible nature of thermo-mechanical strain.

6. Numerical Studies

The presented thermo-mechanical model for concrete is implemented into a 3D finite element (FE) code. The implementation is first verified on two examples from the literature. Subsequently, the influence of the high temperature on the performance of a headed stud anchor that is pulled out from a concrete block is studied.

6.1. VERIFICATION

In the first example, transient test data reported by Thelandersson (1987) are reproduced using the presented model. The concrete cylinder was loaded by different levels of sustained compressive loads and heated by constant heating rate. The specimen geometry (cylinder) was discretized by eight node solid finite elements. The results of the numerical analysis are shown in Fig. 8. As can be seen, the numerical prediction fits the experimental data for all load histories very well.

In the second example a concrete specimen with in-plane fully restrained ends and with only two restrained ends, respectively, were exposed to the constant heating rate (Ehm, 1986). The numerical analysis for both boundary conditions was performed by the use of the eight node solid finite elements. The results of the analysis are shown in Fig. 9. The comparison between numerical and experimental results shows again good trend.

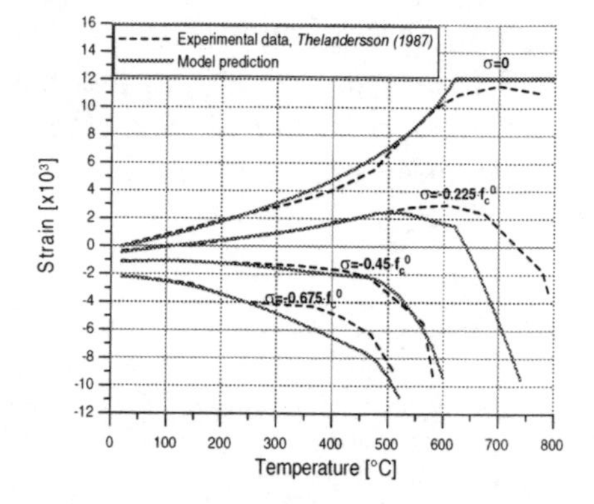

Figure 8. Total strains versus temp-erature.

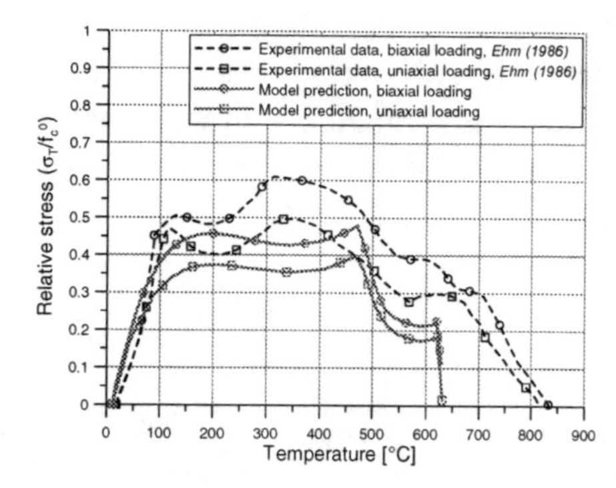

Figure 9. Thermal induced stresses versus temperature.

6.2. PULL-OUT OF HEADED STUD ANCHOR FROM A CONCRETE BLOCK

The performance of a headed stud anchors exposed to high temperature is numerically investigated. A concrete block with a single headed anchor (see Fig. 10) was exposed to fire at its upper side (anchor side). The analysis consists of two parts. In the first part a 3D transient thermal FE analysis was carried out. The resulting thermal distribution is then used in the second part of the analysis in which the above presented thermo-mechanical model was applied.

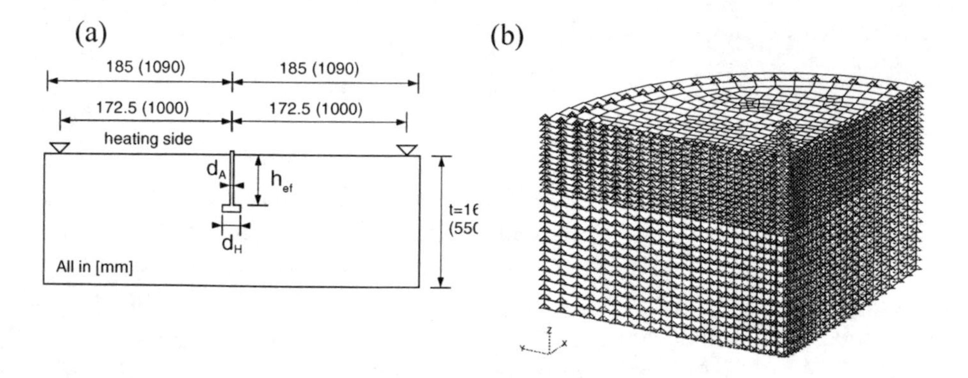

Figure 10. Pull-out of the headed stud anchor from a concrete block: (a) Geometry of the specimen (b). Typical finite element discreti-zation of the concrete specimen.

Table 1. Mechanical and thermal properties of steel and concrete used in the FE analysis.

	Concrete	Steel
Young's modulus E [MPa]	28000	200000
Poisson's ratio ν	0.18	0.34
Tensile strength f_t [MPa]	2.5	
Uniaxial compressive strength f_c [MPa]	21.25	
Fracture energy G_F [Nmm/mm^2]	0.07	
Conductivity λ [W/(mK)]	2.0	53.0
Heat capacity c [J/(kgK)]	900	470
Density ρ [kg/m^3]	2300	7850
Transfer coefficient α [W/(m^{2}K)]	8.0	99.0

The investigated geometry was principally the same as the one tested by Reick (2001). However, in the experiment the concrete block was relatively large. To save computer time, a concrete member of a diameter d_S = 370 mm and thickness t = 160 mm was used for the analysis of anchors with embedment depths h_{ef} = 25 and 50 mm. For anchors with h_{ef} = 150 mm the dimensions of the concrete member were take as d_S = 2180 mm and t = 550 mm (see Fig. 10a). The typical finite element mesh, in which a four node solid elements were used, is shown in Fig. 10b. One fourth of the geometry was discretized, i.e. double symmetry was utilized. To approximately meet experimental boundary conditions, two outer vertical rows of the finite elements around the concrete block were taken as linear elastic. The upper (heated) side of the specimen was supported in vertical direction. The diameter of the supporting ring was 345 (2000) mm. The thermal and mechanical properties of concrete used in the analyses are summarized in Table 1. To prevent anchor failure, the behaviour of steel is assumed to be linear elastic.

The FE analysis was carried out for three different embedment depths h_{ef} = 25 mm (d_A = 8 mm and d_H = 13 mm – see Fig. 10a), h_{ef} = 50 mm (d_A = 10 mm and d_H = 20 mm) and h_{ef} = 150 mm (d_A = 10 mm and d_H = 26 mm). For all analysed embedment depths, anchor was first pre-loaded by design load at temperature of T = 20°C (h_{ef} = 25 mm, P_D = 1.5 kN; h_{ef} = 50 mm, P_D = 5.7 kN; h_{ef} = 150 mm, P_D = 30.0 kN). In the next step the fire was applied at the anchor side of the specimen. The air heating temperature at the upper specimen side was taken according to ISO 833 (equivalent to DIN 4102, part 2):

$$T_{Air}(t) - T_{Air}(t_0) = 345\ log(8t + 1) \tag{19}$$

where $T_{Air}(t_0)$ is the air starting temperature (in our case room temperature of 20°C) and t is time in minutes measured from the start of the fire. In the case of cooling, it was assumed that the air temperature is linearly decreasing from the start of cooling back to the room temperature of 20°C (see Fig. 11a). The temperature at the bottom side of the specimen was assumed to be constant during the entire thermal loading history and equal to 20°C.

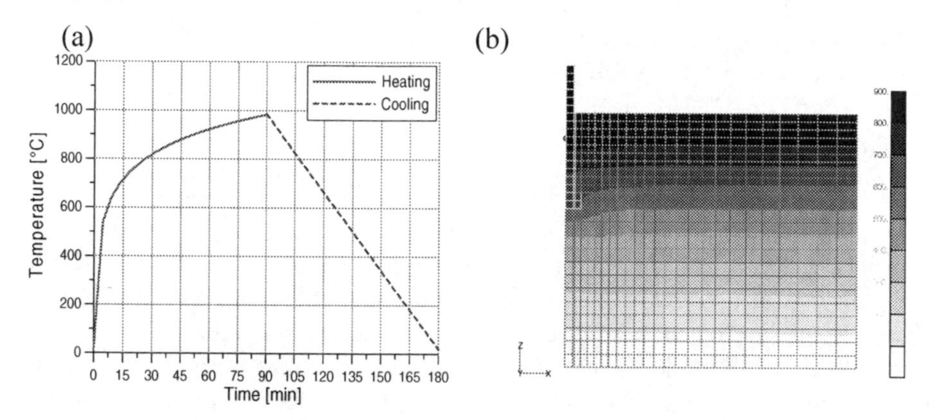

Figure 11. Thermal loading: (a) Assumed temperature increase of air as a function of time (b) Calculated temperature distribution in °C after t = 90 min.

Figure 11b shows distribution of temperature in the concrete specimen (h_{ef} = 50 mm), 90 minutes after start of fire. As one would expect, due to the relatively high conductivity of steel, temperature of concrete around the anchor is higher than in the rest of the specimen. The numerical analysis shows that after cooling of the upper concrete surface down to the room temperature (180 minutes after start of fire), in the mid of the concrete specimen the temperature was still around 300°C.

To investigate the pull-out resistance, the anchor was pulled out from a concrete block for the following thermal loading histories: (i) before heating, (ii) 30 minutes after start of heating, (iii) 90 minutes after start of heating and (iv) after 90 minutes of heating followed by 90 minutes of cooling back to the air temperature of 20°C. Calculated load-displacement curves for h_{ef} = 25 and 150 mm and for four thermal loading histories are plotted in Fig. 12. The relative resistance is calculated as the ratio of the calculated temperature dependent resistance to the calculated pull-out resistance of unheated concrete. Figure 13 shows the relative anchor resistance as a function of embedment depth (Fig. 13a) and the relative anchor resistance as a function of time (Fig. 13b) for all three embedment depths and for all thermal loading histories. For the third loading history, Fig. 13a shows the available test data (Reick 2001). It can be seen that the numerical results fit well the experimental results.

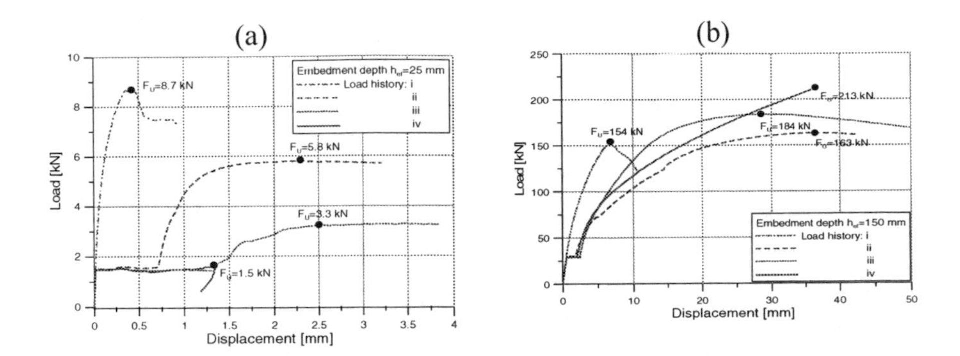

Figure 12. Typical calculated load-displacement curves for all four thermal loading histories with embedment depths: (a) h_{ef} = 25 mm (b) h_{ef} = 150 mm (solid dots indicate maximum load).

As expected, due to damage caused by thermal loading, the pull-out resistance of a headed stud anchors is significantly reduced. It can be seen that with increase of temperature the peak load and stiffness of anchors decrease. Moreover, displacement at peak load significantly increases if the concrete member was exposed to fire. Compared to the initial resistance at $t = 0$ and $T_{Air} = 20°C$, the largest reduction of the ultimate load was obtained for the smallest embedment depth (h_{ef} = 25 mm) and for the forth thermal loading history (90 min. heating followed by 90 min. cooling). However, this does not hold for all embedment depths. The pull-out behaviour of anchors with h_{ef} = 150 mm shows somewhat different response from the one for anchors with smaller embedment depth. Namely, for the second and third thermal loading history there is no reduction of the pull-out capacity. Moreover, for cooling the pull-out resistance is even greater than the pull-out resistance at room temperature.

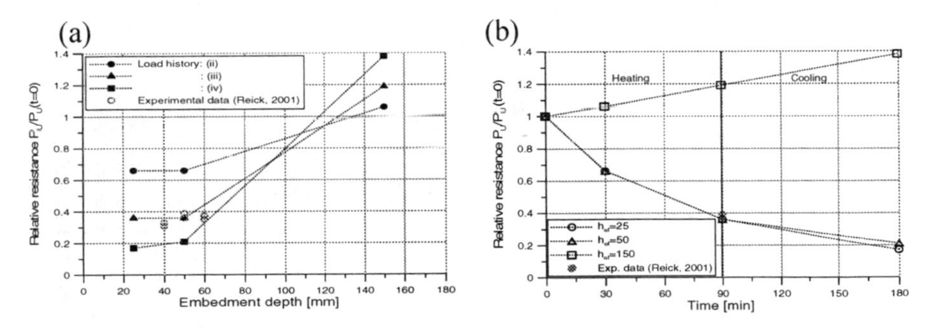

Figure 13. Relative pull-out resistance: (a). As a function of embedment depth (b). As a function of temperature.

For relatively small embedment depths the anchor lies over the entire length in the zone of very high temperature in which the concrete is almost completely destroyed. Extreme cases are observed for embedment depths $h_{ef} = 25$ and 50 *mm* and for the forth loading history. For these cases the ultimate pull-out capacity is even smaller than the initially applied design load (see Figs. 12a), i.e. the anchors failed during cooling. The reason is the existence of large thermal strains, which are partly irreversible, and which in restrained concrete member generate tensile stresses and damage. This damage, together with the degradation of concrete mechanical properties at high temperature, causes significant reduction of the pull-out resistance. To illustrate the effect of the thermal strains, Figure 14a shows damage (dark zones) of the concrete block with $h_{ef} = 25$ mm, which was heated and than cooled down according to the forth thermal loading history, through the maximal principal strains of the mechanical strain tensor. The vertical cracks in concrete, which appear as a consequence of cooling, can be clearly recognized.

For larger anchors and especially if the embedment depth is large compared to the thickness of a concrete member, the head of the stud lies in the zone of lower temperature. In this zone the concrete is less damaged. Moreover, due to the restraining conditions it is possible that the head of the anchor comes into the zone in which the stresses perpendicular to the axis of the anchor are compressive. Such conditions contribute to the relatively low reduction of the pull-out resistance or, as in the present case, to an increase of it. To illustrate this, Fig. 14b shows distribution of horizontal stresses (σ_{yy}) for the embedment depth of 150 mm after cooling of the concrete member ($t = 190$ min). As can be seen, the head of the stud comes into the compressive zone. This explains why the resistance of the anchor in case of the fourth loading history becomes large than the resistance of the anchor pulled out from unheated concrete member.

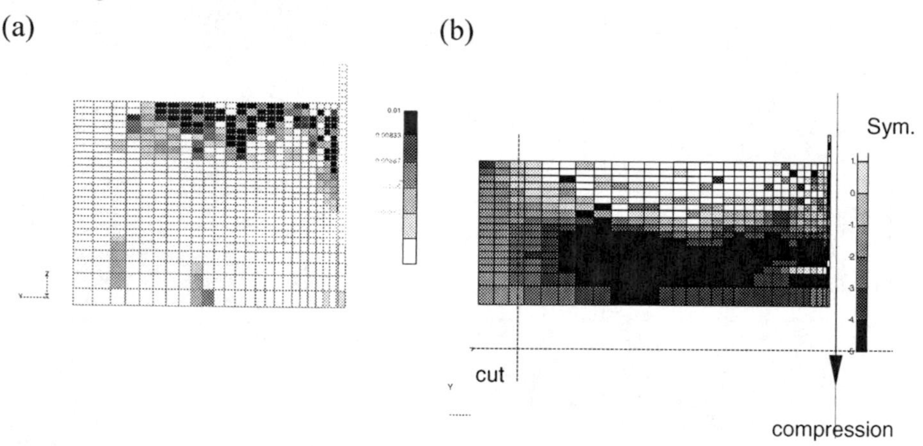

Figure. 14 Results of the thermal analysis: (a) Damaged zone in the concrete member ($h_{ef} = 50$ mm) after heating and cooling – maximal principal strains of the mechanical strain tensor (b) Distribution of horizontal stresses σ_{yy} after heating and cooling ($h_{ef} = 150$ mm).

Figure 15a shows a typical crack pattern (maximal principal strains of the mechanical strain tensor) for unheated concrete member (h_{ef} = 50 mm). Figure 15b shows a typical crack pattern for the anchor (h_{ef} = 50 mm) pulled out from heated concrete member (third thermal loading history). As can be seen, for unheated specimen a typical concrete cone forms. The crack starts from the head of the stud and propagates under an average angle of 35° to the horizontal plane. On the contrary to this, for the heated specimen the angle of the crack propagation close to the anchor head is rather steep. As it approaches the surface of the concrete member, where the concrete is completely destroyed, the crack becomes almost horizontal. Typical failure cone is shown in Figure 15c. This observation is in a very good agreement with experimental evidence.

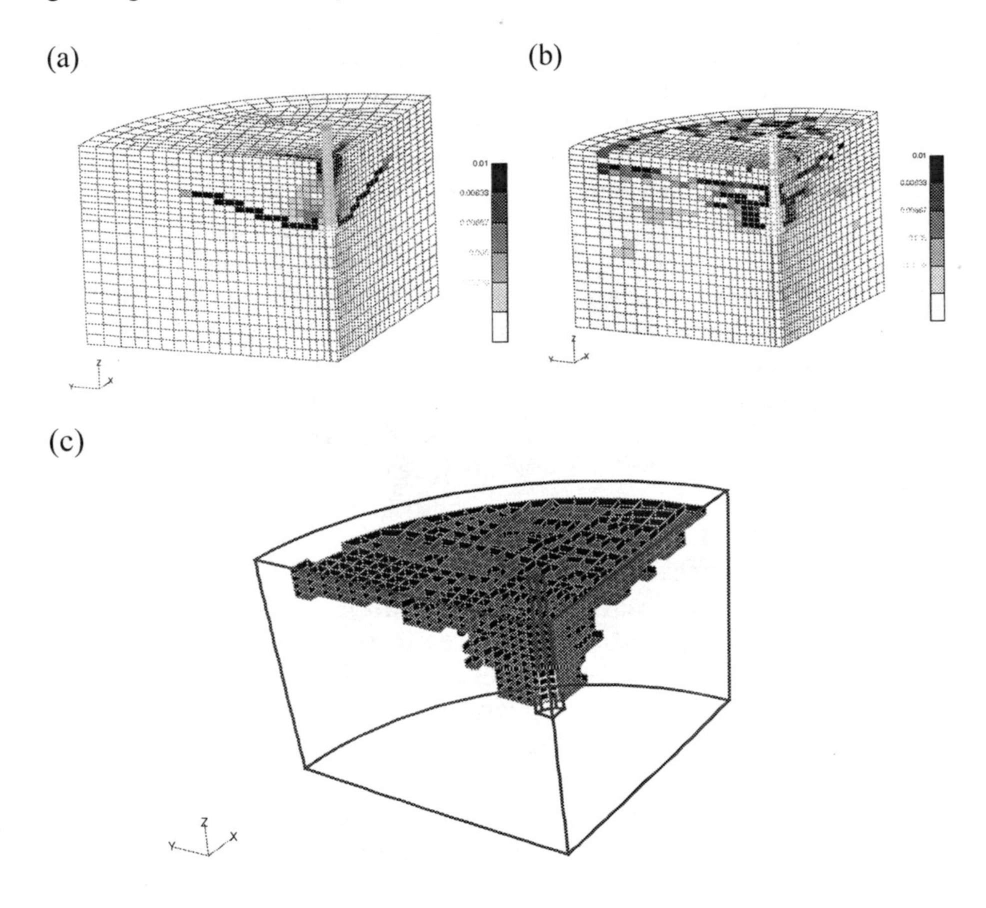

Figure 15. Typical crack patterns for h_{ef} = 50 *mm*: (a) Unheated specimen (b) Specimen exposed 90 min. to fire (loading history (iii)) (c) The concrete cone for the case shown in (b).

7. Conclusions

In the present paper a transient three-dimensional thermo-mechanical model for concrete is presented. For a given boundary conditions, temperature distribution is calculated by the three-dimensional transient thermal finite element analysis using direct integration method. The thermal properties of concrete are assumed to be constant and independent of the stress-strain distribution. In the thermo-mechanical model the total strain is decomposed into pure mechanical strain, free thermal strain and load induced thermal strain. The mechanical strain is calculated based on the temperature dependent microplane model for concrete (Ožbolt et al. 2001). The dependency of the macroscopic concrete properties (Young's modulus, tensile & compressive strength and fracture energy) on temperature is taken from the available experimental database. The free thermal strain, which is stress independent, is calculated by following the proposal of Nielsen et al. (2001). The bi-parabolic model proposed by Nielsen et al. (2004) is used for the prediction of the load induced thermal strain. It is assumed that the total load induced thermal strain is irrecoverable, i.e. creep component is neglected. The model is implemented into a three-dimensional FE code.

To check the model and its implementation two examples taken from the literature were analyzed. In the first the load independent strain and the load dependent thermal strain were predicted. In the second example the thermal induced stresses were calculated. In both cases the model prediction agrees well with the experimental data. Subsequently, the performance of headed stud anchors under fire was investigated. For a given geometry of the concrete member with a constant concrete properties, three-dimensional transient thermal FE analysis was carried out for three embedment depths and for four thermal loading histories. The analysis shows that the resistance of anchors with relatively small embedment depth can be significantly reduced if they are exposed to fire. This is especially true if the concrete member is first exposed to fire and than cooled down. In such a case for small embedment depth the anchor resistance is even smaller than design load. These results agree well with the experiments. For $h_{ef} \geq 150$ mm a slight increase of the pull-out resistance was obtained.

Further studies are needed to investigate the pull-out problem in more detail. Finally, it can be concluded that the presented, relatively simple, model is a powerful numerical tool which can be used to clarify the behavior of a number of structures and structural components exposed to fire.

8. References

Abrams, M.S. (1971), "Compressive strength of concrete at temperatures to 1600F", *ACI SP 25, Temperature and Concrete*, American Concrete Institute, Detroit.

Batdorf, S.B. and Budianski, B. (1949), "A mathematical theory of plasticity based on the concept of slip", *Technical Note No. 1871*, National Advisory Committee for Aeronautics, Washington D.C.

Bažant, Z.P. and Chern, J.C. (1987), "Stress-induced thermal and shrinkage strains in Concrete", *J. of Eng. Mech.*, **113**(10), 1493-1511.

Bažant, Z.P. and Prat, P.C. (1988), "Microplane model for brittle-plastic material - parts I and II", *J. of Eng. Mech.*, **114**(10), 1672-1702.

Bažant, Z.P. and Ožbolt, J. (1990), "Nonlocal microplane model for fracture, damage and size effect in structures", *J. of Eng. Mech.*, **116**(11), 2485-2504.

Bažant, Z.P. and Kaplan, M.F. (1996), *Concrete at High Temperatures: Material Properties and Mathematical Models*, Harlow, Longman.

Carol, I., Jirásek, M., and Bažant, Z.P. (2001), "New thermodynamically consistent approach to microplane theory: Part I - Free energy and consistent microplane stress", *Int. J. of Sol. and Struct.*, **38**(17), 2921-2931.

Belytschko, T., Liu, W.K. and Moran, B. (2001), *Nonlinear Finite Elements for Continua and Structures,* John Wiley & Sons Ltd.

Cook, R.D., Malkus, D.S., Plesha, M.E. and Witt, R.J. (2002), *Concepts and Applications of Finite Element Analysis*, 4th edition, John Wiley & Sons Inc.

Ehm, C. (1986), "Versuche zur Festigkeit und Verformung von Beton unter zweiaxialer Beanspruchung und hohen Temperaturen", *PhD thesis*, Heft 71, TU Braunschweig, Braunschweig.

Gawin, D., Majorana, C.E. and Schrefler, B.A. (1999), "Numerical Analisys of hygro-thermal behaviour and damage of concrete at high temperatures", *Mech. Cohes.-Frict. Mater.*, **4**(1), 37-74.

Khoury, G.A.,Grainger, B.N. and Sullivan, P.J.E. (1985a), "Transient thermal strain of concrete: literature review, conditions within specimens and behaviour of individual constituents", *Mag. of Conc. Res.*, **37**(132), 131-144.

Khoury, G.A., Grainger, B.N. and Sullivan, P.J.E. (1985b), "Strain of concrete during first heating to 600°C under load", *Mag. of Conc. Res.*, **37**(133), 195-215.

Nielsen, C.V., Pearce, C.J. and Bićanić, N. (2001), "Theoretical model of high temperature effects on uniaxial concrete member under elastic restraint", *Mag. of Conc. Res.*, **54**(4), 239-249

Nielsen, C.V., Pearce, C.J. and Bićanić, N. (2004), "Improved phenomenological modelling of transient thermal strains for concrete at high temperatures", *Computers and Concrete,* in press.

Ožbolt, J., Li, Y.-J. and Kožar, I. (2001), "Microplane model for concrete with relaxed kinematic constraint", *Int. J. of Sol. and Struct.*, **38**(16), 2683-2711.

Pearce, C.J., Bićanić, N. and Nielsen, C.V. (2003), "A transient thermal creep model for concrete", *Computational Modeling of Concrete Structures,* Sweets & Zeitlinger, Lisse.

Reick, M. (2001), "Brandverhalten von Befestigungen mit großem Randabstand in Beton bei zentrischer Zugbeanspruchung", *Mitteilungen des Institut für Werkstoffe im Bauwesen*, Band 2001/4, IWB, Universität Stuttgart, Stuttgart.

Schneider, U. (1986), *Properties of Materials at High Temperatures, Concrete*, 2nd. Edition, RILEM Technical Comitee 44-PHT, Technical University of Kassel, Kassel.

Schneider, U. (1988), "Concrete at High Temperatures – A General Review", *Fire Safety Journal,* **13**(1), 55-68

Stabler, J. (2000), "Computational modelling of thermomechanical damage and plasticity in concrete", *PhD thesis*, The University of Queensland, Brisbane.

Taylor, G.I. (1938), "Plastic strain in metals", *J. of the Inst. of Metals.*, **62**, 307-324.

Terro, M.J. (1998), "Numerical modelling of the behaviour of concrete structures in fire", *ACI Struct. J.*, **95**(2), 183-193.

Thelandersson, S. (1983), "On the multiaxial behaviour of concrete exposed to high temperature", *Nucl. Eng. and Design*, **75**(2), 271-282.

Thelandersson, S. (1987), "Modelling of combined thermal and mechanical action in concrete", *J. of Eng. Mech.*, **113**(6), 893-906.

Thienel, K.-C. (1993), "Festigkeit und Verformung von Beton bei hoher Temperatur und biaxialer Beanspruchung – Versuche und Modellbildung", *PhD thesis*, Heft 10, IBMB, TU Braunschweig, Braunschweig.

Thienel, K.-C. and Rostassy, F.S. (1996), "Transient creep of concrete under biaxial stress and high temperature", *Cem. and Conc. Res.*, **26**(9), 1409-1422.

Zhang, B. and Bićanić, N. (2002), "Residual Fracture Toughness of Normal- and High-Strength Gravel Concrete after Heating to 600°C", *ACI Mat. J.,* **99**(3), 217-226.

Part IV: FLUID FLOW INDUCED EXTREME LOADING CONDITIONS

DYNAMICS OF TSUNAMI WAVES

Frédéric Dias (dias@cmla.ens-cachan.fr), Denys Dutykh
Ecole Normale Supérieure de Cachan
CMLA, 61 avenue du président Wilson, 94235 Cachan, France

Abstract. The life of a tsunami is usually divided into three phases: the generation (tsunami source), the propagation and the inundation. Each phase is complex and often described separately. A brief description of each phase is given. Model problems are identified. Their formulation is given. While some of these problems can be solved analytically, most require numerical techniques. The inundation phase is less documented than the other phases. It is shown that methods based on Smoothed Particle Hydrodynamics (SPH) are particularly well-suited for the inundation phase. Directions for future research are outlined.

Key words: dislocations, tsunamis, shallow-water equations, Boussinesq equations, breaking waves, bores

1. Introduction

Given the broadness of the topic of tsunamis, our purpose here is to recall some of the basics of tsunami modeling and to emphasize some general aspects, which are sometimes overlooked. The life of a tsunami is usually divided into three phases: the generation (tsunami source), the propagation and the inundation. The third and most difficult phase of the dynamics of tsunami waves deals with their breaking as they approach the shore. This phase depends greatly on the bottom bathymetry and on the coastline type. The breaking can be progressive. Then the inundation process is relatively slow and can last for several minutes. Structural damages are mainly caused by inundation. The breaking can also be explosive and lead to the formation of a plunging jet. The impact on the coast is then very rapid. In very shallow water, the amplitude of tsunami waves grows to such an extent that typically an undulation appears on the long wave, which develops into a progressive bore (Chanson, 2005). This turbulent front, similar to the wave that occurs when a dam breaks, can be quite high and travel onto the beach at great speed. Then the front and the turbulent current behind it move onto the shore, past buildings and vegetation until they are finally stopped by rising ground. The water level can rise rapidly, typically from 0 to 3 meters in 90 seconds.

The trajectory of these currents and their velocity are quite unpredictible, especially in the final stages because they are sensitive to small changes in the

A. Ibrahimbegovic and I. Kozar (eds.),Extreme Man-Made and Natural Hazards in Dynamics of Structures, 201–224.

topography, and to the stochastic patterns of the collapse of buildings, and to the accumulation of debris such as trees, cars, logs, furniture. The dynamics of this final stage of tsunami waves is somewhat similar to the dynamics of flood waves caused by dam breaking, dyke breaking or overtopping of dykes (cf. the recent tragedy of hurricane Katrina in August 2005). Hence research on flooding events and measures to deal with them may be able to contribute to improved warning and damage reduction systems for tsunami waves in the areas of the world where these waves are likely to occur as shallow surge waves (cf. the recent tragedy of the Indian Ocean tsunami in December 2004).

Civil engineers who visited the damage area following the Boxing day tsunami came up with several basic conclusions. Buildings that had been constructed to satisfy modern safety standards offered a satisfactory resistance, in particular those with reinforced concrete beams properly integrated in the frame structure. These were able to withstand pressure associated with the leading front of the order of 1 atmosphere (recall that an equivalent pressure p is obtained with a windspeed U of about 450 m/s, since $p = \rho_{\text{air}}U^2/2$). By contrast brick buildings collapsed and were washed away. Highly porous or open structures survived. Buildings further away from the beach survived the front in some cases, but they were then destroyed by the erosion of the ground around the buildings by the water currents (Hunt and Burgers, 2005).

Section 2 provides a description of the tsunami source when the source is an earthquake. In Section 3, we review the equations that are often used for tsunami propagation. Section 4 provides a short discussion on the energy of tsunamis. Section 5 is devoted to the run-up and inundation of tsunamis. Finally directions for future research are outlined.

2. Tsunami Induced by Near-shore Earthquake

The inversion of seismic data allows one to reconstruct the permanent deformations of the sea bottom following earthquakes. In spite of the complexity of the seismic source and of the internal structure of the Earth, scientists have been relatively successful in using simple models for the source (Okada, 1985). A description of Okada's model follows.

2.1. INTRODUCTION

The fracture zones, along which the foci of earthquakes are to be found, have been described in various papers. For example, it has been suggested that Volterra's theory of dislocations might be the proper tool for a quantitative description of these fracture zones (Steketee, 1958). This suggestion was made for the following reason. If the mechanism involved in earthquakes and the fracture zones is indeed one of fracture, discontinuities in the displacement components across the fractured surface will exist. As dislocation theory may be described as that part of the theory of elasticity dealing with surfaces across which the displacement field is discontinuous, the suggestion seems reasonable.

As commonly done in mathematical physics, it is necessary for simplicity's sake to make some assumptions. Here we neglect the curvature of the earth, its gravity, temperature, magnetism, non-homogeneity, and consider a semi-infinite medium, which is homogeneous and isotropic. We further assume that the laws of classical linear elasticity theory hold.

Several studies showed that the effect of earth curvature is negligible for shallow events at distances of less than 20° (McGinley, 1969; Ben-Menahem et al., 1969; Ben-Mehanem et al., 1970; Smylie and Mansinha, 1971). The sensitivity to earth topography, homogeneity, isotropy and half-space assumptions was studied and discussed recently (Masterlark, 2003). The author used a commercially available code, ABACUS, which is based on a finite element model (FEM). Six FEMs were constructed to test the sensitivity of deformation predictions to each assumption. The main conclusion is that the vertical layering of lateral inhomogeneity can sometimes cause considerable effects on the deformation fields.

The usual boundary conditions for dealing with earth's problems require that the surface S of the elastic medium (the earth) shall be free from forces. The resulting mixed boundary-value problem was solved a century ago (Volterra, 1907). Later, Steketee proposed an alternative method to solve this problem using Green's functions (Steketee, 1958).

2.2. VOLTERRA'S THEORY OF DISLOCATIONS

In order to introduce the concept of dislocation and for simplicity's sake, this section is devoted to the case of an entire elastic space. The second reason is that in his original paper Volterra solved the problem in this case (Volterra, 1907).

Let O be the origin of a Cartesian coordinate system in an infinite elastic medium, x_i the Cartesian coordinates ($i = 1, 2, 3$), and $\mathbf{e}_i$ a unit vector in the positive x_i-direction. A force $\mathbf{F} = F\mathbf{e}_k$ at O generates a displacement field $u_i^k(P, O)$ at point P, which is determined by the well-known Somigliana tensor

$$u_i^k(P, O) = \frac{F}{8\pi\mu}(\delta_{ik} r,_{nn} - \alpha r,_{ik}), \quad \text{with } \alpha = \frac{\lambda + \mu}{\lambda + 2\mu}. \tag{1}$$

In this relation δ_{ik} is the Kronecker delta, λ and μ are Lamé's constants, and r is the distance from P to O. The coefficient α can be rewritten as $\alpha = 1/2(1 - \nu)$, where ν is Poisson's ratio. Later we will also use Young's modulus E, which is defined as

$$E = \frac{\mu\,(3\lambda + 2\mu)}{\lambda + \mu}.$$

The notation $r,_i$ means $\partial r/\partial x_i$ and the summation convention applies.

The stresses due to the displacement field (1) are easily computed from Hooke's law:

$$\sigma_{ij} = \lambda \delta_{ij} u_{k,k} + \mu(u_{i,j} + u_{j,i}). \tag{2}$$

We find

$$\sigma_{ij}^k(P,O) = -\frac{\alpha F}{4\pi}\left(3\frac{x_i x_j x_k}{r^5} + \frac{\mu}{\lambda+\mu}\frac{\delta_{ki}x_j + \delta_{kj}x_i - \delta_{ij}x_k}{r^3}\right).$$

The components of the force per unit area on a surface element are denoted as follows:

$$T_i^k = \sigma_{ij}^k \cdot \nu_j,$$

where the ν_j's are the direction cosines of the normal to the surface element (Sokolnikoff and Specht, 1946). A Volterra dislocation is defined as a surface Σ in the elastic medium across which there is a discontinuity Δu_i in the displacement fields of the type

$$\Delta u_i = u_i^+ - u_i^- = U_i + \Omega_{ij} x_j, \tag{3}$$

$$\Omega_{ij} = -\Omega_{ji}. \tag{4}$$

Equation (3) in which U_i and Ω_{ij} are constants is the well-known Weingarten relation which states that the discontinuity Δu_i should be of the type of a rigid body displacement, thereby maintaining continuity of the components of stress and strain across Σ.

The displacement field in an infinite elastic medium due to a dislocation of type (1) is then determined by Volterra's formula (Volterra, 1907)

$$u_k(y_1, y_2, y_3) := u_k(y_l) = \frac{1}{F}\iint_{\Sigma} \Delta u_i T_i^k dS. \tag{5}$$

Once the surface Σ is given, the dislocation is essentially determined by the six constants U_i and Ω_{ij}. Therefore we also write

$$u_k(y_l) = \frac{U_i}{F}\iint_{\Sigma} \sigma_{ij}^k(P,Q)\nu_j dS + \frac{\Omega_{ij}}{F}\iint_{\Sigma} \{x_j \sigma_{il}^k(P,Q) - x_i \sigma_{jl}^k(P,Q)\}\nu_l dS, \tag{6}$$

where Ω_{ij} takes only the values $\Omega_{12}, \Omega_{23}, \Omega_{31}$. Following Volterra (Volterra, 1907) and Love (Love, 1944) we call each of the six integrals in (6) an elementary dislocation.

It is clear from (5) and (6) that the computation of the displacement field $u_k(Q)$ is performed as follows: A force $F\mathbf{e}_k$ is applied at Q, and the stresses $\sigma_{ij}^k(P,Q)$ that this force generates are computed at the points $P(x_i)$ on Σ. In particular the components of the force on Σ are computed. After multiplication with prescribed weights of magnitude Δu_i these forces are integrated over Σ to give the displacement component in Q due to the dislocation on Σ.

2.3. DISLOCATIONS IN ELASTIC HALF-SPACE

When the case of an elastic half-space is considered, equation (5) remains valid, but we have to replace σ_{ij}^k by another tensor ω_{ij}^k. This can be explained by the fact that the elementary solutions for a half-space are different from Somigliana solution (1).

The ω_{ij}^k can be obtained from the displacements corresponding to nuclei of strain in a half-space through relation (2). Steketee showed a method of obtaining the six ω_{ij}^k fields by using a Green's function and derived ω_{12}^k, which is relevant to a vertical strike-slip fault. Maruyama derived the remaining five functions (Maruyama, 1964).

It is interesting to mention here that historically these solutions were first derived in a straightforward manner by Mindlin (Mindlin, 1936; Mindlin and Cheng, 1950), who gave explicit expressions of the displacement and stress fields for half-space nuclei of strain consisting of single forces with and without moment. It is only necessary to write the single force results since the other forms can be obtained by taking appropriate derivatives. Their method consists in finding the displacement field in Westergaard's form of the Galerkin vector (Westergaard, 1935). This vector is then determined by taking a linear combination of some biharmonic elementary solutions. The coefficients are chosen to satisfy boundary and equilibrium conditions. These solutions were also derived by Press in a slightly different manner (Press, 1965).

Here, we take the Cartesian coordinate system shown in Figure 1. The elastic medium occupies the region $z \leq 0$ and the $x-$axis is taken to be parallel to the strike direction of the fault. In this coordinate system, $u_i^j(x_1, x_2, x_3; \xi_1, \xi_2, \xi_3)$ is the ith component of the displacement at (x_1, x_2, x_3) due to the jth direction point force of magnitude F at (ξ_1, ξ_2, ξ_3). It can be expressed as follows (Mindlin, 1936; Press, 1965; Okada, 1985; Okada, 1992):

$$u_i^j(x_1, x_2, x_3) = u_{iA}^j(x_1, x_2, -x_3) - u_{iA}^j(x_1, x_2, x_3) + u_{iB}^j(x_1, x_2, x_3) + x_3 u_{iC}^j(x_1, x_2, x_3), \tag{7}$$

where

$$u_{iA}^j = \frac{F}{8\pi\mu}\left((2-\alpha)\frac{\delta_{ij}}{R} + \alpha\frac{R_i R_j}{R^3}\right),$$

$$u_{iB}^j = \frac{F}{4\pi\mu}\left(\frac{\delta_{ij}}{R} + \frac{R_i R_j}{R^3} + \frac{1-\alpha}{\alpha}\left[\frac{\delta_{ij}}{R+R_3} + \right.\right.$$

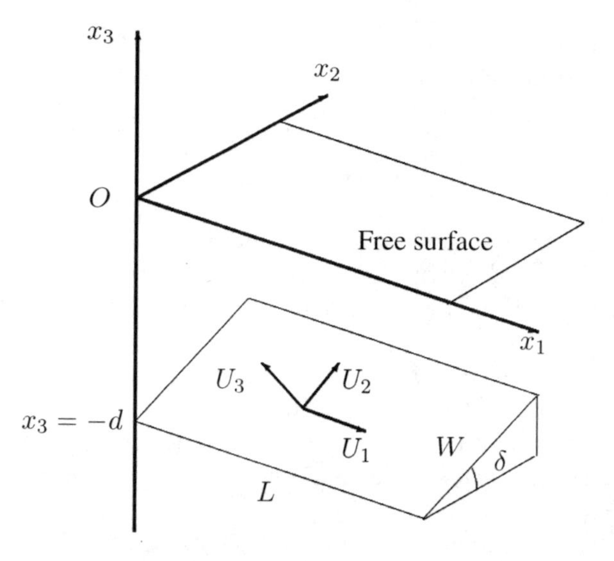

Figure 1. Coordinate system adopted in this study and geometry of the source model.

$$+\frac{R_i\delta_{j3} - R_j\delta_{i3}(1-\delta_{j3})}{R(R+R_3)} - \frac{R_iR_j}{R(R+R_3)^2}(1-\delta_{i3})(1-\delta_{j3})\Big]\Big),$$

$$u^j_{iC} = \frac{F}{4\pi\mu}(1-2\delta_{i3})\left((2-\alpha)\frac{R_i\delta_{j3} - R_j\delta_{i3}}{R^3} + \alpha\xi_3\left[\frac{\delta_{ij}}{R^3} - 3\frac{R_iR_j}{R^5}\right]\right).$$

In these expressions $R_1 = x_1 - \xi_1$, $R_2 = x_2 - \xi_2$, $R_3 = x_3 - \xi_3$, $R^2 = R_1^2 + R_2^2 + R_3^2$.

The first term in equation (7), $u^j_{iA}(x_1, x_2, x_3)$, is the well-known Somigliana tensor, which represents the displacement field due to a single force placed at (ξ_1, ξ_2, ξ_3) in an infinite medium (Love, 1944). The second term also looks like a Somigliana tensor. This term corresponds to a contribution from an image source of the given point force placed at $(\xi_1, \xi_2, -\xi_3)$ in the infinite medium. The third term, $u^j_{iB}(x_1, x_2, x_3)$, and $u^j_{iC}(x_1, x_2, x_3)$ in the fourth term are naturally depth dependent. When x_3 is set equal to zero in equation (7), the first and the second terms cancel each other, and the fourth term vanishes. The remaining term, $u^j_{iB}(x_1, x_2, 0)$ reduces to the formula for the surface displacement field due to a point force in a half-space (Okada, 1985):

$$\begin{cases} u^1_1 = \frac{F}{4\pi\mu}\left(\frac{1}{R} + \frac{(x_1-\xi_1)^2}{R^3} + \frac{\mu}{\lambda+\mu}\left[\frac{1}{R-\xi_3} - \frac{(x_1-\xi_1)^2}{R(R-\xi_3)^2}\right]\right), \\ u^1_2 = \frac{F}{4\pi\mu}(x_1-\xi_1)(x_2-\xi_2)\left(\frac{1}{R^3} - \frac{\mu}{\lambda+\mu}\frac{1}{R(R-\xi_3)^2}\right), \\ u^1_3 = \frac{F}{4\pi\mu}(x_1-\xi_1)\left(-\frac{\xi_3}{R^3} - \frac{\mu}{\lambda+\mu}\frac{1}{R(R-\xi_3)}\right), \end{cases}$$

$$\begin{cases} u_1^2 = \frac{F}{4\pi\mu}(x_1-\xi_1)(x_2-\xi_2)\left(\frac{1}{R^3} - \frac{\mu}{\lambda+\mu}\frac{1}{R(R-\xi_3)^2}\right), \\ u_2^2 = \frac{F}{4\pi\mu}\left(\frac{1}{R} + \frac{(x_2-\xi_2)^2}{R^3} + \frac{\mu}{\lambda+\mu}\left[\frac{1}{R-\xi_3} - \frac{(x_2-\xi_2)^2}{R(R-\xi_3)^2}\right]\right), \\ u_3^2 = \frac{F}{4\pi\mu}(x_2-\xi_2)\left(-\frac{\xi_3}{R^3} - \frac{\mu}{\lambda+\mu}\frac{1}{R(R-\xi_3)}\right), \end{cases}$$

$$\begin{cases} u_1^3 = \frac{F}{4\pi\mu}(x_1-\xi_1)\left(-\frac{\xi_3}{R^3} + \frac{\mu}{\lambda+\mu}\frac{1}{R(R-\xi_3)}\right), \\ u_2^3 = \frac{F}{4\pi\mu}(x_2-\xi_2)\left(-\frac{\xi_3}{R^3} + \frac{\mu}{\lambda+\mu}\frac{1}{R(R-\xi_3)}\right), \\ u_3^3 = \frac{F}{4\pi\mu}\left(\frac{1}{R} + \frac{\xi_3^2}{R^3} + \frac{\mu}{\lambda+\mu}\frac{1}{R}\right). \end{cases}$$

In these formulas $R^2 = (x_1-\xi_1)^2 + (x_2-\xi_2)^2 + \xi_3^2$.

In order to obtain the displacements due to the dislocation we need to calculate the corresponding ξ_k-derivatives of the point force solution (7) and to put it in Steketee-Volterra formula (5)

$$u_i = \frac{1}{F}\iint\limits_{\Sigma} \Delta u_j \left[\lambda\delta_{jk}\frac{\partial u_i^n}{\partial \xi_n} + \mu\left(\frac{\partial u_i^j}{\partial \xi_k} + \frac{\partial u_i^k}{\partial \xi_j}\right)\right]\nu_k dS.$$

It is expressed as follows:

$$\begin{aligned} \frac{\partial u_i^j}{\partial \xi_k}(x_1,x_2,x_3) &= \frac{\partial u_{iA}^j}{\partial \xi_k}(x_1,x_2,-x_3) - \frac{\partial u_{iA}^j}{\partial \xi_k}(x_1,x_2,x_3) + \\ &+ \frac{\partial u_{iB}^j}{\partial \xi_k}(x_1,x_2,x_3) + x_3\frac{\partial u_{iC}^j}{\partial \xi_k}(x_1,x_2,x_3), \end{aligned}$$

with

$$\begin{aligned} \frac{\partial u_{iA}^j}{\partial \xi_k} &= \frac{F}{8\pi\mu}\left((2-\alpha)\frac{R_k}{R^3}\delta_{ij} - \alpha\frac{R_i\delta_{jk} + R_j\delta_{ik}}{R^3} + 3\alpha\frac{R_iR_jR_k}{R^5}\right), \\ \frac{\partial u_{iB}^j}{\partial \xi_k} &= \frac{F}{4\pi\mu}\Bigg(-\frac{R_i\delta_{jk} + R_j\delta_{ik} - R_k\delta_{ij}}{R^3} + 3\frac{R_iR_jR_k}{R^5} + \\ &+ \frac{1-\alpha}{\alpha}\Big[\frac{\delta_{3k}R + R_k}{R(R+R_3)^2}\delta_{ij} - \frac{\delta_{ik}\delta_{j3} - \delta_{jk}\delta_{i3}(1-\delta_{j3})}{R(R+R_3)} + \\ &+ (R_i\delta_{j3} - R_j\delta_{i3}(1-\delta_{j3}))\frac{\delta_{3k}R^2 + R_k(2R+R_3)}{R^3(R+R_3)^2} + \\ &+ (1-\delta_{i3})(1-\delta_{j3})\Big(\frac{R_i\delta_{jk} + R_j\delta_{ik}}{R(R+R_3)^2} - R_iR_j\frac{2\delta_{3k}R^2 + R_k(3R+R_3)}{R^3(R+R_3)^3}\Big)\Big]\Bigg) \\ \frac{\partial u_{iC}^j}{\partial \xi_k} &= \frac{F}{4\pi\mu}(1-2\delta_{i3})\Bigg((2-\alpha)\Big[\frac{\delta_{jk}\delta_{i3} - \delta_{ik}\delta_{j3}}{R^3} + \frac{3R_k(R_i\delta_{j3} - R_j\delta_{i3})}{R^5}\Big] + \\ &+ \alpha\delta_{3k}\Big[\frac{\delta_{ij}}{R^3} - \frac{3R_iR_j}{R^5}\Big] + 3\alpha\xi_3\Big[\frac{R_i\delta_{jk} + R_j\delta_{ik} + R_k\delta_{ij}}{R^5} - \frac{5R_iR_jR_k}{R^7}\Big]\Bigg). \end{aligned}$$

2.4. FINITE RECTANGULAR SOURCE

Now, let us consider a more practical problem. We define the elementary dislocations U_1, U_2, and U_3, corresponding to the strike-slip, dip-slip, and tensile components of an arbitrary dislocation. In Figure 1 each vector represents the direction of the elementary faults. The vector $\mathbf{D}$ is the so-called Burger's vector, which shows how both sides of the fault are spread out: $\mathbf{D} = \mathbf{u}^+ - \mathbf{u}^-$.

A general dislocation can be determined by three angles: the dip angle δ of the fault, the slip angle θ, and the angle ϕ between the fault plane and Burger's vector $\mathbf{D}$. This situation is schematically described in Figure 2.

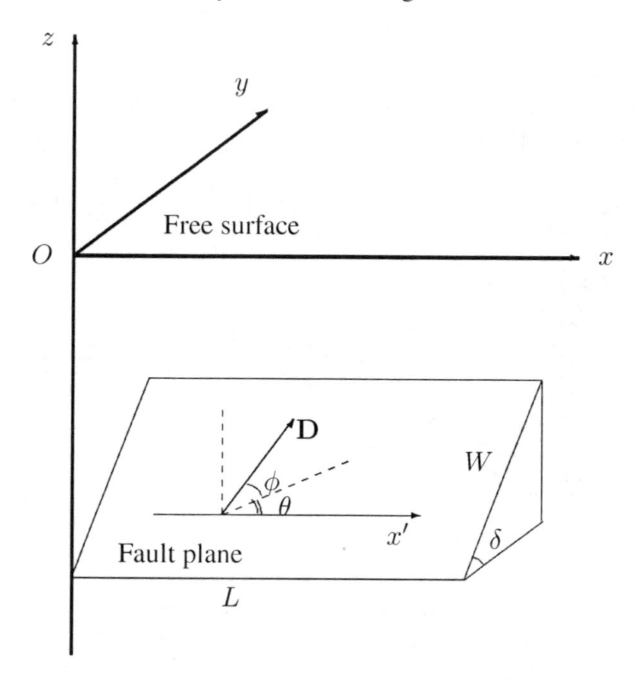

Figure 2. Geometry of the source model and orientation of Burger's vector **D**.

For a finite rectangular fault with length L and width W occurring at depth d (Figure 2), the deformation field can be evaluated analytically by changing the variables and performing integration over the rectangle. This was done by several authors (Chinnery, 1963; Sato and Matsu'ura, 1974; Iwasaki and Sato, 1979; Okada, 1985; Okada, 1992). Here we give the results of their computations. The final results represented in compact form are listed below using Chinnery's notation $\|$ to represent the substitution

$$f(\xi, \eta)\| = f(x, p) - f(x, p - W) - f(x - L, p) + f(x - L, p - W).$$

Let us introduce the following notation:

$$p = y\cos\delta + d\sin\delta, \quad q = y\sin\delta - d\cos\delta,$$

$$\tilde{y} = \eta\cos\delta + q\sin\delta, \quad \tilde{d} = \eta\sin\delta - q\cos\delta,$$

$$R^2 = \xi^2 + \eta^2 + q^2 = \xi^2 + \tilde{y}^2 + \tilde{d}^2, \quad X^2 = \xi^2 + q^2.$$

The quantities U_1, U_2 and U_3 are linked to Burger's vector through the identities

$$U_1 = |\mathbf{D}|\cos\phi\cos\theta, \quad U_2 = |\mathbf{D}|\cos\phi\sin\theta, \quad U_3 = |\mathbf{D}|\sin\phi.$$

For a strike-slip dislocation, one has

$$\begin{aligned}
u_1 &= -\frac{U_1}{2\pi}\left(\frac{\xi q}{R(R+\eta)} + \arctan\frac{\xi\eta}{qR} + I_1\sin\delta\right)\Bigg\|, \\
u_2 &= -\frac{U_1}{2\pi}\left(\frac{\tilde{y}q}{R(R+\eta)} + \frac{q\cos\delta}{R+\eta} + I_2\sin\delta\right)\Bigg\|, \\
u_3 &= -\frac{U_1}{2\pi}\left(\frac{\tilde{d}q}{R(R+\eta)} + \frac{q\sin\delta}{R+\eta} + I_4\sin\delta\right)\Bigg\|.
\end{aligned}$$

For a dip-slip dislocation, one has

$$\begin{aligned}
u_1 &= -\frac{U_2}{2\pi}\left(\frac{q}{R} - I_3\sin\delta\cos\delta\right)\Bigg\|, \\
u_2 &= -\frac{U_2}{2\pi}\left(\frac{\tilde{y}q}{R(R+\xi)} + \cos\delta\arctan\frac{\xi\eta}{qR} - I_1\sin\delta\cos\delta\right)\Bigg\|, \\
u_3 &= -\frac{U_2}{2\pi}\left(\frac{\tilde{d}q}{R(R+\xi)} + \sin\delta\arctan\frac{\xi\eta}{qR} - I_5\sin\delta\cos\delta\right)\Bigg\|.
\end{aligned}$$

For a tensile fault dislocation, one has

$$\begin{aligned}
u_1 &= \frac{U_3}{2\pi}\left(\frac{q^2}{R(R+\eta)} - I_3\sin^2\delta\right)\Bigg\|, \\
u_2 &= \frac{U_3}{2\pi}\left(\frac{-\tilde{d}q}{R(R+\xi)} - \sin\delta\left[\frac{\xi q}{R(R+\eta)} - \arctan\frac{\xi\eta}{qR}\right] - I_1\sin^2\delta\right)\Bigg\|, \\
u_3 &= \frac{U_3}{2\pi}\left(\frac{\tilde{y}q}{R(R+\xi)} + \cos\delta\left[\frac{\xi q}{R(R+\eta)} - \arctan\frac{\xi\eta}{qR}\right] - I_5\sin^2\delta\right)\Bigg\|.
\end{aligned}$$

The terms $I_1, \ldots, I_5$ are given by

$$I_1 = -\frac{\mu}{\lambda+\mu}\frac{\xi}{(R+\tilde{d})\cos\delta} - \tan\delta I_5,$$

Table 1. Parameter set used in Figures 3, 4, and 5.

Parameter	*Value*
Dip angle δ	$13°$
Fault depth d, km	25
Fault length L, km	220
Fault width W, km	90
U_i, m	30
Young modulus E, GPa	9.5
Poisson's ratio ν	0.23

$$
\begin{aligned}
I_2 &= -\frac{\mu}{\lambda+\mu}\log(R+\eta) - I_3,\\
I_3 &= \frac{\mu}{\lambda+\mu}\left[\frac{1}{\cos\delta}\frac{\tilde{y}}{R+\tilde{d}} - \log(R+\eta)\right] + \tan\delta I_4,\\
I_4 &= \frac{\mu}{\mu+\lambda}\frac{1}{\cos\delta}\Big(\log(R+\tilde{d}) - \sin\delta\log(R+\eta)\Big),\\
I_5 &= \frac{\mu}{\lambda+\mu}\frac{2}{\cos\delta}\arctan\frac{\eta(X+q\cos\delta)+X(R+X)\sin\delta}{\xi(R+X)\cos\delta},
\end{aligned}
$$

and if $\cos\delta = 0$,

$$
\begin{aligned}
I_1 &= -\frac{\mu}{2(\lambda+\mu)}\frac{\xi q}{(R+\tilde{d})^2},\\
I_3 &= \frac{\mu}{2(\lambda+\mu)}\left[\frac{\eta}{R+\tilde{d}} + \frac{\tilde{y}q}{(R+\tilde{d})^2} - \log(R+\eta)\right],\\
I_4 &= -\frac{\mu}{\lambda+\mu}\frac{q}{R+\tilde{d}},\\
I_5 &= -\frac{\mu}{\lambda+\mu}\frac{\xi\sin\delta}{R+\tilde{d}}.
\end{aligned}
$$

Figures 3, 4, and 5 show the free-surface deformation after three elementary dislocations. The values of the parameters are given in Table 1.

The traditional approach for hydrodynamic modelers is indeed to use elastic models similar to the model just described with the seismic parameters as input to evaluate details of the seafloor deformation. Then this deformation is translated to the initial condition of the evolution problem described in the next section. A few authors have solved the linearized water wave equations in the presence of a moving bottom (Hammack, 1973; Todorovska and Trifunac, 2001).

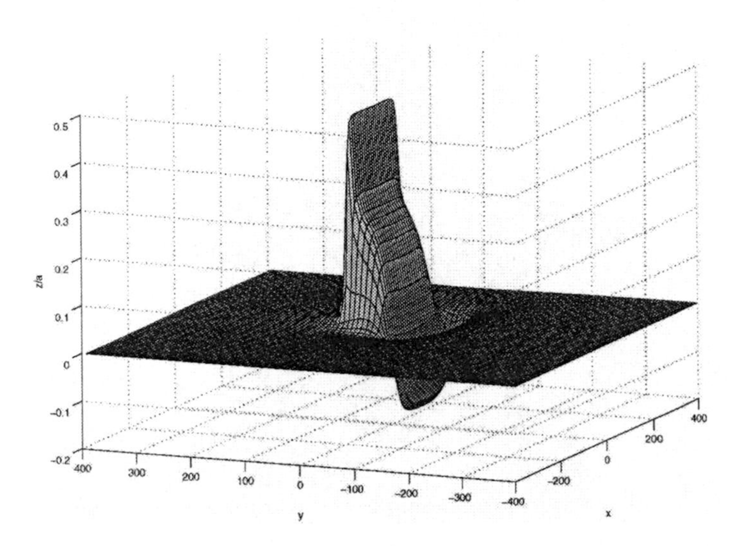

Figure 3. Dimensionless free-surface deformation z/a after dip-slip fault. Here a is $|\mathbf{D}|$ (30 m in the present application).

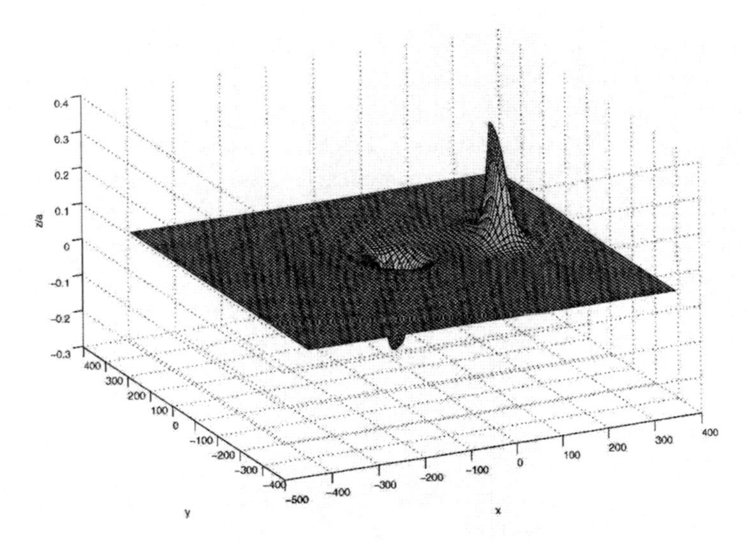

Figure 4. Dimensionless free-surface deformation z/a after strike-slip fault. Here a is $|\mathbf{D}|$ (30 m in the present application).

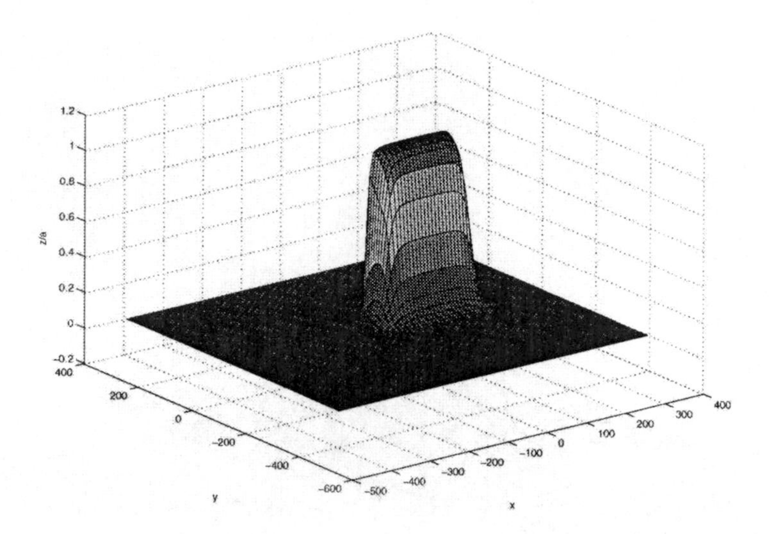

Figure 5. Dimensionless free-surface deformation z/a after tensile fault. Here $\mathbf{D} = (0, 0, U_3)$ and $a = U_3$.

3. Propagation of Tsunamis

The problem of tsunami propagation is a special case of the general water-wave problem. The study of water waves relies on several common assumptions. Some are obvious while some others are questionable under certain circumstances. The water is assumed to be incompressible. Dissipation is not often included. However there are three main sources of dissipation for water waves: bottom friction, surface dissipation and body dissipation. For tsunamis, bottom friction is the most important one, especially in the later stages, and is sometimes included in the computations in an ad-hoc way. In most theoretical analyses, it is not included.

A brief description of the common mathematical model used to study water waves follows. The horizontal coordinates are denoted by x and y, and the vertical coordinate by z. The horizontal gradient is denoted by

$$\nabla := \left(\frac{\partial}{\partial x}, \frac{\partial}{\partial y} \right).$$

The horizontal velocity is denoted by

$$\mathbf{u}(x, y, z, t) = (u, v)$$

and the vertical velocity by $w(x, y, z, t)$. The three-dimensional flow of an inviscid and incompressible fluid is governed by the conservation of mass

$$\nabla \cdot \mathbf{u} + \frac{\partial w}{\partial z} = 0 \tag{8}$$

and by the conservation of momentum

$$\rho \frac{\mathrm{D}\mathbf{u}}{\mathrm{D}t} = -\nabla p, \quad \rho \frac{\mathrm{D}w}{\mathrm{D}t} = -\rho g - \frac{\partial p}{\partial z}. \tag{9}$$

In (9), ρ is the density of water (assumed to be constant throughout the fluid domain), g is the acceleration due to gravity and $p(x, y, z, t)$ the pressure field.

The assumption that the flow is irrotational is commonly made to analyze surface waves. Then there exists a scalar function $\phi(x, y, z, t)$ (the velocity potential) such that

$$\mathbf{u} = \nabla \phi, \quad w = \frac{\partial \phi}{\partial z}.$$

The continuity equation (8) becomes

$$\nabla^2 \phi + \frac{\partial^2 \phi}{\partial z^2} = 0\,. \tag{10}$$

The equation of momentum conservation (9) can be integrated into Bernoulli's equation

$$\frac{\partial \phi}{\partial t} + \frac{1}{2}|\nabla \phi|^2 + \frac{1}{2}\left(\frac{\partial \phi}{\partial z}\right)^2 + gz + \frac{p - p_0}{\rho} = 0\,, \tag{11}$$

which is valid everywhere in the fluid. The constant p_0 is a pressure of reference, for example the atmospheric pressure. The effects of surface tension are not important for tsunami propagation.

3.1. CLASSICAL FORMULATION

The surface wave problem consists in solving Laplace's equation (10) in a domain $\Omega(t)$ bounded above by a moving free surface (the interface between air and water) and below by a fixed solid boundary (the bottom).[1] The free surface is represented by $F(x, y, z, t) := \eta(x, y, t) - z = 0$. The shape of the bottom is given by $z = -h(x, y)$. The main driving force is gravity.

The free surface must be found as part of the solution. Two boundary conditions are required. The first one is the kinematic condition. It can be stated as $\mathrm{D}F/\mathrm{D}t = 0$ (the material derivative of F vanishes), which leads to

$$\eta_t + \nabla \phi \cdot \nabla \eta - \phi_z = 0 \quad \text{at} \quad z = \eta(x, y, t)\,. \tag{12}$$

The second boundary condition is the dynamic condition which states that the normal stresses must be in balance at the free surface. The normal stress at the free surface is given by the difference in pressure. Bernoulli's equation (11) evaluated on the free surface $z = \eta$ gives

$$\phi_t + \tfrac{1}{2}|\nabla \phi|^2 + \tfrac{1}{2}\phi_z^2 + g\eta = 0 \quad \text{at} \quad z = \eta(x, y, t)\,. \tag{13}$$

Finally, the boundary condition at the bottom is

$$\nabla\phi \cdot \nabla h + \phi_z = 0 \quad \text{at} \;\; z = -h(x, y)\,. \tag{14}$$

To summarize, the goal is to solve the set of equations (10), (12), (13) and (14) for $\eta(x, y, t)$ and $\phi(x, y, z, t)$. When the initial value problem is integrated, the fields $\eta(x, y, 0)$ and $\phi(x, y, z, 0)$ must be specified at $t = 0$. The conservation of momentum equation (9) is not required in the solution procedure; it is used *a posteriori* to find the pressure p once η and ϕ have been found.

In the following subsections, we will consider various approximations of the full water-wave equations. One is the system of Boussinesq equations, that retains nonlinearity and dispersion up to a certain order. Another one is the system of nonlinear shallow-water equations that retains nonlinearity but no dispersion. The simplest one is the system of linear shallow-water equations. The concept of shallow water is based on the smallness of the ratio between water depth and wave length. In the case of tsunamis propagating on the surface of deep oceans, one can consider that shallow-water theory is appropriate because the water depth (typically several kilometers) is much smaller than the wave length (typically several hundred kilometers).

3.2. DIMENSIONLESS FORMULATION

The derivation of shallow-water type equations is a classical topic. Two dimensionless numbers, which are supposed to be small, are introduced:

$$\alpha = \frac{a}{d} \ll 1, \quad \beta = \frac{d^2}{\ell^2} \ll 1, \tag{15}$$

where d is a typical water depth, a a typical wave amplitude and ℓ a typical wavelength. The assumptions on the smallness of these two numbers are satisfied for the Indian Ocean tsunami. Indeed the satellite altimetry observations of the tsunami waves obtained by two satellites that passed over the Indian Ocean a couple of hours after the rupture occurred give an amplitude a of roughly 60 cm in the open ocean. The typical wavelength estimated from the width of the segments that experienced slip is between 160 and 240 km (Lay et al., 2005). The water depth ranges from 4 km towards the west of the rupture to 1 km towards the east. These values give the following ranges for the two dimensionless numbers:

$$1.5 \times 10^{-4} < \alpha < 6 \times 10^{-4}, \quad 1.7 \times 10^{-5} < \beta < 6.25 \times 10^{-4}. \tag{16}$$

[1] The surface wave problem can be easily extended to the case of a moving bottom. This extension may be needed to model tsunami generation if the bottom deformation is relatively slow.

The equations are more transparent when written in dimensionless variables. The new independent variables are

$$x = \ell\tilde{x}, \quad y = \ell\tilde{y}, \quad z = d\tilde{z}, \quad t = \ell\tilde{t}/c_0, \tag{17}$$

where $c_0 = \sqrt{gd}$, the famous speed of propagation of tsunamis in the open ocean ranging from 356 km/h for a 1 km water depth to 712 km/h for a 4 km water depth. The new dependent variables are

$$\eta = a\tilde{\eta}, \quad h = d\tilde{h}, \quad \phi = ga\ell\tilde{\phi}/c_0. \tag{18}$$

In dimensionless form, and after dropping the tildes, the equations become

$$\begin{aligned}
\beta\nabla^2\phi + \phi_{zz} &= 0, & (19)\\
\beta\nabla\phi\cdot\nabla h + \phi_z &= 0 \quad \text{at } z = -h(x,y), & (20)\\
\beta\eta_t + \alpha\beta\nabla\phi\cdot\nabla\eta &= \phi_z \quad \text{at } z = \alpha\eta(x,y,t), & (21)\\
\beta\phi_t + \tfrac{1}{2}\alpha\beta|\nabla\phi|^2 + \tfrac{1}{2}\alpha\phi_z^2 + \beta\eta &= 0 \quad \text{at } z = \alpha\eta(x,y,t). & (22)
\end{aligned}$$

So far, no approximation has been made. In particular, we have not used the fact that the numbers α and β are small.

3.3. SHALLOW-WATER EQUATIONS

When β is small, the water is considered to be shallow. The linearized theory of water waves is recovered by letting α go to zero. For the shallow water-wave theory, one assumes that β is small and expand ϕ in terms of β:

$$\phi = \phi_0 + \beta\phi_1 + \beta^2\phi_2 + \cdots.$$

This expansion is substituted into the governing equation and the boundary conditions. The lowest-order term in Laplace's equation is

$$\phi_{0zz} = 0. \tag{23}$$

The boundary conditions imply that $\phi_0 = \phi_0(x,y,t)$. Thus the vertical velocity component is zero and the horizontal velocity components are independent of the vertical coordinate z at lowest order. Let $\phi_{0x} = u(x,y,t)$ and $\phi_{0y} = v(x,y,t)$. Assume now for simplicity that the water depth is constant ($h = 1$). Solving Laplace's equation and taking into account the bottom kinematic condition yields the following expressions for ϕ_1 and ϕ_2:

$$\begin{aligned}
\phi_1(x,y,z,t) &= -\tfrac{1}{2}(1+z)^2(u_x + v_y), & (24)\\
\phi_2(x,y,z,t) &= \tfrac{1}{24}(1+z)^4[(\nabla^2 u)_x + (\nabla^2 v)_y]. & (25)
\end{aligned}$$

The next step consists in retaining terms of requested order in the free-surface boundary conditions. Powers of α will appear when expanding in Taylor series the free-surface conditions around $z = 0$. For example, if one keeps terms of order $\alpha\beta$ and β^2 in the dynamic boundary condition (22) and in the kinematic boundary condition (21), one obtains

$$\beta\phi_{0t} - \tfrac{1}{2}\beta^2(u_{tx} + v_{ty}) + \beta\eta + \tfrac{1}{2}\alpha\beta(u^2 + v^2) = 0, \quad (26)$$

$$\beta[\eta_t + \alpha(u\eta_x + v\eta_y) + (1 + \alpha\eta)(u_x + v_y)] = \tfrac{1}{6}\beta^2[(\nabla^2 u)_x + (\nabla^2 v)_y] \quad (27)$$

Differentiating (26) first with respect to x and then to respect to y gives a set of two equations:

$$u_t + \alpha(uu_x + vv_x) + \eta_x - \tfrac{1}{2}\beta(u_{txx} + v_{txy}) = 0, \quad (28)$$

$$v_t + \alpha(uu_y + vv_y) + \eta_y - \tfrac{1}{2}\beta(u_{txy} + v_{tyy}) = 0. \quad (29)$$

The kinematic condition (27) can be rewritten as

$$\eta_t + [u(1 + \alpha\eta)]_x + [v(1 + \alpha\eta)]_y = \tfrac{1}{6}\beta[(\nabla^2 u)_x + (\nabla^2 v)_y]. \quad (30)$$

Equations (28)–(30) contain in fact various shallow-water models. The so-called fundamental shallow-water equations are obtained by neglecting the terms of order β:

$$u_t + \alpha(uu_x + vu_y) + \eta_x = 0, \quad (31)$$

$$v_t + \alpha(uv_x + vv_y) + \eta_y = 0, \quad (32)$$

$$\eta_t + [u(1 + \alpha\eta)]_x + [v(1 + \alpha\eta)]_y = 0. \quad (33)$$

Recall that we assumed h to be constant for the derivation. Going back to an arbitrary water depth and to dimensional variables, the system of nonlinear shallow water equations reads

$$u_t + uu_x + vu_y + g\eta_x = 0, \quad (34)$$

$$v_t + uv_x + vv_y + g\eta_y = 0, \quad (35)$$

$$\eta_t + [u(h + \eta)]_x + [v(h + \eta)]_y = 0. \quad (36)$$

This system of equations has been used for example by Titov and Synolakis for the numerical computation of tidal wave run-up (Titov and Synolakis, 1998). Note that this model does not include any bottom friction terms. To solve the problem of tsunami generation caused by bottom displacement, the motion of the seafloor obtained from seismological models (Okada, 1985) and described in Section 3 can be prescribed during a time t_0. Usually t_0 is assumed to be small, so that the bottom displacement is considered as an instantaneous vertical displacement. This assumption may not be appropriate for slow events.

The satellite altimetry observations of the Indian Ocean tsunami clearly show dispersive effects. The question of dispersive effects in tsunamis is open. Most propagation codes ignore dispersion. A few propagation codes that include dispersion have been developed (Dalrymple et al., 2006). A well-known code is FUNWAVE, developed at the University of Delaware over the past ten years (Kirby et al., 1998). Dispersive shallow water-wave models are presented next.

3.4. BOUSSINESQ EQUATIONS

An additional dimensionless number, sometimes called the Stokes number, is introduced:

$$S = \frac{\alpha}{\beta} \approx 1. \tag{37}$$

For the Indian Ocean tsunami, one finds

$$0.24 < S < 46. \tag{38}$$

Therefore the additional assumption that $S \approx 1$ may be realistic.

In this subsection, we provide the guidelines to derive Boussinesq-type systems of equations (Bona et al., 2002). Of course, the variation of bathymetry is essential for the propagation of tsunamis, but for the derivation the water depth will be assumed to be constant. Some notation is introduced. The potential evaluated along the free surface is denoted by $\Phi(x, y, t) := \phi(x, y, \eta, t)$. The derivatives of the velocity potential evaluated on the free surface are denoted by $\Phi_{(*)}(x, y, t) := \phi_*(x, y, \eta, t)$, where the star stands for x, y, z or t. Consequently, Φ_* (defined for $* \neq z$) and $\Phi_{(*)}$ have different meanings. They are however related since

$$\Phi_* = \Phi_{(*)} + \Phi_{(z)}\eta_* \, .$$

The vertical velocity at the free surface is denoted by $W(x, y, t) := \phi_z(x, y, \eta, t)$.

The boundary conditions on the free surface (12) and (13) become

$$\eta_t + \nabla\Phi \cdot \nabla\eta - W(1 + \nabla\eta \cdot \nabla\eta) = 0, \tag{39}$$

$$\Phi_t + g\eta + \tfrac{1}{2}|\nabla\Phi|^2 - \tfrac{1}{2}W^2(1 + \nabla\eta \cdot \nabla\eta) = 0. \tag{40}$$

These two nonlinear equations provide time-stepping for η and Φ. In addition, Laplace's equation as well as the kinematic condition on the bottom must be satisfied. In order to relate the free-surface variables with the bottom variables, one must solve Laplace's equation in the whole water column. In Boussinesq-type models, the velocity potential is represented as a formal expansion,

$$\phi(x, y, z, t) = \sum_{n=0}^{\infty} \phi^{(n)}(x, y, t)\, z^n. \tag{41}$$

Here the expansion is about $z = 0$, which is the location of the free surface at rest. Demanding that ϕ formally satisfy Laplace's equation leads to a recurrence relation between $\phi^{(n)}$ and $\phi^{(n+2)}$. Let ϕ_o denote the velocity potential at $z = 0$, $\mathbf{u}_o$ the horizontal velocity at $z = 0$, and w_o the vertical velocity at $z = 0$. Note that ϕ_o and w_o are nothing else than $\phi^{(0)}$ and $\phi^{(1)}$. The potential ϕ can be expressed in terms of ϕ_o and w_o only. Finally, one obtains the velocity field in the whole water column $(-h \leq z \leq \eta)$ (Madsen et al., 2003):

$$\mathbf{u}(x, y, z, t) = \cos(z\nabla)\mathbf{u}_o + \sin(z\nabla)w_o, \tag{42}$$

$$w(x, y, z, t) = \cos(z\nabla)w_o - \sin(z\nabla)\mathbf{u}_o. \tag{43}$$

Here the cosine and sine operators are infinite Taylor series operators defined by

$$\cos(z\nabla) = \sum_{n=0}^{\infty}(-1)^n \frac{z^{2n}}{(2n)!}\nabla^{2n}, \quad \sin(z\nabla) = \sum_{n=0}^{\infty}(-1)^n \frac{z^{2n+1}}{(2n+1)!}\nabla^{2n+1}.$$

Then one can substitute the representation (42)-(43) into the kinematic bottom condition and use successive approximations to obtain an explicit recursive expression for w_o in terms of $\mathbf{u}_o$ to infinite order in $h\nabla$.

A wide variety of Boussinesq systems can been derived (Madsen et al., 2003). One can generalize the expansions to an arbitrary $z-$level, instead of the $z = 0$ level. The Taylor series for the cosine and sine operators can be truncated, Padé approximants can be used in operators at $z = -h$ and/or at $z = 0$.

The classical Boussinesq equations are more transparent when written in the dimensionless variables used in the previous subsection. We further assume that h is constant, drop the tildes, and write the equations for one spatial dimension (x). Performing the expansion about $z = 0$ leads to the vanishing of the odd terms in the velocity potential. Substituting the expression for ϕ into the free-surface boundary conditions evaluated at $z = 1 + \alpha\eta(x, t)$ leads to two equations in η and ϕ_o with terms of various order in α and β. The small parameters α and β are of the same order, while η and ϕ_o as well as their partial derivatives are of order one.

3.5. CLASSICAL BOUSSINESQ EQUATIONS

The classical Boussinesq equations are obtained by keeping all terms that are at most linear in α or β. In the derivation of the fundamental nonlinear shallow-water equations (31)–(33), the terms in β were neglected. It is therefore implicitly assumed that the Stokes number is large. Since the cube of the water depth appears in the denominator of the Stokes number ($S = \alpha/\beta = a\ell^2/d^3$), it means that the Stokes number is 64 times larger in a 1 km depth than in a 4 km depth!

Based on these arguments, dispersion is more important to the west of the rupture. Considering the Stokes number to be of order one leads to the following system in dimensional form[2]:

$$u_t + uu_x + g\eta_x - \tfrac{1}{2}h^2 u_{txx} = 0, \quad (44)$$

$$\eta_t + [u(h+\eta)]_x - \tfrac{1}{6}h^3 u_{xxx} = 0. \quad (45)$$

The classical Boussinesq equations are in fact slightly different. They are obtained by replacing u with the depth averaged velocity

$$\frac{1}{h}\int_{-h}^{\eta} u\, dz.$$

They read

$$u_t + uu_x + g\eta_x - \tfrac{1}{3}h^2 u_{txx} = 0, \quad (46)$$

$$\eta_t + [u(h+\eta)]_x = 0. \quad (47)$$

A number of variants of the classical Boussinesq system were studied by Bona et al., who in particular showed that depending on the modeling of dispersion the linearization about the rest state may or may not be well-posed (Bona et al., 2002).

3.6. KORTEWEG–DE VRIES EQUATION

The previous system allows the propagation of waves in both the positive and negative $x-$directions. Seeking solutions travelling in only one direction, for example the positive $x-$direction, leads to a single equation for η, the Korteweg–de Vries equation:

$$\eta_t + c_0\left(1 + \frac{3\eta}{2d}\right)\eta_x + \frac{1}{6}c_0 d^2 \eta_{xxx} = 0, \quad (48)$$

where d is the water depth. It admits solitary wave solutions travelling at speed V in the form

$$\eta(x,t) = a\,\mathrm{sech}^2\left(\sqrt{\frac{3a}{4d^3}}(x - Vt)\right), \quad \text{with } V = c_0\left(1 + \frac{a}{2d}\right).$$

The solitary wave solutions of the Korteweg–de Vries equation are of elevation ($a > 0$) and travel faster than c_0. Their speed increases with amplitude. Note that a natural length scale appears:

$$\ell = \sqrt{\frac{4d^3}{3a}}.$$

[2] Equations (44) and (45) could have been obtained from equations (28) and (30).

For the Indian Ocean tsunami, it gives roughly $\ell = 377$ km. It is of the order of magnitude of the wavelength estimated from the width of the segments that experienced slip.

4. Energy of a Tsunami

The energy of the earthquake is measured via the strain energy released by the faulting. The part of the energy transmitted to the tsunami wave is less than one percent (Lay et al., 2005). They estimate the tsunami energy to be 4.2×10^{15} J. They do not give details on how they obtained this estimate. However, a simple calculation based on considering the tsunami as a soliton

$$\eta(x) = a\,\mathrm{sech}^2\left(\frac{x}{\ell}\right), \quad u(x) = \alpha c_0\,\mathrm{sech}^2\left(\frac{x}{\ell}\right),$$

gives for the energy

$$E = \frac{1}{\sqrt{3}}\alpha^{3/2}\rho d^2(c_0^2 + gd)\int_{-\infty}^{\infty}\mathrm{sech}^4 x\,dx + O(\alpha^2).$$

The value for the integral is $4/3$. The numerical estimate for E is close to that of Lay et al. (2005). Incidently, at this level of approximation, there is equipartition between kinetic and potential energy. It is also important to point out that a tsunami being a shallow water wave, the whole water column is moving as the wave propagates. For the parameter values used so far, the maximum horizontal current is 3 cm/s. However, as the water depth decreases, the current increases and becomes important when the depth becomes less than 500 m. Additional properties of solitary waves can be found for example in (Longuet-Higgins, 1974).

5. Tsunami Run-up

The last phase of a tsunami is its run-up and inundation. Although in some cases it may be important to consider the coupling between fluid and structures, we restrict ourselves to the description of the fluid flow. The problem of waves climbing a beach is a classical one (Carrier and Greenspan, 1958). The transformations used by Carrier and Greenspan are still used nowadays. The basis of their analysis is the one-dimensional counterpart of the system (34)–(36). In addition, they assume the depth to be of uniform slope: $h = -x\tan\theta$. Introduce the following dimensionless quantities, where ℓ is a characteristic length[3]:

$$x = \ell\tilde{x}, \;\; \eta = \ell\tilde{\eta}, \;\; u = \sqrt{g\ell}\,\tilde{u}, \;\; t = \sqrt{\ell/g}\,\tilde{t}, \;\; c^2 = (h+\eta)/\ell.$$

After dropping the tildes, the dimensionless system of equations (34)-(36) becomes

$$\begin{aligned} u_t + uu_x + \eta_x &= 0, \\ \eta_t + [u(-x\tan\theta + \eta)]_x &= 0. \end{aligned}$$

In terms of the variable c, these equations become

$$\begin{aligned} u_t + uu_x + 2cc_x + \tan\theta &= 0, \\ 2c_t + cu_x + 2uc_x &= 0. \end{aligned}$$

The equations written in characteristic form are

$$\begin{aligned} \left[\frac{\partial}{\partial t} + (u+c)\frac{\partial}{\partial x}\right](u + 2c + t\tan\theta) &= 0, \\ \left[\frac{\partial}{\partial t} + (u-c)\frac{\partial}{\partial x}\right](u - 2c + t\tan\theta) &= 0. \end{aligned}$$

The characteristic curves C^+ and C^- as well as the Riemann invariants are

$$\begin{aligned} C^+ &: \frac{dx}{dt} = u + c, \quad u + 2c + t\tan\theta = r, \\ C^- &: \frac{dx}{dt} = u - c, \quad u - 2c + t\tan\theta = s. \end{aligned}$$

Next one can rewrite the hyperbolic equations in terms of the new variables λ and σ defined as follows:

$$\begin{aligned} \frac{\lambda}{2} &= \frac{1}{2}(r+s) = u + t\tan\theta, \\ \frac{\sigma}{4} &= \frac{1}{4}(r-s) = c. \end{aligned}$$

One obtains

$$\begin{aligned} x_s - \left[\frac{1}{4}(3r+s) - t\tan\theta\right] t_s &= 0, \\ x_r - \left[\frac{1}{4}(r+3s) - t\tan\theta\right] t_r &= 0. \end{aligned}$$

The elimination of x results in the *linear* second-order equation for t

$$\sigma(t_{\lambda\lambda} - t_{\sigma\sigma}) - 3t_\sigma = 0. \tag{49}$$

---

[3] In fact there is no obvious characteristic length in this idealized problem. Some authors simply say at this point that ℓ is specific to the problem under consideration.

Since $u+t\tan\theta = \lambda/2$, u must also satisfy (49). Introducing the potential $\phi(\sigma,\lambda)$ such that

$$u = \frac{\phi_\sigma}{\sigma},$$

one obtains the equation

$$(\sigma\phi_\sigma)_\sigma - \sigma\phi_{\lambda\lambda} = 0$$

after integrating once. Two major simplifications have been obtained. The nonlinear set of equations have been reduced to a linear equation for u or ϕ and the free boundary is now the fixed line $\sigma = 0$ in the $(\sigma, \lambda)-$plane. The free boundary is the instantaneous shoreline $c = 0$, which moves as a wave climbs a beach.

The above formulation has been used by several authors to study the run-up of various types of waves on sloping beaches (Tadepalli and Synolakis, 1994; Carrier et al., 2003; Tinti and Tonini, 2005). For example, it has been shown that leading depression N-waves run-up higher than leading elevation N-waves, suggesting that perhaps the solitary wave model may not be adequate for predicting an upper limit for the run-up of near-shore generated tsunamis.

There is a rule of thumb that says that the run-up does not usually exceed twice the fault slip. Since run-ups of 30 meters were observed in Sumatra during the Boxing Day tsunami, the slip might have been of 15 meters or even more.

Analytical models are useful, especially to perform parametric studies. However, the breaking of tsunami waves as well as the subsequent floodings must be studied numerically. The most natural methods that can be used are the free surface capturing methods based on a finite volume discretisation, such as the Volume of Fluid (VOF) or the Level Set methods, and the family of Smoothed Particle Hydrodynamics methods (SPH), applied to free-surface flow problems (Monaghan, 1994; Gomez-Gesteira and Dalrymple, 2004; Gomez-Gesteira et al., 2005). Such methods allow a study of flood wave dynamics, of wave breaking on the land behind beaches, and of the flow over rising ground with and without the presence of obstacles. This task is an essential part of tsunami modelling, since it allows the determination of the level of risk due to major flooding, the prediction of the resulting water levels in the flooded areas, the determination of security zones. It also provides some help in the conception and validation of protection systems in the most exposed areas.

6. Direction for Future Research

A useful direction for future research in the dynamics of tsunami waves is the three-dimensional (3D) simulation of tsunami breaking along a coast. For this purpose, different validation steps are necessary. First more simulations of a two-dimensional (2D) tsunami interacting with a sloping beach ought to be performed. Then these simulations should be extended to the case of a 2D tsunami interacting with a sloping beach in the presence of obstacles. An important output of these computations will be the hydrodynamic loading on obstacles. The nonlinear inelastic behaviour of the obstacles may be accounted for using damage or plasticity models. The development of Boussinesq type models coupled with structure interactions is also a promising task. Finally there is a need for 3D numerical simulations of a tsunami interacting with a beach of complex bathymetry, with or without obstacles. These simulations will hopefully demonstrate the usefulness of numerical simulations for the definition of protecting devices or security zones. An important challenge in that respect is to make the numerical methods capable of handling interaction problems involving different scales: the fine scale needed for representing the damage of a flexible obstacle and a coarse scale needed to quantify the tsunami propagation.

References

Ben-Mehanem, A., Singh, S. J., and Solomon, F. (1970) Deformation of an homogeneous earth model finite by dislocations, *Rev. Geophys. Space Phys.* **8**, 591–632.

Ben-Menahem, A., Singh, S. J., and Solomon, F. (1969) Static deformation of a spherical earth model by internal dislocations, *Bull. Seism. Soc. Am.* **59**, 813–853.

Bona, J. L., Chen, M., and Saut, J.-C. (2002) Boussinesq equations and other systems for small-amplitude long waves in nonlinear dispersive media. I: Derivation and linear theory, *Journal of Nonlinear Science* **12**, 283–318.

Carrier, G. F. and Greenspan, H. P. (1958) Water waves of finite amplitude on a sloping beach, *Journal of Fluid Mechanics* **2**, 97–109.

Carrier, G. F., Wu, T. T., and Yeh, H. (2003) Tsunami run-up and draw-down on a plane beach, *Journal of Fluid Mechanics* **475**, 79–99.

Chanson, H. (2005) Le tsunami du 26 décembre 2004: un phénomène hydraulique d'ampleur internationale. Premiers constats, *La Houille Blanche* **2**, 25–32.

Chinnery, M. A. (1963) The stress changes that accompany strike-slip faulting, *Bull. Seism. Soc. Am.* **53**, 921–932.

Dalrymple, R. A., Grilli, S. T., and Kirby, J. T. (2006) Tsunamis and challenges for accurate modeling, *Oceanography* **19**, 142–151.

Gomez-Gesteira, M., Cerqueiro, D., Crespo, C., and Dalrymple, R. A. (2005) Green water overtopping analyzed with a SPH model, *Ocean Engineering* **32**, 223–238.

Gomez-Gesteira, M. and Dalrymple, R. A. (2004) Using SPH for wave impact on a tall structure, *Journal of Waterways, Port, Coastal, and Ocean Engineering* **130**, 63–69.

Hammack, J. (1973) A note on tsunamis: their generation and propagation in an ocean of uniform depth, *Journal of Fluid Mechanics* **60**, 769–799.

Hunt, J. and Burgers, J. M. (2005) Tsunami waves and coastal flooding, *Mathematics TODAY* pp. 144–146.

Iwasaki, T. and Sato, R. (1979) Strain field in a semi-infinite medium due to an inclined rectangular fault, *J. Phys. Earth* **27**, 285–314.

Kirby, J. T., Wei, G., Chen, Q., Kennedy, A. B., and Dalrymple, R. A. (1998) FUNWAVE 1.0, Fully nonlinear Boussinesq wave model documentation and user's manual, Research Report No. CACR-98-06.

Lay, T., Kanamori, H., Ammon, C. J., Nettles, M., Ward, S. N., Aster, R. C., Beck, S. L., Bilek, S. L., Brudzinski, M. R., Butler, R., DeShon, H. R., Ekstrom, G., Satake, K., and Sipkin, S. (2005) The great Sumatra-Andaman earthquake of 26 December 2004, *Science* **308**, 1127–1133.

Longuet-Higgins, M. S. (1974) On the mass, momentum, energy and circulation of a solitary wave, *Proc. R. Soc. Lond. A* **337**, 1–13.

Love, A. E. H. (1944) *A treatise on the mathematical theory of elasticity*, Dover Publications, New York.

Madsen, P. A., Bingham, H. B., and Schaffer, H. A. (2003) Boussinesq-type formulations for fully nonlinear and extremely dispersive water waves: derivation and analysis, *Proc. R. Soc. Lond. A* **459**, 1075–1104.

Maruyama, T. (1964) Statical elastic dislocations in an infinite and semi-infinite medium, *Bull. Earthquake Res. Inst., Tokyo Univ.* **42**, 289–368.

Masterlark, T. (2003) Finite element model predictions of static deformation from dislocation sources in a subduction zone: Sensivities to homogeneous, isotropic, Poisson-solid, and half-space assumptions, *J. Geophys. Res.* **108**, 2540.

McGinley, J. R. (1969) A comparison of observed permanent tilts and strains due to earthquakes with those calculated from displacement dislocations in elastic earth models, Ph.D. thesis, California Institute of Technology, Pasadena, California.

Mei, Z., Roberts, A. J., and Li, Z. (2002) Modelling the dynamics of turbulent floods, *SIAM Journal on Applied Mathematics* **63**, 423–458.

Mindlin, R. D. (1936) Force at a point in the interior of a semi-infinite medium, *Physics* **7**, 195–202.

Mindlin, R. D. and Cheng, D. H. (1950) Nuclei of strain in the semi-infinite solid, *J. Appl. Phys.* **21**, 926–930.

Monaghan, J. J. (1994) Simulating free surface flows with SPH, *Physica D* **110**, 399–406.

Okada, Y. (1985) Surface deformation due to shear and tensile faults in a half-space, *Bull. Seism. Soc. Am.* **75**, 1135–1154.

Okada, Y. (1992) Internal deformation due to shear and tensile faults in a half-space, *Bull. Seism. Soc. Am.* **82**, 1018–1040.

Press, F. (1965) Displacements, strains and tilts at tele-seismic distances, *J. Geophys. Res.* **70**, 2395–2412.

Sato, R. and Matsu'ura, M. (1974) Strains and tilts on the surface of a semi-infinite medium, *J. Phys. Earth* **22**, 213–221.

Smylie, D. E. and Mansinha, L. (1971) The elasticity theory of dislocations in real earth models and changes in the rotation of the earth, *Geophys. J.R. Astr. Soc.* **23**, 329–354.

Sokolnikoff, I. S. and Specht, R. D. (1946) *Mathematical theory of elasticity*, McGraw-Hill, New York.

Steketee, J. A. (1958) On Volterra's dislocation in a semi-infinite elastic medium, *Can. J. Phys.* **36**, 192–205.

Tadepalli, S. and Synolakis, C. E. (1994) The run-up of N-waves on sloping beaches, *Proc. R. Soc. Lond. A* **445**, 99–112.

Tinti, S. and Tonini, R. (2005) Analytical evolution of tsunamis induced by near-shore earthquakes on a constant-slope ocean, *Journal of Fluid Mechanics* **535**, 33–64.

Titov, V. V. and Synolakis, C. E. (1998) Numerical modeling of tidal wave runup, *J. Waterway, Port, Coastal, and Ocean Engineering* **124**, 157–171.

Todorovska, M. I. and Trifunac, M. D. (2001) Generation of tsunamis by a slowly spreading uplift of the seafloor, *Soil Dynamics and Earthquake Engineering* **21**, 151–167.

Volterra, V. (1907) Sur l'équilibre des corps élastiques multiplement connexes, *Annales Scientifiques de l'Ecole Normale Supérieure* **24**, 401–517.

Westergaard, H. M. (1935), *Bull. Amer. Math. Soc.* **41**, 695.

MODELLING FLUID-INDUCED STRUCTURAL VIBRATIONS: REDUCING THE STRUCTURAL RISK FOR STORMY WINDS

D. Perić (d.peric@swansea.ac.uk), W. Dettmer and P.H. Saksono
School of Engineering, University of Wales Swansea

Abstract. The objective of this work is the modelling of the interaction between fluid flow and solid bodies, either rigid or flexible. The fluid flow is governed by the incompressible Navier-Stokes equations and modelled by using stabilised low order velocity-pressure finite elements. The motion of the fluid domain is accounted for by an arbitrary Lagrangian-Eulerian (ALE) strategy. The flexible structure is represented by means of appropriate standard finite element formulations while the motion of the rigid body is described by rigid body dynamics. For temporal discretisation of both fluid and solid bodies, the discrete implicit generalised-α method is employed. The resulting strongly coupled set of nonlinear equations is solved by means of a novel partitioned solution procedure, which is based on the Newton-Raphson methodology and incorporates full linearisation of the overall incremental problem. The strong coupling is resolved and optimal convergence of the residuals is achieved. Numerical examples are presented to demonstrate the robustness and efficiency of the methodology.

Key words: Finite element, fluid-structure interaction, partitioned approach, Newton-Raphson solution method

1. Introduction

Many important problems encountered in the broad area of engineering involve interaction between fluid and structures. Complexity of such problems has for a long time presented a significant challenge to designers, and, historically fluid-structure interaction has been the main cause behind a number of spectacular failures of engineering structures. In this context it is worth mentioning the failure of the Tacoma Narrows Bridge, which at the time of completion in July 1940 was the third longest suspension bridge in the world. The bridge collapsed spectacularly some four months later during the storm on 7 November, with torsional instability due to rotational galloping being commonly considered as the most dominant cause of failure. In November 1965, during high gusty winds, three out of a group of eight cooling towers at Ferrybridge C Power Station collapsed, while the remaining towers sustained severe structural damage. Underestimated wind pressure loads and failure to account for complex wind loading due to interaction

A. Ibrahimbegovic and I. Kozar (eds.),Extreme Man-Made and Natural Hazards in Dynamics of Structures, 225–254.

of air-flow with a group of several large towers have been considered as the main causes of collapse.

In recent years, numerical modelling of fluid-structure interaction has become a focus of major research activity (see. *e.g.* (Ohayon and Felippa, 2001; Matthies and Ohayon, 2005). However, the optimal choice of numerical strategy for the discretisation of the fluid, solid and time domains, and for the modelling of the fluid-structure interface still remains open.

In this work, we propose a specific combination of discretisation strategies and then present a novel solution procedure for the coupled sets of discrete non-linear equations. In particular, we employ a finite difference type time integration scheme as opposed to the space-time strategies (Johnson et al., 1984; Hughes et al., 1989). A major contribution of the present work is the development of a novel solution procedure based on the exact linearisation of the overall problem, which results in optimal rates of convergence of the residuals. The computational ingredients of the adopted strategy are as follows:

For the modelling of the incompressible Newtonian fluid flow we employ the SUPG/PSPG finite element formulation adapted to a moving domain, which allows for computationally convenient linear equal order velocity-pressure interpolations. This stabilised methodology is based on the extensive work by Hughes and co-workers and has been further developed by Tezduyar and others (Tezduyar et al., 1992b).

An arbitrary Lagrangian-Eulerian (ALE) description is used to account for the deformation of the fluid domain which arises from the displacement and deformation of the solid structure.

The solid structure, depending on the physical problem under consideration, may be modelled by an appropriate standard finite element technique, involving membrane, beam, shell and/or continuum elements. In some cases it is sufficient to consider the solid structure as a rigid body.

The strategy adopted in this work allows the employment of non-matching fluid and structural finite element meshes. Hence, the fluid and the structural domains can be discretised independently. A simple interpolation strategy based on a finite element type discretisation of the interface is employed to strongly enforce the kinematic constraints between fluid and solid. The incorporation of these constraints into the weak form of the overall problem leads to a straightforward transfer of the traction fields in terms of the nodal finite element forces.

The generalised-α method is employed for the integration in time (Jansen et al., 2000; Dettmer and Perić, 2003). This method belongs to the class of discrete and implicit single step integration schemes. For linear problems the scheme can be shown to be second order accurate and unconditionally stable. It allows user controlled damping of high unresolved frequencies.

The fully discretised model consists of coupled sets of nonlinear equations. The coupling can be described briefly as follows: The deformation of the structure is driven by the traction forces exerted by the fluid at the fluid-solid interface. The structural displacements, on the other hand, define the geometry and the geometry changes of the fluid domain. In a discrete FE setting this situation may be regarded as a coupled three field problem, involving the fluid flow, the motion of the fluid mesh and the structural dynamics.

The final numerical ingredient required is a robust and efficient solution procedure to compute the complete set of unknowns at the current time instant. In this work, a novel partitioned solution scheme has been developed relying on the Newton-Raphson procedure, which incorporates the full linearisation of the incremental problem and hence exhibits asymptotically quadratic convergence of the solution for all problem unknowns. Alternative strategies, some of which resemble Quasi-Newton procedures or employ linearisations obtained by numerical differentiation, have been suggested by Tezduyar, Matthies and others (see (Dettmer and Perić, 2005b; Dettmer and Perić, 2005a) and references therein).

A detailed description of the overall methodology is given in (Dettmer and Perić, 2005b; Dettmer and Perić, 2005a).

2. Governing Equations

2.1. FLUID MECHANICS ON A MOVING DOMAIN

An essential feature of the fluid-structure interaction is the motion of the boundary of the fluid domain. The geometry of the fluid domain may change substantially during the time domain of interest. Therefore, it is convenient to formulate the problem in the ALE description relying on a moving reference frame, in which the conservation laws are expressed.

In this context it is well-established (see *e.g.* (Belytschko et al., 2000; Donea and Huerta, 2003)) that the time derivative of the velocity $\mathbf{u}$ of the fluid particle which traverses through the coordinate $\hat{\mathbf{x}}$ of the reference frame at a specific time instant can be written as

$$\frac{D\mathbf{u}}{Dt} = \nabla_{\hat{\mathbf{x}}}\mathbf{u}\,(\mathbf{u} - \hat{\mathbf{v}}) + \dot{\mathbf{u}}, \tag{1}$$

where $\hat{\mathbf{v}} = \partial\hat{\mathbf{x}}/\partial t$ is the velocity of the reference point. The operator $\nabla_{\hat{\mathbf{x}}}(\bullet)$ denotes the derivatives with respect to the current referential coordinates $\hat{\mathbf{x}}$. The expression $\dot{\mathbf{u}}$ corresponds to the change of the material particle velocity, which is noted by an observer travelling with a point on the reference frame. The velocity $\mathbf{u} - \hat{\mathbf{v}}$ is denoted as the *convective velocity*. In the framework of the finite element method, the moving reference frame is identified with the finite element mesh. The Eulerian or Lagrangian representations of the material time derivative of $\mathbf{u}$ are easily recovered from Equation (1) by setting $\hat{\mathbf{v}} = \mathbf{0}$ or $\hat{\mathbf{v}} = \mathbf{u}$, respectively.

Let Ω_f denotes the current fluid referential domain, the momentum conservation law and the continuity equation for incompressible flow are formulated in the referential description as

$$\rho_f\,(\dot{\mathbf{u}} + ((\nabla_{\hat{\mathbf{x}}}\mathbf{u})(\mathbf{u} - \hat{\mathbf{v}}) - \mathbf{f}_f) - \nabla_{\hat{\mathbf{x}}} \cdot \boldsymbol{\sigma}_f \;=\; \mathbf{0} \quad \text{in} \quad \Omega_f \times I \tag{2}$$

$$\nabla_{\hat{\mathbf{x}}} \cdot \mathbf{u} \;=\; 0 \quad \text{in} \quad \Omega_f \times, I \tag{3}$$

where $\rho_f, \mathbf{f}_f$ and σ_f represent, respectively, the fluid density, the volume force vector and the Cauchy stress tensor acting on the fluid. The constitutive equation for the Newtonian fluid is given in a standard way as

$$\boldsymbol{\sigma}_f = -p\mathbf{I} + 2\mu\nabla^s_{\hat{\mathbf{x}}}\mathbf{u}, \tag{4}$$

where p represents the pressure, $\mathbf{I}$ is the second order identity tensor, μ denotes the fluid viscosity and $\nabla^s_{\hat{\mathbf{x}}}(\bullet)$ is the symmetric gradient operator with respect to the current referential coordinates $\hat{\mathbf{x}}$.

2.2. SOLID MECHANICS

In a standard Lagrangian description, the conservation of momentum of a solid continuum may be expressed in spatial description as

$$\rho_s(\ddot{\mathbf{d}} - \mathbf{b}_s) - \nabla \cdot \boldsymbol{\sigma}_s = \mathbf{0} \quad \text{in} \quad \Omega_s \times I, \tag{5}$$

where ρ_s is the current density of the deformed solid and the vector $\mathbf{d}$ represents the displacement field, whereas the body forces are given by the vector $\mathbf{b}_s$. The symmetric second order $\boldsymbol{\sigma}_s$ denotes the Cauchy stress acting in the solid that is related to appropriate local strain measures by a constitutive relation suitable to model the behaviour of the solid material.

If the solid body is considered as rigid then Equation (5), in two dimensions, is simply replaced by

$$\mathbf{M}\ddot{\mathbf{d}} + \mathbf{C}\dot{\mathbf{d}} + \mathbf{K}\mathbf{d} \;=\; \mathbf{f}_r \tag{6}$$

$$I_\theta\ddot{\theta} + C_\theta\dot{\theta} + K_\theta\theta \;=\; T \tag{7}$$

where $\mathbf{d} = [d_x, d_y]^T$ represents the rigid body translation, θ describes the rigid body rotation with respect to the center of gravity c_g of the rigid body. The quantities $\mathbf{M}$, $\mathbf{C}$, $\mathbf{K}$, I_θ, C_θ and K_θ denote the mass, the damping and the stiffness moduli for translational and rotational degree of freedom, respectively, whereas

$\mathbf{f}_r$ is the force vector acting on the rigid body and T the component along the out of plane basis vector $\hat{\mathbf{e}}_z$ of the moment $\mathbf{T}$.

2.3. BOUNDARY CONDITIONS

Let the boundary of the fluid domain be denoted by $\Gamma_f = \Gamma_f^g \cup \Gamma_f^h \cup \Gamma^i$ and the boundary of the solid domain be represented by $\Gamma_s = \Gamma_s^g \cup \Gamma_s^h \cup \Gamma^i$. Then the following boundary conditions have to be satisfied

$$\mathbf{u} - \mathbf{g}_f = \mathbf{0} \quad \text{on} \quad \Gamma_f^g \times I \tag{8}$$

$$\boldsymbol{\sigma}_f \hat{\mathbf{n}} - \mathbf{h}_f = \mathbf{0} \quad \text{on} \quad \Gamma_f^h \times I \tag{9}$$

$$\mathbf{d} - \mathbf{g}_s = \mathbf{0} \quad \text{on} \quad \Gamma_s^g \times I \tag{10}$$

$$\boldsymbol{\sigma}_s \hat{\mathbf{n}} - \mathbf{h}_s = \mathbf{0} \quad \text{on} \quad \Gamma_s^h \times I \tag{11}$$

$$\mathbf{u} - \dot{\mathbf{d}} = \mathbf{0} \quad \text{on} \quad \Gamma^i \times I \tag{12}$$

$$(\mathbf{u} - \hat{\mathbf{v}})\hat{\mathbf{n}} = 0 \quad \text{on} \quad \Gamma^i \times I \tag{13}$$

$$\boldsymbol{\sigma}_f \hat{\mathbf{n}} - \boldsymbol{\sigma}_s \hat{\mathbf{n}} = \mathbf{0} \quad \text{on} \quad \Gamma^i \times I \tag{14}$$

where Γ^g and Γ^h denote the part of the boundary where the velocity vector and traction vector are prescribed, respectively, and Γ^i is a part of boundary where the interaction between the fluid and the solid bodies is taking place. The quantities $\mathbf{g}$, $\mathbf{h}$, and $\hat{\mathbf{n}}$ denote, respectively, the prescribed velocity and traction vectors and the current outward normal unit vector on the boundary. Here the subscripts f and s denote the association with fluid or solid body.

Equations (8),(10),and(13) determine the motion of the part of the boundary Γ^h for both fluid and solid and the interface boundary Γ^i. The movement of the internal finite element nodes of the fluid domain should be chosen such that the mesh quality does not deteriorate as the displacements of the solid body become large. For this purpose several algorithms have been suggested in the literature (see *e.g.*, (Tezduyar et al., 1992a; Bar-Yoseph et al., 2001; Degand and Farhat, 2002; Dettmer, 2004)). In this work we use either the pseudo-elastic technique or technique based on the optimisation of mesh quality as described in more detail in Section 4.

3. Stabilised Fnite Element Formulation for the Fuid Fow

Let $\mathcal{S}^h, \mathcal{V}^h, \mathcal{P}^h$ be the appropriate finite element spaces of continuous piecewise linear functions on Ω^h, where $\Omega^h = \cup_{e=1}^{n_{\text{el}}} \Omega^e$ is a standard discretisation of the fluid domain Ω with n_{el} finite elements. A stabilised velocity-pressure finite element formulation of the fluid flow then reads:

For any $t \in I$, find $\mathbf{u}^h \in \mathcal{S}^h$ and $p^h \in \mathcal{P}^h$ such that the following weak form is satisfied for any admissible $\delta\,\mathbf{u}^h \in \mathcal{V}^h$ and $\delta\mathbf{p}^h \in \mathcal{P}^h$,

$$G^{\mathrm{f}}(\mathbf{u}^h, p^h; \delta\mathbf{u}^h, \delta p^h) = G_{\mathrm{Gal}}(\mathbf{u}^h, p^h; \delta\mathbf{u}^h, \delta p^h) + G_{\mathrm{stab}}(\mathbf{u}^h, p^h; \delta\mathbf{u}^h, \delta p^h) = 0\,. \quad (15)$$

The variational form (15) consists of the standard Galerkin terms summarised in G_{Gal}, to which a stabilisation term G_{stab} of the momentum equation has been added.

In order to simplify the notation in the remainder of this section, which is dedicated to the detailed presentation of G_{Gal} and G_{stab}, the coupling of Equation (15) with the deformation of the solid structure is not explicitly included. This issue, together with the appropriate notation, is elaborated in detail in Section 7.

The Galerkin terms, which can be obtained from Equations (2–9) by the standard procedure, read

$$G_{\mathrm{Gal}}(\mathbf{u}^h, p^h; \delta\mathbf{u}^h, \delta p^h) = \int_{\Omega^h} \Big(\delta\rho\, \mathbf{u}^h \cdot \big(\dot{\mathbf{u}}^h + (\nabla_{\hat{\mathbf{x}}^h} \mathbf{u}^h)(\mathbf{u}^h - \hat{\mathbf{v}}^h) - \mathbf{f}_f \big) + \\ + \nabla_{\hat{\mathbf{x}}^h} \delta\mathbf{u}^h : \boldsymbol{\sigma}(\mathbf{u}^h, p^h) + \delta p^h (\nabla_{\hat{\mathbf{x}}^h} \cdot \mathbf{u}^h) \Big)\, \mathrm{d}v - \int_{\Gamma^h_h} \delta\mathbf{u}^h \cdot \mathbf{h}^h\, \mathrm{d}a, \quad (16)$$

where the vector field $\hat{\mathbf{v}}^h$ denotes the mesh motion, which is based on the same finite element interpolation as $\mathbf{u}^h$.

The stabilisation term used in this work is similar to the one employed by the authors in (Tezduyar et al., 1992b), but has been extended to incorporate modifications required in the ALE framework. In (Tezduyar et al., 1992b), the stabilisation technique is referred to as a combination of the *streamline-upwind/-* and the *pressure-stabilising/Petrov-Galerkin* schemes (SUPG/PSPG).

The stabilisation serves two purposes: First, it provides stability to the velocity field $\mathbf{u}^h$ in convection dominated regions of the domain. Second, it circumvents the Babuška-Brezzi condition, which standard mixed Galerkin methods are required to satisfy. Thus, it effectively renders a smooth pressure field without jeopardising the weak enforcement of the continuity condition.

The stabilisation term employed here reads

$$G_{\mathrm{stab}}(\mathbf{u}^h, p^h; \delta\mathbf{u}^h, \delta p^h) = \sum_{e=1}^{n_{\mathrm{el}}} \int_{\Omega^e} \Big[\tau_u\, \rho\, (\nabla_{\hat{\mathbf{x}}^h} \delta\mathbf{u}^h)(\mathbf{u}^h - \hat{\mathbf{v}}^h) + \tau_p \nabla_{\hat{\mathbf{x}}^h} \delta p^h \Big] \\ \cdot \Big[\rho \big(\dot{\mathbf{u}}^h + (\nabla_{\hat{\mathbf{x}}^h} \mathbf{u}^h)(\mathbf{u}^h - \hat{\mathbf{v}}^h) - \mathbf{f}_f \big) + \nabla_{\hat{\mathbf{x}}^h} p^h \Big]\, \mathrm{d}v. \quad (17)$$

We note that due to the absence of the viscous term in the second pair of brackets in (17), the stabilisation term does not vanish as the spatial discretisation is refined. Consequently, the choice of the weighting parameter τ and its limit behaviour are essential for the success of the methodology.

In the experience of the authors, it has proved useful (see also *e.g.* (Tezduyar et al., 1992b)) to employ two parameters, here denoted as τ_u and τ_p, and hence treat the two stabilisation purposes separately. Both stabilisation parameters τ_u and τ_p are defined as follows

$$\tau = \frac{h^e}{2\|\mathbf{u}^e - \hat{\mathbf{v}}^e\|\,\rho}\, z, \quad z = \frac{\beta_1}{\sqrt{1 + \left(\dfrac{\beta_1}{\beta_2\, Re^e}\right)^2}}, \quad Re^e = \frac{\|\mathbf{u}^e - \hat{\mathbf{v}}^e\|\, h^e\, \rho}{2\mu}, \tag{18}$$

but they have different scaling parameters β_1 and β_2, which may be set independently. The characteristic element size, the convective velocity in the element centroid and the element Reynolds number are represented by h, $\mathbf{u}^e - \hat{\mathbf{v}}^e$, and Re^e, respectively. Thus, τ_u and τ_p are constant within every element and, hence, the stabilisation terms are discontinuous across the inter element boundaries, which explains the summation of integrals in (17). The parameters β_1 and β_2 define the limit of z as $Re^e \to \infty$ and the derivative $\mathrm{d}z/\mathrm{d}Re^e$ at $Re^e = 0$, respectively. The examples in Section 9 have all been obtained with ($\beta_1 = 1,\ \beta_2 = \frac{1}{2}$) for τ_u and ($\beta_1 = 30,\ \beta_2 = \frac{1}{2}$) for τ_p. This specific choice makes the parameter τ_u identical to the expression which yields nearly exact solutions for the one dimensional advection-diffusion equation (see (Brooks and Hughes, 1982) and references therein). In this work, the characteristic element size h^e is defined as the diameter of the circle, the area of which corresponds to the finite element e.

The choice of the formula in (18) is rather heuristic, and many different expressions have been introduced in literature (see *e.g.* (Tezduyar et al., 1992b; Tezduyar and Osawa, 2000; Dettmer and Perić, 2003; Oñate, 2000)). In some early publications (*e.g.* (Tezduyar et al., 1992b)) it was suggested that τ should vanish as the discretisation in time is refined. In the context of a finite difference time integration scheme, the expressions (18), which do not depend on the time increment, have been shown to yield a robust method, allowing, to a wide extent, independent refinement of the discretisations of space and time (see (Dettmer and Perić, 2003) for a more detailed discussion and numerical verification).

4. Motion of the Fuid Fnite Element Mesh

At this stage, the motion of the fluid mesh is arbitrary except for its outline: On the interface boundary $\Gamma^h_{\mathrm{f-s}}$, Equation (13) has to be satisfied. On the other parts of Γ^h, the user prescribes at least the motion of the nodes normal to the current configuration of the boundary. Note that, eventually, there may be regions of Ω^h, which are not required to adapt to a new geometry, since they are far away from the moving rigid body. In such regions we set $\hat{\mathbf{v}}^h = \mathbf{0}$, and the flow problem becomes purely Eulerian.

4.1. MOTION OF INTERNAL NODES

The movement of the internal finite element nodes should be chosen such that the mesh quality does not deteriorate as the displacements of the solid structure become large. For this purpose, many different algorithms have been suggested in literature (see *e.g.* (Tezduyar et al., 1992a; Bar-Yoseph et al., 2001; Degand and Farhat, 2002; Dettmer, 2004) and references therein).
In this work the following techniques are used:

- *Pseudo-elastic technique.* In this approach, the mesh is simply assumed to represent an elastic solid body. A standard Lagrangian finite element technique typically employed in solid mechanics can then be used to adapt the mesh to the new geometry of the domain.
  For small distortions of the geometry the linear elastic model is sufficient. In the presence of large deformations of the fluid domain a hyperelastic model may be more suitable. Note that the mesh need not necessarily represent an elastic continuum. In literature, alternative methodologies have been suggested in which the mesh is, for example, assumed to be a network of elastic springs (see *e.g.* (Degand and Farhat, 2002) and references therein). In the pseudo-elastic approach used in this work, the mesh is treated as a simple hyperelastic Neo-Hookean continuum, with two parameters μ^{m} and K^{m}, representing the shear and bulk modulus, respectively. For the two dimensional situation the plane strain condition is employed.
- *Optimisation of mesh quality.* A simple strategy to compute the mesh motion can be defined by enforcing the condition, which requires that the mesh quality, with respect to a certain criteria, is optimal at all times $t \in I$. In this work, the chosen criteria is the ratio of the inner and outer circles of the triangular or tetrahedral finite element. Thus, the mesh movement satisfies

$$W = \sum_{e=1}^{n_{\mathrm{el}}} \left(\frac{r^e_{\mathrm{out}}}{r^e_{\mathrm{in}}} \right) \Rightarrow MIN \,, \tag{19}$$

  which is a simplified version of the expression used in (Braess and Wriggers, 2000). The quantities r^e_{in} and r^e_{out} denote the inner and the outer radii of a triangular or tetrahedral finite element. The equations, which determine the nodal positions $\hat{\mathbf{x}}^{\mathrm{f}}_i$, then read

$$\frac{\partial W}{\partial \hat{\mathbf{x}}_i^{\mathrm{f}}} = \mathbf{0}, \qquad i = 1, 2, .., N^{\mathrm{f}}\,, \tag{20}$$

where N^{f} denotes the number of nodes in the mesh interior. In the authors' experience this methodology renders acceptable meshes even for very distorted geometries. Note that also the initial mesh should satisfy (19).

Both strategies can be fully linearised and thus allow the employment of the Newton-Raphson procedure to solve for the new nodal positions. This is of particular importance with respect to the overall solution procedure described in Section 8.

The employment of large time steps in problems involving severe deformations of the domain often requires the adaptation of the mesh to substantial changes of the geometry within one time step. In such cases the Newton-Raphson procedure may fail to converge. In this work, this problem is overcome by increment cutting within the mesh update procedure, *i.e.*, the new displacement of the boundary is applied in increments if necessary.

4.2. MOTION OF NODES ON THE INTERFACE BOUNDARY

Provided that the initial boundary of the fluid mesh appropriately resolves the surface of the solid structure, there is normally no need to allow for any tangential movement of the fluid nodes along the interface. The fluid mesh boundary, similarly to the fluid particles, can then be required to "stick" to the surface of the structure. Thus, we satisfy Equations (11) and (13) by employing a purely Lagrangian description of the interface, *i.e.*

$$\hat{\mathbf{v}}^h = \mathbf{u}^h = \mathcal{I}\left(\dot{\mathbf{d}}^h\right) \qquad \forall\, (\hat{\mathbf{x}}^h, t) \in \Gamma^h_{\mathrm{f-s}} \times I\,. \tag{21}$$

where the vector $\dot{\mathbf{d}}^h$ denotes the finite element approximation of the structural velocity field $\dot{\mathbf{d}}$. In order to allow for non-matching fluid and structural meshes, it is necessary to define an appropriate interpolation operator $\mathcal{I}(\bullet)$.

The current configuration of the interface boundary $\Gamma^h_{\mathrm{f-s}}$ is then described by

$$\hat{\mathbf{x}}^h = \mathcal{I}\left(\mathbf{x}^h_{\mathrm{s}0} + \mathbf{d}^h\right) \qquad \forall\, (\hat{\mathbf{x}}^h, t) \in \Gamma^h_{\mathrm{f-s}} \times I\,, \tag{22}$$

where $\mathbf{x}^h_{\mathrm{s}0}$ denotes the discretisation of the initial configuration of the solid structure.

5. Finite Element Formulation for the Solid Structure

The fluid-structure interaction solution methodology adopted in this work does not impose any restriction on the specific choice of structural element to be used. Importantly, the structural mesh at the interface is not required to match the fluid finite element mesh. Thus, any appropriate standard finite element method may be used for the discretisation of the solid structure.

The starting point of a structural finite element method is the balance of momentum as given by (5) or by an appropriate equivalent representation. A standard finite element formulation of (5) reads as follows: For any $t \in I$, find $\mathbf{d}^h \in \mathcal{S}^h$ such that for any $\delta\mathbf{d} \in \mathcal{V}^h$

$$G^{\mathrm{s}}(\mathbf{d}^h; \delta\mathbf{d}^h) = \int_{\Omega^h} \left(\delta\mathbf{d}^h \rho \left(\ddot{\mathbf{d}}^h - \mathbf{f}\right) + \nabla\delta\mathbf{d}^h : \boldsymbol{\sigma}(\mathbf{d}^h)\right) dv - \int_{\Gamma_h^h} \delta\mathbf{d}^h \cdot \mathbf{h}^h \, da = 0 \quad (23)$$

where $\mathcal{S}^h$ and $\mathcal{V}^h$ are the appropriate finite element spaces and Ω^h is a finite element discretisation of the solid domain.

Introducing the vector of the nodal displacements $\mathbf{d}$, the formulation (23) can be rewritten in an equivalent matrix form as

$$\mathsf{M}\ddot{\mathsf{d}} + \mathsf{C}\dot{\mathsf{d}} + \mathsf{K}\mathsf{d} = \mathsf{P}\,, \quad (24)$$

where the matrices M, C and K are denoted, respectively, as the mass, damping and stiffness matrices. It should be noted that some structural finite elements, such as beam and shell elements, also include the rotational degrees of freedom in addition to the translational displacements. We assume such rotations to be included in the vector d.

6. Integration in Time

In order to complete the discretisation of the problem, it remains to apply a numerical time integration scheme to the variational form given by (15). The most popular choices are standard *discrete* time stepping schemes and the so-called *time finite element methods*. Both approaches have been extensively discussed in recent publications (see *e.g.* (Ramaswamy, 1990; Braess and Wriggers, 2000) for discrete and (Hughes et al., 1989; Masud and Hughes, 1997) for time finite element methods).

In (Dettmer and Perić, 2003; Dettmer, 2004), we have provided a detailed comparison of implicit time integration schemes with respect to incompressible Newtonian fluid flow in the Eulerian framework. As a result, the generalised-α

method has been suggested as a very efficient and robust alternative to the more expensive time finite element methods. The generalised-α method has originally been developed in (Chung and Hulbert, 1993) for the second order differential equation arising in solid dynamics, but has been adapted to the first order problem of fluid mechanics in (Jansen et al., 2000). For linear problems the scheme can be shown to be unconditionally stable and second order accurate (see (Chung and Hulbert, 1993; Jansen et al., 2000; Dettmer and Perić, 2003)). Furthermore, it enables user controlled high frequency damping, which is desirable especially for coarse discretisation in space and time. This is achieved by specifying the single integration parameter, which, for linear problems, can be identified with the spectral radius ρ_∞ associated with very large time steps. In the following the application of the generalised-α method to Equations (15) and (24) is described.

First, the time interval $I = [0, T]$ is replaced by a sequence of discrete time instants t_n, $n = 0, 1, 2, \ldots, N$ with $t_0 = 0$ and $t_N = T$. The time step size $\Delta t = t_{n+1} - t_n$ is allowed to vary.

6.1. FLUID SOLVER

The generalised-α method is now used to express $\mathbf{u}^h$ and its time derivative $\dot{\mathbf{u}}^h$ in (15) in terms of $\mathbf{u}^h$ and $\dot{\mathbf{u}}^h$ at the discrete time instants t_n and t_{n+1}. These values are henceforth denoted as $\mathbf{u}^h_n$, $\mathbf{u}^h_{n+1}$, $\dot{\mathbf{u}}^h_n$ and $\dot{\mathbf{u}}^h_{n+1}$.

In (Jansen et al., 2000), the generalised-α method is given as

$$\mathbf{u}^h_{n+1} = \mathbf{u}^h_n + \Delta t\,(1-\gamma^{\mathrm{f}})\,\dot{\mathbf{u}}^h_n + \Delta t\gamma^{\mathrm{f}}\,\dot{\mathbf{u}}^h_{n+1} \tag{25}$$

$$\mathbf{u}^h_{n+\alpha^{\mathrm{f}}_f} = (1-\alpha^{\mathrm{f}}_f)\,\mathbf{u}^h_n + \alpha^{\mathrm{f}}_f\,\mathbf{u}^h_{n+1} \tag{26}$$

$$\dot{\mathbf{u}}^h_{n+\alpha^{\mathrm{f}}_m} = (1-\alpha^{\mathrm{f}}_m)\,\dot{\mathbf{u}}^h_n + \alpha^{\mathrm{f}}_m\,\dot{\mathbf{u}}^h_{n+1}\,, \tag{27}$$

where γ^{f}, α^{f}_m and α^{f}_f are integration parameters. These equations can be rewritten as

$$\mathbf{u}^h_{n+\alpha^{\mathrm{f}}_f} = (1-\alpha^{\mathrm{f}}_f)\,\mathbf{u}^h_n + \alpha^{\mathrm{f}}_f\,\mathbf{u}^h_{n+1} \tag{28}$$

$$\dot{\mathbf{u}}^h_{n+\alpha^{\mathrm{f}}_m} = \left(1-\frac{\alpha^{\mathrm{f}}_m}{\gamma^{\mathrm{f}}}\right)\dot{\mathbf{u}}^h_n + \frac{\alpha^{\mathrm{f}}_m}{\Delta t\gamma^{\mathrm{f}}}\,(\mathbf{u}^h_{n+1}-\mathbf{u}^h_n) \tag{29}$$

$$\dot{\mathbf{u}}^h_{n+1} = \frac{1}{\Delta t\gamma^{\mathrm{f}}}\,(\mathbf{u}^h_{n+1}-\mathbf{u}^h_n) - \frac{1-\gamma^{\mathrm{f}}}{\gamma^{\mathrm{f}}}\,\dot{\mathbf{u}}^h_n\,. \tag{30}$$

In G_{Gal} and G_{stab}, given, respectively, by the relations (16) and (17), the expressions $\mathbf{u}^h$ and $\dot{\mathbf{u}}^h$ are now replaced by $\mathbf{u}^h_{n+\alpha^{\mathrm{f}}_f}$ and $\dot{\mathbf{u}}^h_{n+\alpha^{\mathrm{f}}_m}$, respectively.

Thus, the fluid velocity and its time derivative in (15) are expressed exclusively in terms of the unknown $\mathbf{u}^h_{n+1}$ and in terms of the quantities $\mathbf{u}^h_n$ and $\dot{\mathbf{u}}^h_n$, which are known from the solution at the previous time instant. Due to its nature as the Lagrangian multiplier, which enforces continuity of the flow, the pressure should not be subjected to a time integration scheme, but be computed independently for each time increment. Once $\mathbf{u}^h_{n+1}$ has been computed, the quantity $\dot{\mathbf{u}}^h_{n+1}$ can be obtained from (30).

The integration parameters are reduced to one independent control variable as follows

$$\gamma^{\mathrm{f}} = \frac{1}{2} + \alpha^{\mathrm{f}}_m - \alpha^{\mathrm{f}}_f\,, \qquad \alpha^{\mathrm{f}}_m = \frac{1}{2}\frac{3-\rho^{\mathrm{f}}_\infty}{1+\rho^{\mathrm{f}}_\infty}\,, \qquad \alpha^{\mathrm{f}}_f = \frac{1}{1+\rho^{\mathrm{f}}_\infty}\,, \tag{31}$$

where ρ^{f}_∞ has to be chosen such that $0 \leq \rho^{\mathrm{f}}_\infty \leq 1$. For $\rho^{\mathrm{f}}_\infty = 1$, the method is identical to the trapezoidal rule, whereas the numerical damping of the method increases with smaller values of ρ^{f}_∞. For a detailed study of the generalised-α method in combination with stabilised finite elements we refer to (Dettmer and Perić, 2003).

6.2. SOLID BODY TIME INTEGRATION

Similarly to Section 6.1, the quantities $\mathbf{d}_n, \dot{\mathbf{d}}_n, \ddot{\mathbf{d}}_n, n = 0, 1, .., N_{\text{time}}\, \theta_n, \dot{\theta}, \ddot{\theta}_n$, are introduced. For the time integration of the momentum conservation equation of the solid body (23), we then employ the generalised-α method as presented in (Chung and Hulbert, 1993) for second order initial value problems. In the following, only the time integration strategy for $\delta\mathbf{d}$ is presented. The rotation θ is integrated analogously.

As suggested in (Chung and Hulbert, 1993), we define

$$\mathbf{d}_{n+1} = \mathbf{d}_n + \Delta t\, \dot{\mathbf{d}}_n + \Delta t^2 \left(\left(\frac{1}{2} - \beta^{\mathrm{s}} \right) \ddot{\mathbf{d}}_n + \beta^{\mathrm{s}}\, \ddot{\mathbf{d}}_{n+1} \right) \tag{32}$$

$$\dot{\mathbf{d}}_{n+1} = \dot{\mathbf{d}}_n + \Delta t \left((1-\gamma^{\mathrm{s}})\, \ddot{\mathbf{d}}_n + \gamma^{\mathrm{s}}\, \ddot{\mathbf{d}}_{n+1} \right) \tag{33}$$

$$\mathbf{d}_{n+\alpha^{\mathrm{s}}_f} = (1-\alpha^{\mathrm{s}}_f)\, \mathbf{d}_n + \alpha^{\mathrm{s}}_f\, \mathbf{d}_{n+1} \tag{34}$$

$$\dot{\mathbf{d}}_{n+\alpha^{\mathrm{s}}_f} = (1-\alpha^{\mathrm{s}}_f)\, \dot{\mathbf{d}}_n + \alpha^{\mathrm{s}}_f\, \dot{\mathbf{d}}_{n+1} \tag{35}$$

$$\ddot{\mathbf{d}}_{n+\alpha^{\mathrm{s}}_m} = (1-\alpha^{\mathrm{s}}_m)\, \ddot{\mathbf{d}}_n + \alpha^{\mathrm{s}}_m\, \ddot{\mathbf{d}}_{n+1}\,, \tag{36}$$

where $\mathbf{d}_{n+\alpha^s_f}$, $\dot{\mathbf{d}}_{n+\alpha^s_f}$, $\ddot{\mathbf{d}}_{n+\alpha^s_m}$, are the quantities to be employed in (23). Equations (32 – 36) can be rewritten as

$$\begin{aligned}
\mathbf{d}_{n+\alpha^s_f} &= \mathbf{d}_n + \frac{\Delta t\, \alpha^s_f\, (\gamma^s - \beta^s)}{\gamma^s}\, \dot{\mathbf{d}}_n + \frac{\Delta t^2\, \alpha^s_f\, (\gamma^s - 2\,\beta^s)}{2\,\gamma^s}\, \ddot{\mathbf{d}}_n \\
&\quad + \frac{\Delta t\, \alpha^s_f\, \beta^s}{\gamma^s}\, \dot{\mathbf{d}}_{n+1} && (37)\\
\dot{\mathbf{d}}_{n+\alpha^s_f} &= (1 - \alpha^s_f)\, \dot{\mathbf{d}}_n + \alpha^s_f\, \dot{\mathbf{d}}_{n+1} && (38)\\
\ddot{\mathbf{d}}_{n+\alpha^s_m} &= \frac{1 - \alpha^s_m}{\gamma^s}\, \ddot{\mathbf{d}}_n + \frac{\alpha^s_m}{\Delta t\, \gamma^s}\, (\dot{\mathbf{d}}_{n+1} - \dot{\mathbf{d}}_n) && (39)\\
\mathbf{d}_{n+1} &= \mathbf{d}_n + \frac{\Delta t\, (\gamma^s - \beta^s)}{\gamma^s}\, \dot{\mathbf{d}}_n + \frac{\Delta t^2\, (\gamma^s - 2\,\beta^s)}{2\,\gamma^s}\, \ddot{\mathbf{d}}_n + \frac{\Delta t\, \beta^s}{\gamma^s}\, \dot{\mathbf{d}}_{n+1} && (40)\\
\ddot{\mathbf{d}}_{n+1} &= -\frac{1 - \gamma^s}{\gamma^s}\, \ddot{\mathbf{d}}_n + \frac{1}{\Delta t\, \gamma^s}\, (\dot{\mathbf{d}}_{n+1} - \dot{\mathbf{d}}_n) && (41)
\end{aligned}$$

Thus, similarly to the time integration strategy applied to the fluid, $\mathbf{d}_{n+\alpha^s_f}$, $\dot{\mathbf{d}}_{n+\alpha^s_f}$ and $\ddot{\mathbf{d}}_{n+\alpha^s_m}$ are expressed in terms of the unknown $\dot{\mathbf{d}}_{n+1}$ and the solution at the previous time instant. Once $\dot{\mathbf{d}}_{n+1}$ has been computed, $\mathbf{d}_{n+1}$ and $\ddot{\mathbf{d}}_{n+1}$ are obtained from (40) and (41). For linear problems, the following formula for the time integration parameters is optimal (Chung and Hulbert, 1993)

$$\begin{aligned}
\beta^s &= \frac{1}{4}\,(1 + \alpha^s_m - \alpha^s_f)^2\,, \qquad & \gamma^s &= \frac{1}{2} + \alpha^s_m - \alpha^s_f \\
\alpha^s_f &= \frac{1}{1 + \rho^s_\infty}\,, \qquad & \alpha^s_m &= \frac{2 - \rho^s_\infty}{1 + \rho^s_\infty}\,.
\end{aligned} \tag{42}$$

Similarly to (31), the scalar ρ^s_∞ can be identified as the spectral radius associated with infinitely large time steps and has to be chosen by the user such that $0 \leq \rho^s_\infty \leq 1$. Note that (42) is not identical to (31).

6.3. MESH UPDATE

It still remains to discretise the movement of the mesh in time. Therefore, the configuration $\hat{\mathbf{x}}^h_n$ and the velocity field $\hat{\mathbf{v}}^h_n$ at the discrete time instants t_n, $n = 0, 1, 2, \ldots, N$ are introduced. In this work, $\hat{\mathbf{x}}^h_n$ and $\hat{\mathbf{v}}^h_n$ are related by a simple generalised midpoint scheme

$$\hat{\mathbf{v}}^h_{n+1} = \frac{1}{\Delta t \gamma^m}\,(\hat{\mathbf{x}}^h_{n+1} - \hat{\mathbf{x}}^h_n) - \frac{1 - \gamma^m}{\gamma^m}\,\hat{\mathbf{v}}^h_n\,, \tag{43}$$

where γ^m is an integration parameter to be chosen such that $\frac{1}{2} \leq \gamma^m \leq 1$. In this work, we set $\gamma^m = \gamma^f$ (see $(31)_1$), and thus make it dependent on ρ^f_∞. The following expressions are then employed in the weak form (15)

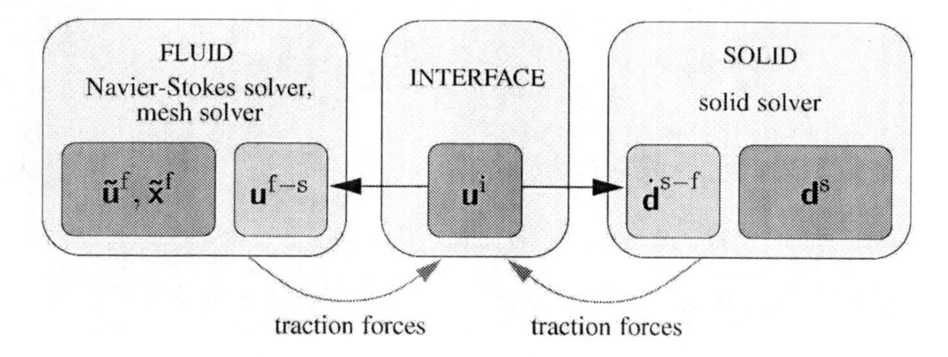

Figure 1. Discrete overall problem: Domain decomposition based on introduction of interface d.o.f. $\mathbf{u}^{\mathrm{i}}$.

$$\hat{\mathbf{x}}^h_{n+\alpha^{\mathrm{f}}_f} = (1-\alpha^{\mathrm{f}}_f)\,\hat{\mathbf{x}}^h_n + \alpha^{\mathrm{f}}_f\,\hat{\mathbf{x}}^h_{n+1} \tag{44}$$

$$\hat{\mathbf{v}}^h_{n+\alpha^{\mathrm{f}}_f} = (1-\alpha^{\mathrm{f}}_f)\,\hat{\mathbf{v}}^h_n + \alpha^{\mathrm{f}}_f\,\hat{\mathbf{v}}^h_{n+1}\,, \tag{45}$$

where α^{f}_f is given by $(31)_3$. The vector field $\hat{\mathbf{x}}^h_{n+\alpha^{\mathrm{f}}_f}$ defines the configuration, which is employed in the computation of the integrals in (15).

Importantly, the mesh update strategy discussed in Section 4.1 is applied to the position of the internal nodes of the finite element mesh at t_{n+1}. Thus, the mesh meets the quality criteria (19) at the time instants t_n, $t_{n+1}, \ldots$. The mesh configuration given by $\hat{\mathbf{x}}^h_{n+\alpha^{\mathrm{f}}_f}$ is therefore an interpolant between two "optimal" configurations.

Remark 6.1. Equation (21) requires that, along the interface boundary, the velocity vectors $\hat{\mathbf{v}}^h$, $\mathbf{u}^h$ and $\dot{\mathbf{d}}_{\mathrm{f-s}}$ coincide. In this work, these velocities are enforced to be identical at the discrete time instants t_n, $t_{n+1}, \ldots$. It then follows from the similarity of (26) and (45), that the interpolated values $\mathbf{u}^h_{n+\alpha^{\mathrm{f}}_f}$ and $\hat{\mathbf{v}}^h_{n+\alpha^{\mathrm{f}}_f}$ also coincide. Yet, due to the different parameters employed in the time integration of the fluid and the solid body, the quantities $\mathbf{u}^h_{n+\alpha^{\mathrm{f}}_f}$ and $\hat{\mathbf{v}}^h_{n+\alpha^{\mathrm{f}}_f}$ are not necessarily identical to the interpolated velocity of the solid body surface, which is associated with $\mathbf{d}_{n+\alpha^{\mathrm{s}}_f}$ and, in the case of rigid body, $\theta_{n+\alpha^{\mathrm{r}}_f}$. However, the authors have not experienced any numerical problems associated with this small inconsistency.

7. Discrete Overall Problem

The degrees of freedom associated with the finite element nodes in mesh interiors are denoted by the vectors $\tilde{\mathbf{u}}^{\mathrm{f}}\,\{\mathbf{u}, \mathbf{p}\}$, $\hat{\mathbf{x}}^{\mathrm{f}}$ and $\mathbf{d}^{\mathrm{s}}$ which represent, respectively, the fluid velocities and pressures, the position of the fluid nodes and the structural displacements. Let the vector $\mathbf{u}^{\mathrm{i}}$ denote the kinematical degrees of freedom of the interface. At the adjacent boundaries of the fluid and the structural finite element meshes, the nodal positions, velocities and accelerations of both the fluid and the solid can be expressed exclusively in terms of the motion of the interface as represented by $\mathbf{u}^{\mathrm{i}}$. This arrangement of kinematic variables is illustrated in Figure 1.

Following the standard finite element discretisation procedure, a compact representation of the overall problem may then be written as

$$\mathbf{g}^{\mathrm{f}}\left(\tilde{\mathbf{u}}^{\mathrm{f}}, \hat{\mathbf{x}}^{\mathrm{f}}, \mathbf{u}^{\mathrm{i}}\right) = \mathbf{0}, \quad n^{\mathrm{f}} \text{ equations} \tag{46}$$

$$\mathbf{g}^{\mathrm{i}}\left(\tilde{\mathbf{u}}^{\mathrm{f}}, \hat{\mathbf{x}}^{\mathrm{f}}, \mathbf{u}^{\mathrm{i}}, \mathbf{d}^{\mathrm{s}}\right) = \mathbf{0}, \quad n^{\mathrm{i}} \text{ equations} \tag{47}$$

$$\mathbf{g}^{\mathrm{s}}\left(\mathbf{d}^{\mathrm{s}}, \mathbf{u}^{\mathrm{i}}\right) = \mathbf{0}, \quad n^{\mathrm{s}} \text{ equations} \tag{48}$$

$$\mathbf{m}\left(\hat{\mathbf{x}}^{\mathrm{f}}, \mathbf{u}^{\mathrm{i}}\right) = \mathbf{0}, \quad n^{\mathrm{m}} \text{ equations}\,, \tag{49}$$

where the numbers of scalar equations n^{f} corresponds to the number of velocity and pressure degrees of freedom of the fluid mesh reduced by the fluid velocities on the interface boundary. The number of interface degrees of freedom is denoted by n^{i}, whereas n^{s} represents the number of structural degrees of freedom not including those on the interface boundary of the structural mesh. Finally, the number of degrees of freedom which describe the motion of the fluid mesh is given as n^{m}. The systems (46) – (49) are strongly coupled and highly nonlinear.

We note that the introduction of the independent interface degrees of freedom, as schematically shown in Figure 1, allows a modular computer implementation and thus facilitates future extensions of the methodology. Such extensions may include, for instance, independent remeshing of the fluid or the solid domains or the employment of a 'mortar'strategy for the data transfer between the adjacent phases. Furthermore, it is pointed out that the computational cost of solving (46) – (49) depends strongly on the number n^{i} of interface degrees of freedom. In many cases, the choice of a coarse interface discretisation may allow a good computational model of the fluid-structure interaction at low computational cost. Finally, the introduction of independent interface degrees of freedom facilitates the application of the computational framework to different physical problems such as fluid-rigid body interaction or free surface flow (see references in (Dettmer and Perić, 2005b)).

If the structures under considerations can be regarded as rigid bodies, Equation (47) becomes $\mathbf{g}^{\mathrm{i}}\left(\tilde{\mathbf{u}}^{\mathrm{f}}, \hat{\mathbf{x}}^{\mathrm{f}}, \mathbf{u}^{\mathrm{i}}\right) = \mathbf{0}$ and set of Equations (48) can be ignored. Furthermore, the interface velocity $\mathbf{u}^{\mathrm{i}}$ becomes $\mathbf{u}^{\mathrm{i}} = \{\dot{\mathbf{d}}, \dot{\theta}\}$ where $\mathbf{d}$ denotes the rigid body translations and θ is the rigid body rotation.

Box1: Solution algorithm for coupled system of equations (46)-(49)

---

1. estimate $\tilde{\mathbf{u}}^{\mathrm{f}}$, $\mathbf{u}^{\mathrm{i}}$
2. fluid mesh solver $\mathbf{m}\left(\hat{\mathbf{x}}^{\mathrm{f}}, \mathbf{u}^{\mathrm{i}}\right) = \mathbf{0} \quad \Rightarrow \quad \hat{\mathbf{x}}^{\mathrm{f}}, \ \dfrac{\partial \hat{\mathbf{x}}^{\mathrm{f}}}{\partial \mathbf{u}^{\mathrm{i}}}$
3. solid solver $\mathbf{g}^{\mathrm{s}}\left(\mathbf{d}^{\mathrm{s}}, \mathbf{u}^{\mathrm{i}}\right) = \mathbf{0} \quad \Rightarrow \quad \mathbf{d}^{\mathrm{s}}, \ \dfrac{\partial \mathbf{d}^{\mathrm{s}}}{\partial \mathbf{u}^{\mathrm{i}}}$
4. if $\mathbf{g}^{\mathrm{f}}\left(\tilde{\mathbf{u}}^{\mathrm{f}}, \hat{\mathbf{x}}^{\mathrm{f}}, \mathbf{u}^{\mathrm{i}}\right)$, $\mathbf{g}^{\mathrm{i}}\left(\mathbf{u}^{\mathrm{f}}, \hat{\mathbf{x}}^{\mathrm{f}}, \mathbf{u}^{\mathrm{i}}, \mathbf{d}^{\mathrm{s}}\right) < tol$, then exit
5. compute derivatives

$$\mathsf{A} = \frac{\partial \mathbf{g}^{\mathrm{f}}}{\partial \tilde{\mathbf{u}}^{\mathrm{f}}}, \quad \mathsf{B} = \frac{\partial \mathbf{g}^{\mathrm{f}}}{\partial \mathbf{u}^{\mathrm{i}}} + \frac{\partial \mathbf{g}^{\mathrm{f}}}{\partial \hat{\mathbf{x}}^{\mathrm{f}}}\frac{\partial \hat{\mathbf{x}}^{\mathrm{f}}}{\partial \mathbf{u}^{\mathrm{i}}}, \quad \mathsf{C} = \frac{\partial \mathbf{g}^{\mathrm{i}}}{\partial \tilde{\mathbf{u}}^{\mathrm{f}}}, \quad \mathsf{D} = \frac{\partial \mathbf{g}^{\mathrm{i}}}{\partial \mathbf{u}^{\mathrm{i}}} + \frac{\partial \mathbf{g}^{\mathrm{i}}}{\partial \hat{\mathbf{x}}^{\mathrm{f}}}\frac{\partial \hat{\mathbf{x}}^{\mathrm{f}}}{\partial \mathbf{u}^{\mathrm{i}}} + \frac{\partial \mathbf{g}^{\mathrm{i}}}{\partial \mathbf{d}^{\mathrm{s}}}\frac{\partial \mathbf{d}^{\mathrm{s}}}{\partial \mathbf{u}^{\mathrm{i}}}$$

6. *combined fluid + interface solver*

$$\begin{bmatrix} \mathsf{A} & \mathsf{B} \\ \mathsf{C} & \mathsf{D} \end{bmatrix} \begin{Bmatrix} \Delta\tilde{\mathbf{u}}^{\mathrm{f}} \\ \Delta\mathbf{u}^{\mathrm{i}} \end{Bmatrix} = - \begin{Bmatrix} \mathbf{g}^{\mathrm{f}} \\ \mathbf{g}^{\mathrm{i}} \end{Bmatrix}, \qquad \begin{Bmatrix} \tilde{\mathbf{u}}^{\mathrm{f}} \\ \mathbf{u}^{\mathrm{i}} \end{Bmatrix} \leftarrow \begin{Bmatrix} \tilde{\mathbf{u}}^{\mathrm{f}} \\ \mathbf{u}^{\mathrm{i}} \end{Bmatrix} + \begin{Bmatrix} \Delta\tilde{\mathbf{u}}^{\mathrm{f}} \\ \Delta\mathbf{u}^{\mathrm{i}} \end{Bmatrix}$$

7. goto 2.

Remark 7.1. At the outset, the derivation of system (46) – (49) was based on identifying the kinematics of the interface with the discretisation of structural surface. However, the coupled system (46) – (49) may also be obtained from a more generic framework based on the introduction of an independent discretisation of the interface, as shown schematically in Figure 2.

8. Solution Algorithm

Given the solution at the time instant t_n, the nonlinear coupled system (46) – (49) has to be solved for $\tilde{\mathbf{u}}^{\mathrm{f}}, \hat{\mathbf{x}}^{\mathrm{f}}, \mathrm{d}^{\mathrm{s}}$ and u^{i} at the next time instant t_{n+1}. The solution strategy adopted in this work is essentially based on the Newton-Raphson method and may be summarised as in Box 1. It can easily be shown that the derivative matrices $\partial \hat{\mathbf{x}}^{\mathrm{f}}/\partial \mathrm{u}^{\mathrm{i}}$ and $\partial \mathrm{d}^{\mathrm{s}}/\partial \mathrm{u}^{\mathrm{i}}$ may be obtained from solving two systems of linear equations for n^{i} different right hand sides. The coefficient matrices to be employed are the stiffness matrices of the converged fluid mesh solver and the solid solver, respectively. The implementation of the derivatives of the fluid nodal forces with respect to the fluid mesh configuration, as required in step 5. of the algorithm, is tedious but straightforward. We note that the computational cost of calculating the submatrices B, C and D is essentially determined by the number

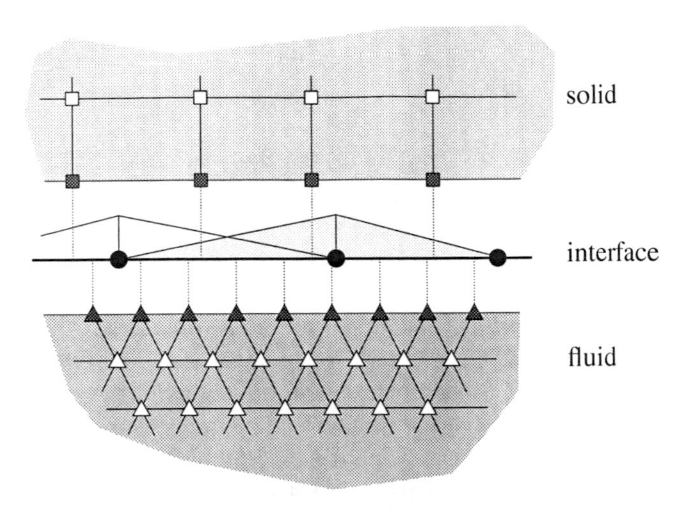

Figure 2. General interface modelling, transfer based on finite element type interpolation of the interface domain.

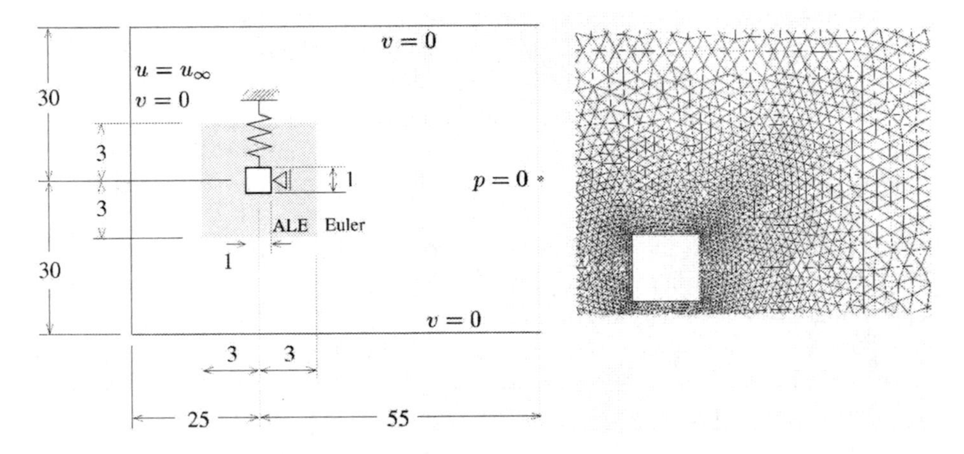

a. Geometry and boundary conditions b. Detail of the finite element mesh

Figure 3. Vortex-induced oscillations and galloping.

of interface degrees of freedom n^{i}. Similarly, the number of non-zero coefficients in the system matrix of the combined fluid + interface solver depends largely on n^{i}.

9. Numerical Examples

9.1. VORTEX-INDUCED OSCILLATIONS AND GALLOPING

Two mechanisms that are often associated with the flow-induced vibration of a structure subject to non-oscillating flow are *vortex-induced oscillation* and *galloping*. Vortex-induced oscillation is associated with synchronisation or "lock-in" of vortex-shedding frequency with the oscillation frequency of the structure, while galloping is driven by the oscillating aero/hydrodynamic forces due to the motion of the structure and has a frequency many times lower than that of vortex-shedding. Both mechanisms can lead to significant oscillation amplitudes, which can potentially result in catastrophic failure of a structure.

As a demonstration of vortex-induced galloping we consider a rigid body with a square cross section exposed to fluid flow. The body is supported by an elastic spring with a small amount of structural damping such that it is free to oscillate perpendicularly to the direction of the flow. The properties of the body and its support are set to $k = 3.08425, c = 0.0581195$ and $m = 20.0$. The length of the sides of the square body is $D = 1.0$. The fluid density and viscosity read $\rho = 1.0$ and $\mu = 0.01$, respectively. The maximum inflow velocity considered here is $u_\infty = 2.5$, which corresponds to the Reynolds number $Re = u_\infty D\rho/\mu = 250$. With the above material parameters the set-up of the problem, at $Re = 250$, corresponds to one of the numerical simulations performed by Robertson et al. (2003). The geometry and the boundary conditions employed here are displayed in Figure 3.a.

The finite element mesh, which is used for the simulations, consists of 8718 elements, and the surface of the rigid body is modelled with 80 fluid element edges. Figure 3.b shows a detail of the mesh. For the rigid body, the time integration parameter is set to $\rho^{\mathrm{r}}_\infty = 0.9$. For the time integration of the fluid, ρ^{f}_∞ is set to 0.5 or 0.8 and the time step size Δt is roughly adapted to the different inflow velocities u_∞ (see Figure 4).

In Figure 4, the frequency f_{o} and the amplitude Y/D of the rigid body oscillations as well as the frequency f_{v} of the vortex shedding are displayed against the Reynolds number. The diagrams in Figure 5 show the evolution of the amplitude Y/D in time. In Figure 6, typical oscillations of the displacement Y/D and the lift force F_y, obtained from two different time step sizes. A typical flow pattern during galloping is shown in Figure 7.

On the basis of the numerical results, the following observations are made:

- The effects of vortex induced oscillations and galloping can clearly be distinguished: In Figure 4, the lock-in region is observed to coincide with the interval of Reynolds numbers $Re \approx 50$ to 55. The maximum amplitude of the vortex induced oscillations is $\max(Y/D) = 0.186$. In the lock-in region, the frequencies f_{o}, f_{v} and f_{n} are almost identical. For $Re \approx 55$ to 150, the

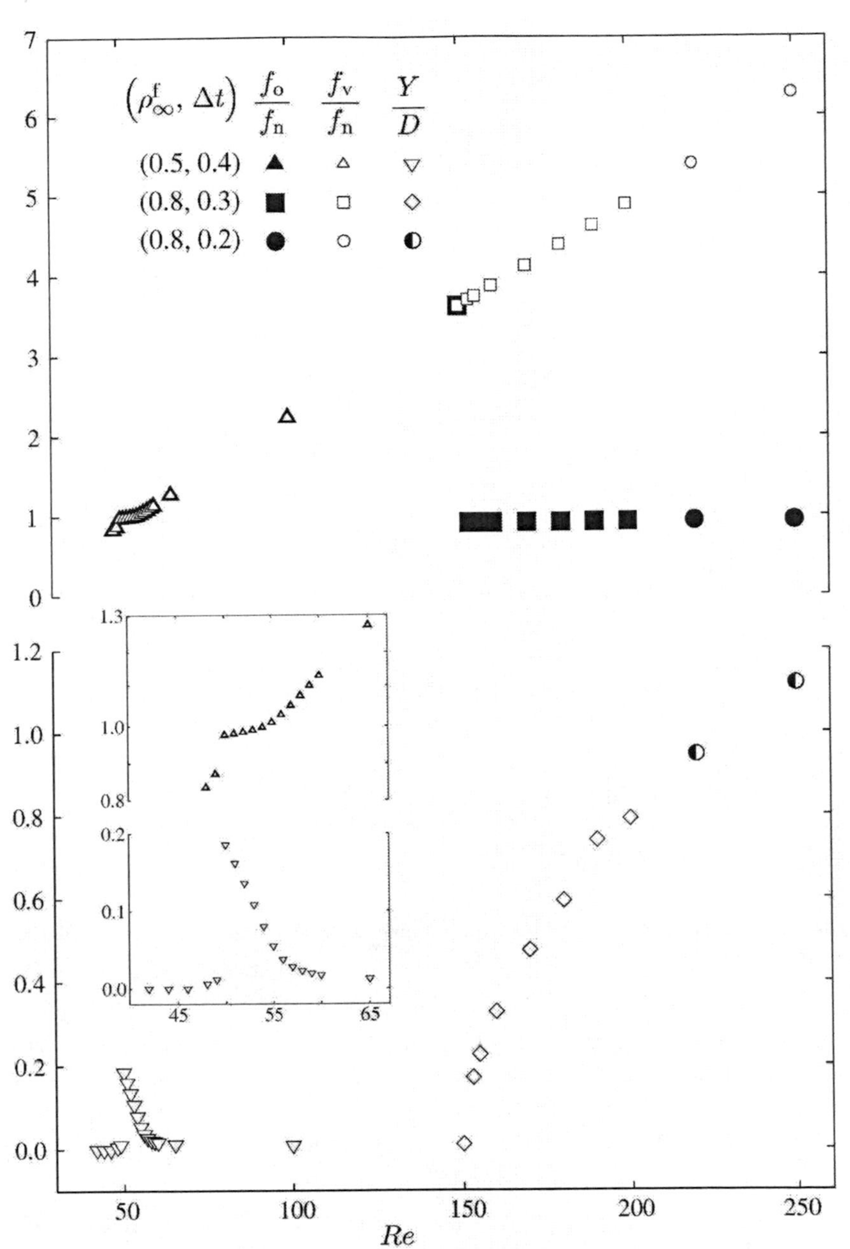

Figure 4. Oscillating square, amplitudes and frequencies; also showing a zoom of the lock-in region; different time integration parameters $(\rho_\infty^{\mathrm{f}}, \Delta t)$.

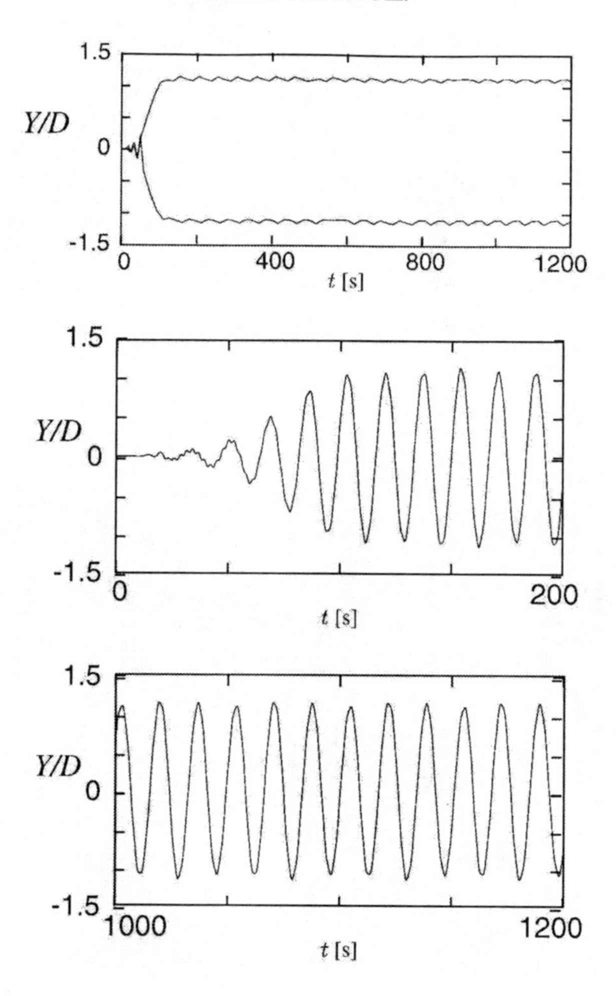

Figure 5. Vortex-induced oscillation and galloping: evolution of the amplitude Y/D; $Re = 250$ with $\rho^{\mathrm{f}}_{\infty} = 0.5, \Delta t = 0.2$.

rigid body oscillations are negligible. At $Re \approx 150$, the rigid body suddenly starts large amplitude oscillations. At $Re = 250$, the amplitude has risen to $\max(Y/D) = 1.117$. The frequency f_{o} coincides with the natural frequency f_{n}, whereas the frequency of the vortex shedding f_{v} is 3.6 to 6.3 times larger than f_{n}.

- At $Re = 250$, the rigid body oscillations ($\max(Y/D) = 1.117$ and $f_{\mathrm{o}} \approx 0.938 f_{\mathrm{n}}$) agree very well with the solution resented by Robertson et al. (2003)($\max(Y/D) \approx 1.15$ and $f_{\mathrm{o}} \approx 0.938 f_{\mathrm{n}}$).
- The independence of the rigid body oscillations at high Reynolds numbers

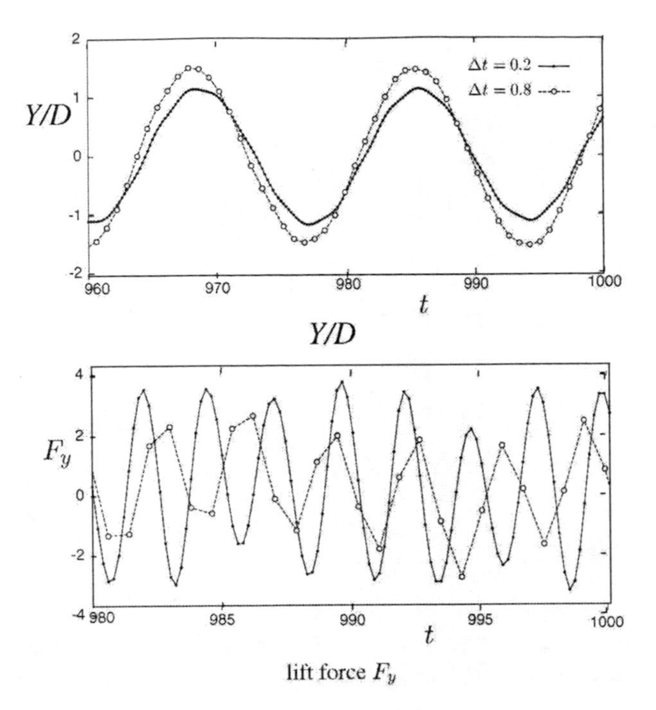

Figure 6. Vortex-induced oscillations and galloping: $Re = 250$, $\rho^{\mathrm{f}}_{\infty} = 0.8$, two different time step sizes Δt.

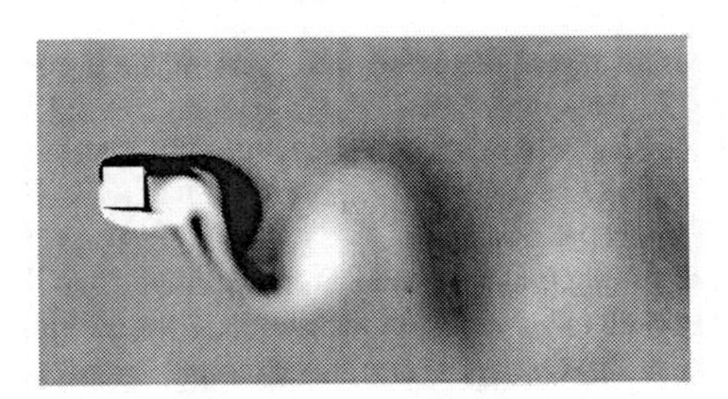

Figure 7. Vortex-induced oscillations and galloping: typical vorticity distribution during galloping, $Re = 250$, $\mathrm{vert}(\mathbf{u}^h) \leq -5 \rightarrow$ black, $\mathrm{vort}(\mathbf{u}^h) \geq +5 \rightarrow$ white.

from the vortex shedding is illustrated in Figure 6. It is shown that the galloping effect is captured rather accurately for very large time step sizes ($\Delta t \approx T_{\mathrm{o}} \,/\, 20, T_{\mathrm{o}} = 1/f_{\mathrm{o}}$),which do not resolve the vortex shedding properly.

- The numerical model and the solution algorithm prove robust and efficient. The convergence of the residuals is observed to be asymptotically quadratic.

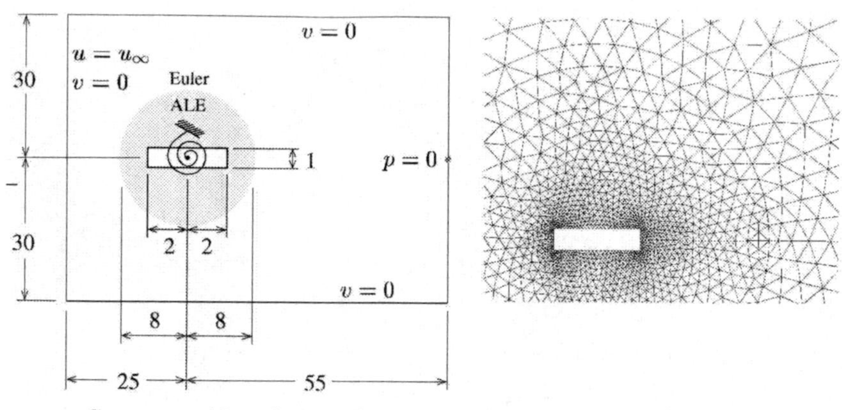

a. Geometry and boundary conditions b. Detail of the finite element mesh

Figure 8. Rotational galloping.

9.2. ROTATIONAL GALLOPING

The set-up of the problem is almost identical to Subsection 9.1, however, the length of the rigid body under consideration in the direction of the flow is now four times longer than its thickness, and the rigid body is free to rotate, but fixed in x- and y-direction. The rotational degree of freedom is associated with an elastic spring and a certain amount of linear damping. The geometry and the boundary conditions of the problem, which are displayed in Figure 8.a, correspond to (Robertson et al., 2003).

The properties of the rigid body-spring system are set to $I_\theta = 400$, $c_\theta = 78.540$ and $k_\theta = 61.685$, thus $f_{\rm n} = \sqrt{k_\theta / I_\theta} / (2\pi) = 0.0625$. The inflow velocity, the diameter of the rectangular body and the fluid density and viscosity are chosen as $u_\infty = 2.5$, $D = 1$, $\rho = 1$ and $\mu = 0.01$, respectively, such that the global Reynolds number becomes $Re = u_\infty D\rho/\mu = 250$.

Two different finite element meshes are considered with 3454 (11696) elements and 84 (164) element edges on the boundary of the rigid body. The detail of the finite element mesh employed in this example is shown in Figure 8.b. The time integration parameters are set to $\rho^{\rm r}_\infty = 0.9$ and $\rho^{\rm f}_\infty = 0.8$. Different time step sizes Δt are used. The sudden application of the inflow velocity at $t = 0$ is overcome by the employment of one large initial time step with $\Delta t = 30$.

Figure 9 illustrates the typical evolution of the rigid body rotation θ in time. Typical flow patterns are displayed in Figure 10.

For the Reynolds number under consideration $(Re = 250)$, the rigid bar is galloping. It is observed that the vortex shedding frequency $f_{\rm v}$ is significantly larger than the frequency $f_{\rm o}$ of the oscillation. A rough estimate of the frequency $f_{\rm v}$, obtained by observing the vortex shedding over a sufficiently long time interval before the onset of galloping, renders $f_{\rm v} \approx 5.2\, f_{\rm n}$, whereas $f_{\rm o} \approx 0.8\, f_{\rm n}$.

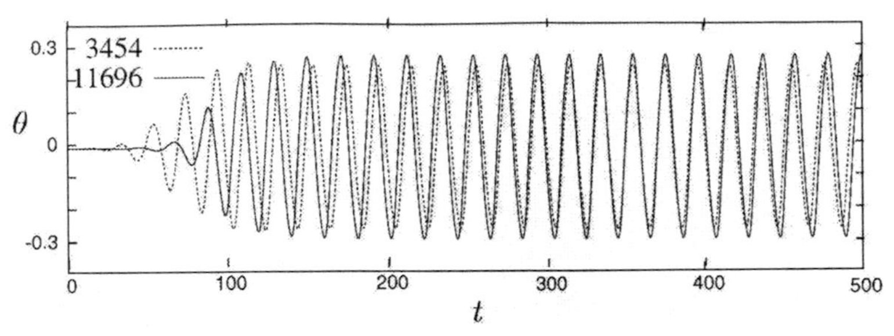

Figure 9. Rotational galloping: evolution of rotation θ for $\Delta t = 0.3$, meshes with 3454 and 11696 elements.

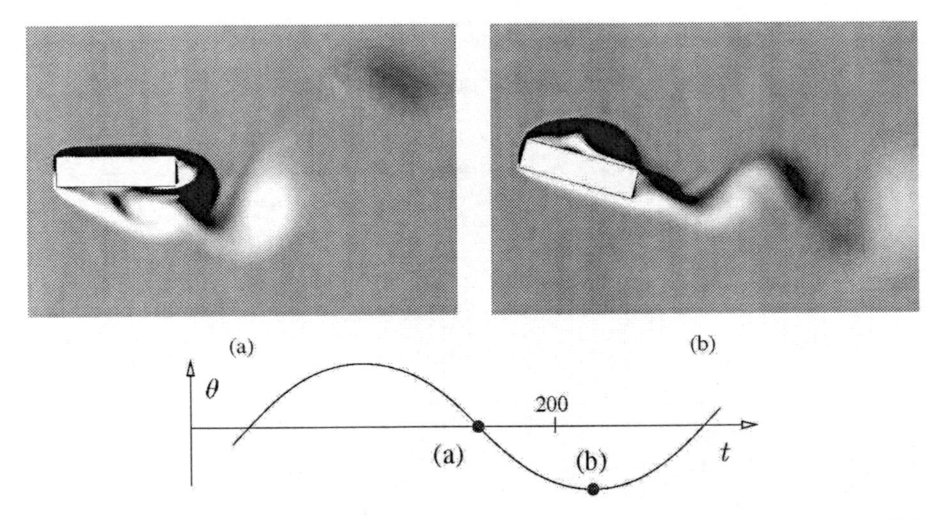

Figure 10. Rotational galloping: typical vorticity distribution, 11696 elements, $\Delta t = 0.2$, $\mathrm{vert}(\mathbf{u}^h) \leq -5 \rightarrow$ black, $\mathrm{vort}(\mathbf{u}^h) \geq +5 \rightarrow$ white.

The amplitude and frequency of the rigid body oscillations agree very well with the solution presented in (Robertson et al., 2003).

9.3. FLUTTER OF A BRIDGE DECK

In this example, a rigid H-profile is considered that is supported by a rotational and a vertical translational linear elastic spring. The horizontal motion is fixed to zero. The profile is exposed to uniform fluid flow in the horizontal direction. If the parameters of the supports are chosen in such that they represent the torsional and vertical stiffness of the superstructure, this model problem may be used to analyse the aerodynamic stability of a suspension or a cable-stayed bridge.

The geometry and material parameters employed here taken from Hübner *et al.*, (Hübner et al., 2001). The geometry and the boundary conditions are displayed

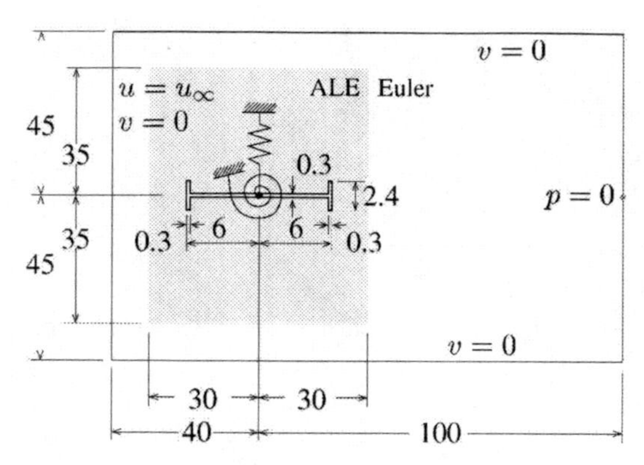

a. Geometry and boundary conditions

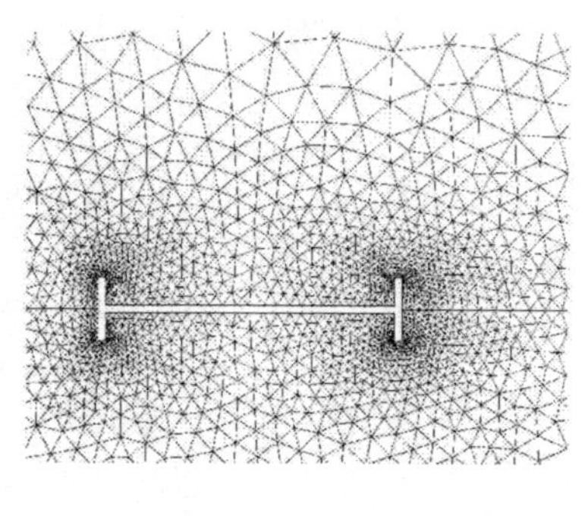

b. Detail of the finite element mesh

Figure 11. Flutter of a bridge deck.

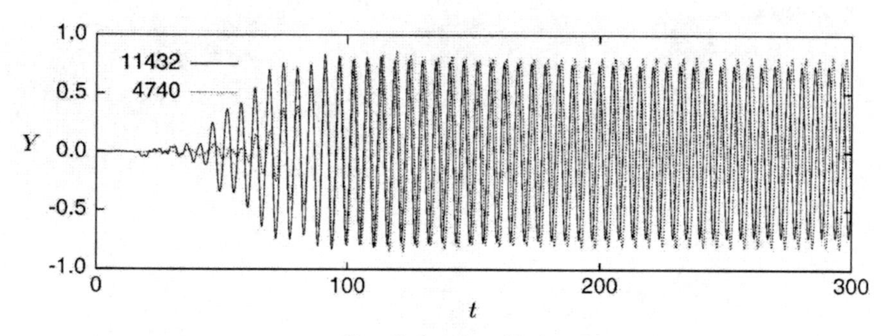

a. Translational oscillations

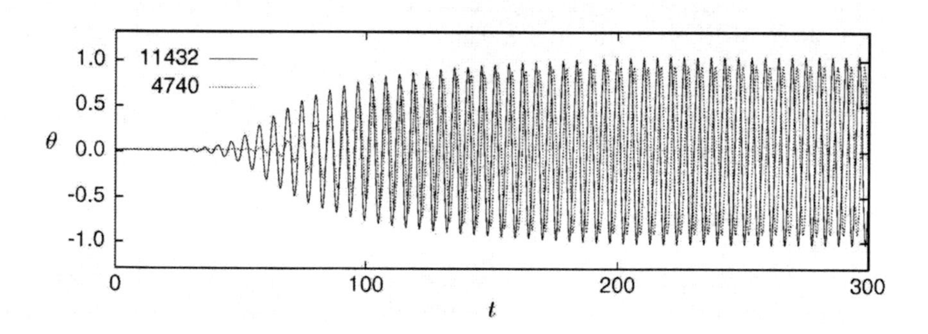

b. Rotational oscillations

Figure 12. Flutter of bridge deck for two different meshes.

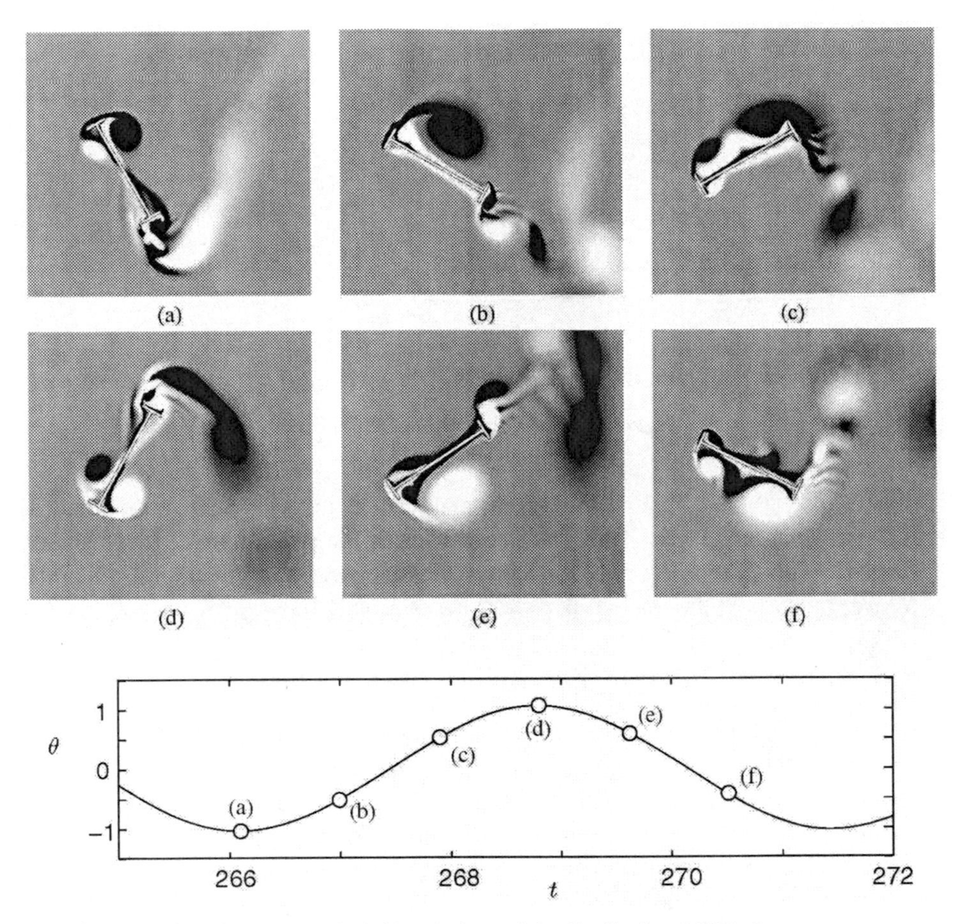

Figure 13. Flutter of a bridge deck, vorticity distribution, 11432 elements.

in Figure 11. The fluid properties are set to $\mu = 0.1$ N s/m^2 and $\rho = 1.25\text{kg}/\text{m}^3$, the inflow velocity is chosen as $u_\infty = 10\,\text{m/s}$ and the rigid body properties read $k_y = 2000\,\text{N/m}$, $m_y = 3000\,\text{kg}$, $k_\theta = 40{,}000\,\text{Nm}$ and $I_\theta = 25,300$ kg m^2. The natural frequencies are thus obtained as $f_{y,\text{n}} = 0.130$ Hz and $f_{\theta,\text{n}} = 0.200$ Hz. Structural damping is ignored. If the Reynolds number is related to the width of the bridge deck $b = 12$ m, it follows that $Re = u_\infty b\rho/\mu = 1500$. As depicted in Figure 11, the Eulerian mesh is used to model the part of the fluid domain away from the solid body. The evolution of the bridge deck degree of freedom Y and θ are displayed in Figures 12.a and 12.b. Figure 13 shows typical flow patterns.

After roughly 70 s, the bridge deck begins to oscillate with large amplitudes. At $t \approx 200$ s the oscillations take a stable pattern. The amplitude of the rota-

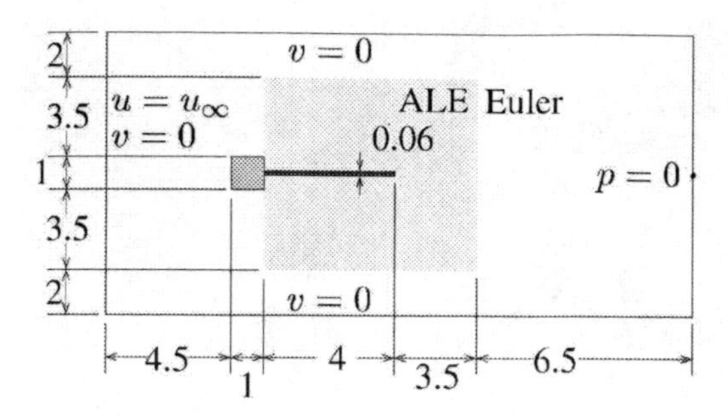

a. Geometry and boundary conditions

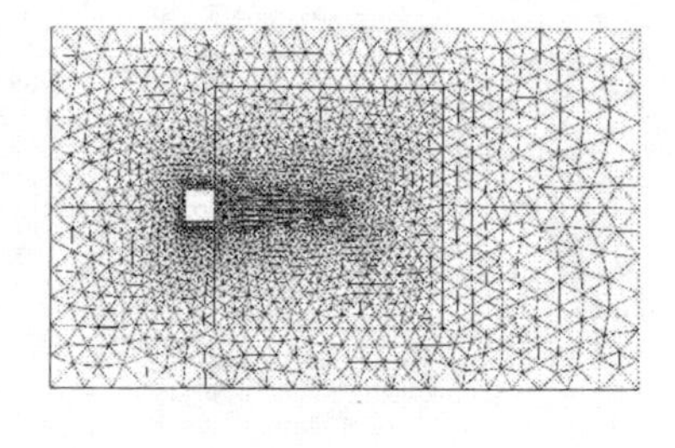

b. Detail of the finite element mesh

Figure 14. Flow around a flexible structure.

tion is $\max(\theta) \approx 57^\circ$ and the maximum vertical displacement is obtained as $0.75\ \mathrm{m} \leq \max(Y) \leq 0.85\ \mathrm{m}$. The frequencies of the rotation and the translation coincide with $f_0 \approx 0.186\ \mathrm{s}^{-1}$. It is noted that this is close to the natural rotational frequency. In fact, the rotation is clearly the dominant motion.

9.4. FLOW AROUND A FLEXIBLE BEAM STRUCTURE

This represents an example of a two dimensional flow past a flexible beam at Reynolds number $Re = 333$. This problem was introduced by Wall and Ramm (Wall and Ramm, 1998) and has become a standard benchmark test. The elastic beam is excited to large amplitude oscillations by the vortex shedding, which results in large periodic deformation of the fluid domain and causes an increase in the complexity of the flow. Simulations have been performed with different spatial and temporal discretisations, whereby coarse fluid meshes (4336 elements) and large time steps ($\Delta t \approx 1/(20f)$ with $f =$ first eigenfrequency of the beam) have rendered good approximations.

The initial geometry and finite element mesh used for analysis given in Figure 14. The material parameters for the density and viscosity of the fluid are $\rho_f = 1.18 \times 10^{-3}$ and $\mu_f = 1.82 \times 10^{-4}$, respectively, while the density, Young's modulus and Poisson's ratio of the solid body are $\rho_s = 0.1$, $E = 2.5 \times 10^6$ and $\nu = 0.35$, respectively. Here, the flexible beam is modelled using 20 nine-noded small strain finite elements. Figure 15 shows the oscillation of the tip of the beam. Typical flow patterns for this example are shown in Figure 16.

The simulation renders an unsteady periodic long term response of the flexible beam. The build-up of the oscillations takes approximately 2 time units. For all discretisations considered, the amplitudes of the oscillating tip displacement d lie between 1.1 and 1.4. The average frequency $\bar{f}$ is obtained between 2.96 and 3.31.

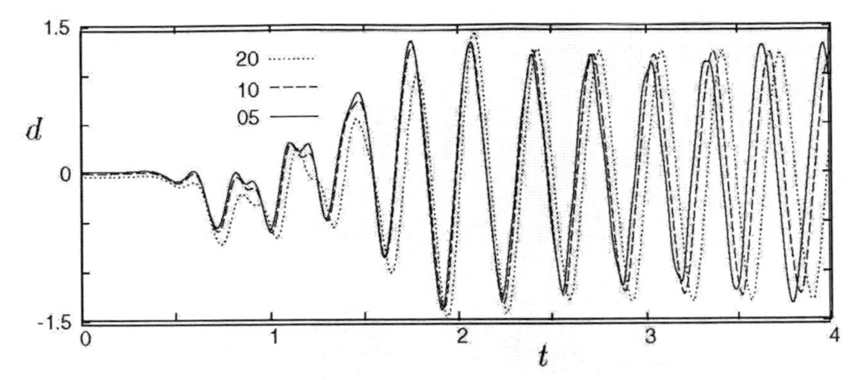

a. The build up of the oscillations for different time step sizes Δt

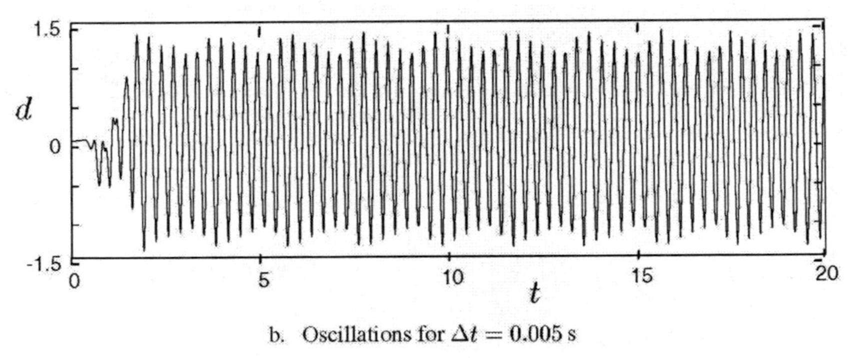

b. Oscillations for $\Delta t = 0.005$ s

Figure 15. Flow around a flexible structure: Vertical displacement of the tip of the structure.

This interval contains the lowest eigenfrequency of the beam $f_1 \approx 3.03$ and also agrees well with the results obtained in (Wall, 1999; Hübner et al., 2001).

10. Conclusions

In this work we have presented a fully implicit computational strategy for the simulation of fluid-structure interaction based on the finite element methodology combined with a discrete time integration scheme. The strong coupling between fluid and solid is resolved with accuracy set by the prescribed tolerance. The consistent linearisation of the overall problem has been achieved, leading to asymptotically quadratic convergence of the residuals. The methodology seems robust and accurate. It is extremely efficient for small numbers n^{i}, and seems competitive for larger problems. The strategy seems to be very suitable for problems with flexible structures experiencing large deformations with possibly large strains.

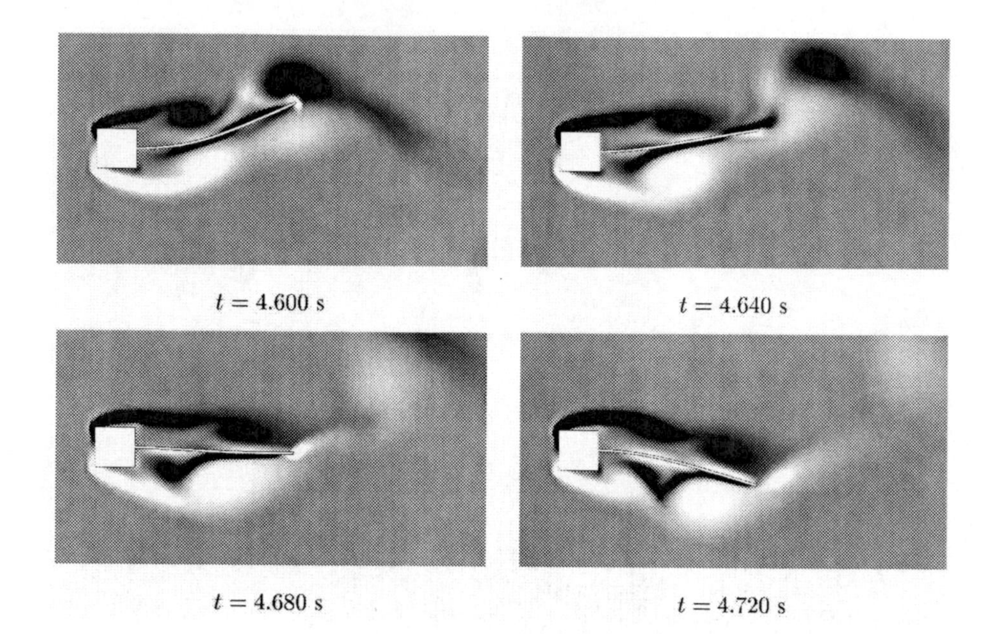

Figure 16. Flow around a flexible structure: Vorticity distribution during periodic long term oscillations.

Acknowledgements

This work was partly funded by EPSRC of UK under grant no, GR/R92318 and this support is gratefully acknowledge.

References

Bar-Yoseph, P. Z., Mereu, S., Chippada, S., and Kalro, V. J. (2001) Automatic Monitoring of element Shape Quality in 2-D and 3-D Computational Mesh Dynamics, *Computational Mechanics* **27**, 378–395.

Belytschko, T., Liu, W. K., and Moran, B. (2000) *Nonlinear Finite Elements for Continua and Structures*, Chicester, UK, John Wiley & Sons.

Braess, H. and Wriggers, P. (2000) Arbitrary Lagrangian Eulerian Finite Element Analysis of Free Surface Flow, *Computer Methods in Applied Mechanics and Engineering* **190**, 95–109.

Brooks, A. N. and Hughes, T. J. R. (1982) Streamline-Upwind/Petrov-Galerkin Formulations for Convection Dominated Flows with Particular Emphasis on the Incompressible Navier-Stokes Equations, *Computer Methods in Applied Mechanics and Engineering* **32**, 199–259.

Chung, J. and Hulbert, G. M. (1993) A Time Integration Algorithm for Structural Dynamics with Improved Numerical Dissipation: The Generalized-α Method, *Journal of Applied Mechanics* **60**, 371–375.

Degand, C. and Farhat, C. (2002) A Three-Dimensional Torsional Spring Analogy Method for Unstructured Dynamic Meshes, *Computers and Structures* **80**, 305–316.

Dettmer, W. G. (2004) Finite Element Modelling of Fluid Flow with Moving Free Surfaces and Interfaces Including Fluid-Solid Interaction, Ph.D. thesis, University of Wales Swansea, United Kingdom.

Dettmer, W. G. and Perić, D. (2003) An Analysis of the Time Integration Algorithms for the Finite Element Solutions of Incompressible Navier-Stokes Equations Based on a Stabilised Formulation, *Computer Methods in Applied Mechanics and Engineering* **192**, 1177–1226.

Dettmer, W. G. and Perić, D. (2005)a A Computational Framework for Fluid-Rigid Body Interaction: Finite Element Formulation and Applications, *Computer Methods in Applied Mechanics and Engineering*, in press.

Dettmer, W. G. and Perić, D. (2005)b A Computational Framework for Fluid-Structure Interaction: Finite Element Formulation and Applications, *Computer Methods in Applied Mechanics and Engineering*, in press.

Dettmer, W. G., Saksono, P. H., and Perić, D. (2005) Efficient Modelling of Fluid-Structure Interaction Based on an ALE Strategy Combined with Mesh Adaptivity, In *Extended Abstracts of the 5th International Conference on Computation of Shell and Spatial Structures*, Salzburg, Austria.

Donea, J. and Huerta, A. (2003) *Finite Element Methods for Flow Problems*, Chicester, UK, John Wiley & Sons.

Hübner, B., Walhorn, E., and Dinkler, D. (2001) Strongly Coupled Analysis of Fluid-Structure Interaction Using Space-Time Finite Elements, In *Proceedings of the European Conference on Computational Mechanics*, Cracow, Poland.

Hughes, T. J. R., Franca, L. P., and Balestra, M. (1986) A New Finite Element Formulation for Computational Fluid Dynamics: V. Circumventing the Babuska-Brezzi Condition: A Stable Petrov-Galerkin Formulation of the Stokes Problem Accomodating Equal-Order Interpolations, *Computer Methods in Applied Mechanics and Engineering* **59**, 85–99.

Hughes, T. J. R., Franca, L. P., and Hulbert, G. M. (1989) A New Finite Element Formulation for Computational Fluid Dynamics: VIII. The Galerkin/Least-Squares Method for Advective-Diffusive Equations, *Computer Methods in Applied Mechanics and Engineering* **73**, 173–189.

Jansen, K. E., Whiting, C. H., and Hulbert, G. M. (2000) A Generalized α-Method for Integrating the Filtered Navier-Stokes Equations with a Stabilized Finite Element Method, *Computer Methods in Applied Mechanics and Engineering* **190**, 305–319.

Johnson, C., Nävert, U., and Pitkäranta, J. (1984) Finite Element Methods for Linear Hyperbolic Problems, *Computer Methods in Applied Mechanics and Engineering* **45**, 285–312.

Masud, A. and Hughes, T. J. R. (1997) A Space Time Galerkin/Least-Squares Finite Element Formulation of the Navier-Stokes Problem, *Computer Methods in Applied Mechanics and Engineering* **146**, 91–126.

Matthies, H. G. and Ohayon, R. (eds.) (2005) Advances in Analysis of Fluid-Structure Interaction, Vol. 83 of *Computers and Structures*, Special Issue.

Ohayon, R. and Felippa, C. (eds.) (2001) Advances in Computational Methods for Fluid-Structure Interaction and Coupled Problems, Vol. 190 of *Computer Methods in Applied Mechanics and Engineering*, Special Issue.

Oñate, E. (2000) A Stabilized Finite Element Method for Incompressible Viscous Flows Using a Finite Increment Calculus Formulation, *Computer Methods in Applied Mechanics and Engineering* **182**, 355–370.

Perić, D., Hochard, C., Dutko, M., and Owen, D. R. J. (1996) Transfer Operator for Evolving Meshes in Small Strain Plasticity, *Computer Methods in Applied Mechanics and Engineering* **137**, 331–334.

Perić, D., Jr, M. V., and Owen, D. R. J. (1999) On Adaptive Strategies for Large Deformations of Elasto-Plastic Solids at Finite Strains: Computational Issues and Industrial Applications, *Computer Methods in Applied Mechanics and Engineering* **176**, 279–312.

Ramaswamy, B. (1990) Numerical Simulation of Unsteady Viscous Free Surface Flow, *Journal of Computational Physics* **90**, 396–430.

Robertson, I., Sherwin, S. J., and Bearman, P. W. (2003) A Numerical Study of Rotational and Transverse Galloping Rectangular Bodies, *Journal of Fluids and Structures* **17**, 681–699.

Rugonyi, S. and Bathe, K. J. (2001) On Finite Element Analysis of Fluid Flows Fully Coupled with Structural Interactions, *Computer Modelling in Engineering and Science* **2**, 195–212.

Tezduyar, T. E., Behr, M., Mittal, S., and Johnson, A. A. (1992)a Computational Unsteady Incompressible flows with the Stabilized Finite Element Methods – Space-Time Formulations, Iterative Strategies and Massively Parallel Implementations, In *New Methods in Transient Analysis*, Vol. 143 of *AMD*, New York, pp. 7–24.

Tezduyar, T. E., Mittal, S., Ray, S. E., and Shih, R. (1992)b Incompressible Flow Computations with Stabilized Bilinear and Linear Equal-Order-Interpolation Velocity-Pressure Elements, *Computer Methods in Applied Mechanics and Engineering* **95**, 221–242.

Tezduyar, T. E. and Osawa, Y. (2000) Finite Element Stabilization Parameters Computed from Element Matrices and Vectors, *Computer Methods in Applied Mechanics and Engineering* **190**, 411–430.

Wall, W. A. (1999) Fluid-Struktur-Interaktion mit Stabilisisierten Finiten Elementen, Ph.D. thesis, Universität Stuttgart.

Wall, W. A. and Ramm, E. (1998) Fluid-Structure Interaction Based upon a Stabilized (ALE) Finite Element Method, In *Proceedings of the 4th World Congress on Computational Mechanics*, Buenos Aires, Argentina.

Part V: EARTHQUAKE INDUCED EXTREME LOADING CONDITIONS

SEISMIC ASSESSMENT OF STRUCTURES BY A PRACTICE-ORIENTED METHOD

*P. Fajfar *(pfajfar@ikpir.fgg.uni-lj.si)*
University of Ljubljana, Faculty of Civil and Geodetic Engineering, Ljubljana, SLOVENIA

Abstract. A relatively simple seismic analysis technique based on the pushover analysis of a multi-degree-of-freedom model and the response spectrum analysis of an equivalent single-degree-of-freedom system, called the N2 method, has been developed at the University of Ljubljana and implemented in the European standard Eurocode 8. The method is formulated in the acceleration –displacement format, which enables the visual interpretation of the procedure and of the relations between the basic quantities controlling the seismic response. Its basic variant was restricted to planar structures. Recently the applicability of the method has been extended to plan-asymmetric buildings, which require a 3D structural model. In the paper, the N2 method is summarized and applied to two test examples.

Key words*:* non-linear seismic analysis; pushover analysis; simplified non-linear method; inelastic spectrum; N2 method

1. Introduction

Seismic assessment of structures requires determination of seismic demand imposed to a structure by earthquake ground motion and of seismic capacity of the structure to resist forces and deformations. Seismic demand is determined with a structural analysis, which is a complex task because (a) the problem is dynamic and usually non-linear, (b) the structural system is usually complex, and (c) input data (structural properties and ground motions) are random and uncertain. In principle, the non-linear time-history analysis is the correct approach. However, such an approach, for the time being, is not practical for everyday design use. It requires additional input data (time-histories of ground motions and detailed hysteretic behaviour of structural members) which cannot

A. Ibrahimbegovic and I. Kozar (eds.),Extreme Man-Made and Natural Hazards in Dynamics of Structures, 257–284.

be reliably predicted. Non-linear dynamic analysis is, at present, appropriate for research and for design of important structures. It represents a long-term trend. On the other hand, the methods applied in the great majority of existing building codes are based on the assumption of linear elastic structural behaviour and do not provide information about real strength, ductility and energy dissipation. They also fail to predict expected damage in quantitative terms. For the time being, the most rational analysis and performance evaluation methods for practical applications, which should be used in addition to the established elastic procedures, seem to be simplified inelastic procedures, which combine the non-linear static (pushover) analysis of a relatively simple multi degree-of-freedom (MDOF) mathematical model and the response spectrum analysis of an equivalent single-degree-of-freedom (SDOF) model. They can be used for a variety of purposes such as design verification for new buildings and bridges, damage assessment for existing structures, determination of basic structural characteristics in direct displacement based design, and rapid evaluation of global structural response to seismic ground motion of different intensities. In recent years, a breakthrough of these procedures has been observed. They have been implemented into the modern guidelines and codes.

The paper discusses one of the simplified non-linear approaches, called the N2 method, developed at the University of Ljubljana. It has been implemented into the Eurocode 8 standard (Annex B of Part 1) (CEN, 2004a).

The development of the N2 method started in the mid-1980s (Fajfar and Fischinger, 1987; Fajfar and Fischinger 1989). The basic idea came from the Q-model developed by Saiidi and Sozen (1981). The method has been gradually developed into a more mature version (Fajfar and Gašperši, 1996). The applicability of the method has been extended to bridges (Fajfar et al., 1997). Later on, following Bertero's (Bertero, 1995) and Reinhorn's idea (Reinhorn, 1997), the N2 method has been formulated in the acceleration – displacement format (Fajfar, 1999; Fajfar, 2000). This version combines the advantages of the visual representation of the capacity spectrum method, developed by Freeman (Freeman et al., 1975; Freeman, 1998), with the sound physical basis of inelastic demand spectra. More recent research has been aimed to extending the applicability of the method to plan-asymmetric buildings (discussed in next paragraphs) and to infilled reinforced concrete frames (Dolšek and Fajfar, 2005). Additional studies have been made also in applications to bridges (Isakovi et al., 2003). Moreover, the incremental N2 (IN2) method has been developed (Dolšek and Fajfar, 2004) as a simple alternative to incremental dynamic analysis (IDA), used in probabilistic seismic assessment.

Originally, the N2 method was limited to planar structural models like all simplified non-linear methods. Only recently, attempts have been made to extend the applicability of simplified methods to asymmetric structures, which require a 3D analysis, e.g. (Kilar and Fajfar, 1997, 2001; Moghadam and Tso, 2001; Ayala and Tavera, 2002; Aydinoglu, 2003; Chopra and Goel, 2004; Fujii et al., 2004).

The extension of the applicability of the N2 method to plan-asymmetric structures is not straightforward. In the N2 method, seismic demand is determined from inelastic spectra and depends on the period of the idealized equivalent SDOF system. The transformation from the MDOF to an equivalent SDOF system is based on the assumption of a time-invariant displacement shape. This assumption represents the major limitation of the applicability of the method. It works well in the case of planar structural models with small influence of higher modes. In the case of asymmetric building structures, represented by a 3D structural model, several modes may substantially contribute to the response and the torsional effects may not be properly taken into account by a straightforward extension of the N2 method to 3D models, used in some earlier publications by the authors (Fajfar, 2002; Fajfar et al., 2002). The results of recent parametric studies suggest that in the majority of cases an upper limit for torsional effects can be estimated by a linear dynamic (spectral) analysis. Based on this observation, it has been proposed that the results obtained by pushover analysis of a 3D structural model be combined with the results of a linear dynamic (spectral) analysis (Fajfar et al., 2005). The former results control the target displacements and the distribution of deformations along the height of the building, whereas the latter results define the torsional amplifications. Any favourable torsional effect on the stiff side, *i.e.* any reduction of displacements compared to the counterpart symmetric building, which may arise from elastic analysis, will probably decrease or may even disappear in the inelastic range, and is thus not taken into account in analysis and design.

In the paper, the extended version of the N2 method is summarized and applied to two test examples. The first test example is a planar four-storey reinforced concrete (RC) frame building and the second test example is an asymmetric three-storey RC frame building (»SPEAR« building, pseudo-dynamically tested in full-scale in ELSA). The results are compared with results of nonlinear dynamic time-history analyses.

2. Description of the N2 Method

In this chapter, the steps of the simple version of the N2 method, extended to asymmetric structures, are described. A simple version of the spectrum for the reduction factor is applied. It should be noted, however, that the suggested procedures used in particular steps of the method can be easily replaced by other available procedures. Additional information on the N2 method can be found in (Fajfar, 2000) (planar version) and (Fajfar, 2002) (extended version).

STEP 1: DATA

A 3-D model of the building structure is used. The floor diaphragms are assumed to be rigid in the horizontal plane. The number of degrees of freedom is three times the number of storeys N. The degrees of freedom are grouped in three sub-vectors, representing displacements at the storey levels in the horizontal directions x and y, and torsional rotations $\mathbf{U}^T = [\mathbf{U}_x^T, \mathbf{U}_y^T, \mathbf{U}_z^T]$.

In addition to the data needed for the usual elastic analysis, the non-linear force–deformation relationships for structural elements under monotonic loading are also required. The most common element model is the beam element with concentrated plasticity at both ends. A bilinear or trilinear moment–rotation relationship is usually used.

Seismic demand is traditionally defined in the form of an elastic (pseudo)-acceleration spectrum S_{ae} ("pseudo" will be omitted in the following text), in which spectral accelerations are given as a function of the natural period of the structure T. In principle, any spectrum can be used. However, the most convenient is a spectrum of the Newmark-Hall type. The specified damping coefficient is taken into account in the spectrum.

STEP 2: SEISMIC DEMAND IN AD FORMAT

Starting from the usual acceleration spectrum (acceleration versus period), inelastic spectra in acceleration – displacement (AD) format can be determined. For an elastic SDOF system, the following relation applies

$$S_{de} = \frac{T^2}{4\pi^2} S_{ae} \tag{1}$$

where S_{ae} and S_{de} are the values in the elastic acceleration and displacement spectrum, respectively, corresponding to the period T and a fixed viscous damping ratio.

For an inelastic SDOF system with a bilinear force – deformation relationship, the acceleration spectrum (S_a) and the displacement spectrum (S_d) can be determined as

$$S_a = \frac{S_{ae}}{R_\mu} \tag{2}$$

$$S_d = \frac{\mu}{R_\mu} S_{de} = \frac{\mu}{R_\mu} \frac{T^2}{4\pi^2} S_{ae} = \mu \frac{T^2}{4\pi^2} S_a \tag{3}$$

where μ is the ductility factor defined as the ratio between the maximum displacement and the yield displacement, and R_μ is the reduction factor due to ductility, i.e., due to the hysteretic energy dissipation of ductile structures. Note that R_μ is not equivalent to the reduction factor R used in seismic codes. The code reduction factor R, which is in Eurocode 8 called behaviour factor q, takes into account both energy dissipation and the so-called overstrength R_s. It can be defined as $R = R_\mu R_s$.

Several proposals have been made for the reduction factor R_μ. In the simple version of the N2 method, we will make use of a bilinear spectrum for the reduction factor R_μ

$$R_\mu = (\mu - 1)\frac{T}{T_C} + 1 \qquad T < T_C \tag{4}$$

$$R_\mu = \mu \qquad T \geq T_C \tag{5}$$

where T_C is the characteristic period of the ground motion. It is typically (e.g. in Eurocode 8) defined as the transition period where the constant acceleration segment of the response spectrum (the short-period range) passes to the constant velocity segment of the spectrum (the medium-period range). Eqs. (3) and (5) suggest that, in the medium – and long-period ranges, the equal displacement rule applies, i.e., the displacement of the inelastic system is equal to the displacement of the corresponding elastic system with the same period.

Starting from the elastic design spectrum, and using Eqs. (3) – (5), the demand spectra for the constant ductility factors μ in AD format can be obtained. They represent inelastic demand spectra. It should be noted that the construction of these spectra is in fact not needed in the computational procedure. They just help for the visualisation of the procedure.

STEP 3: PUSHOVER ANALYSIS

Using a pushover analysis, a characteristic non-linear force –displacement relationship of the MDOF system can be determined. In principle, any force and displacement can be chosen. Usually, base shear and roof (top) displacement are used as representative of force and displacement, respectively. The selection of an appropriate lateral load distribution is an important step within the pushover analysis. A unique solution does not exist. Fortunately, the range of reasonable assumptions is usually relatively narrow and, within this range, different assumptions produce similar results. One practical possibility is to use two different displacement shapes (load patterns) and to envelope the results.

Lateral loads are applied in mass centres of different storeys. The vector of the lateral loads **P**, which generally consists of components in three directions (forces in the x and y direction and torsional moments) is determined as

$$\mathbf{P} = p\,\mathbf{\Psi} = p\,\mathbf{M}\,\mathbf{\Phi} \tag{6}$$

where **M** is the mass matrix. The magnitude of the lateral loads is controlled by p. The distribution of lateral loads $\mathbf{\Psi}$ is related to the assumed displacement shape $\mathbf{\Phi}$. (Note that the displacement shape $\mathbf{\Phi}$ is needed only for the transformation from the MDOF to the equivalent SDOF system in Step 4). Consequently, the assumed load and displacement shapes are not mutually independent as in the majority of other pushover analysis approaches. The procedure can start either by assuming displacement shape $\mathbf{\Phi}$ and determining lateral load distribution $\mathbf{\Psi}$ according to Eq. (6), or by assuming lateral load distribution $\mathbf{\Psi}$ and determining displacement shape $\mathbf{\Phi}$ from Eq. (6). Note that Eq. (6) does not present any restriction regarding the distribution of lateral loads.

Generally, $\mathbf{\Phi}$ can consist of non-zero components in three directions (two horizontal directions and of torsional rotation). In such a case (coupled displacement shape) lateral loads also consist of components in three directions. The procedure can be substantially simplified if lateral loads are applied in one direction only. This is a special case that requires that also the assumed displacement shape has non-zero components in one direction only, e.g.

$$\mathbf{\Phi}^T = [\mathbf{\Phi}_x^T, \mathbf{0}^T, \mathbf{0}^T] \tag{7}$$

This special case is used in the proposed extended version of the N2 method. It should be noted, however, that even in this special case of uncoupled assumed displacement shape, the resulting displacements, determined by a pushover analysis of an asymmetric structure, will be coupled, i.e. they will have components in three directions.

From Eqs. (6) – (7) it follows that the lateral force in the x-direction at the i-th level is proportional to the component $\Phi_{x,i}$ of the assumed displacement shape $\mathbf{\Phi}_x$, weighted by the storey mass m_i

$$P_{x,i} = p\, m_i\, \Phi_{x,i} \tag{8}$$

Such a relation has a physical background: if the assumed displacement shape was equal to the mode shape and constant during ground shaking, i.e. if the structural behaviour was elastic, then the distribution of lateral forces would be equal to the distribution of effective earthquake forces and Eq. (6) was "exact". In inelastic range, the displacement shape changes with time and Eq. (6)

represents an approximation. Nevertheless, by assuming related lateral forces and displacements according to Eq. (6), the transformation from the MDOF to the equivalent SDOF system and vice-versa (Steps 4 and 6) follows from simple mathematics not only in elastic but also in inelastic range. No additional approximations are required, as in the case of some other simplified procedures.

In the proposed method, lateral loading, determined according to Eqs. (6) – (7), is applied independently in two horizontal directions, in each direction with + and –sign.

STEP 4: EQUIVALENT SDOF MODEL AND CAPACITY CURVE

In the N2 method, seismic demand is determined by using response spectra. Inelastic behaviour is taken into account explicitly. Consequently, the structure should, in principle, be modelled as a SDOF system. Different procedures have been used to determine the characteristics of an equivalent SDOF system. One of them, used in the current version of the N2 method, is summarized below.

The starting point is the equation of motion of a 3D structural model (with 3N degrees of freedom) representing a multi-storey building (damping is not taken into account because it will be included in the spectrum)

$$\mathbf{M}\,\ddot{\mathbf{U}} + \mathbf{R} = -\mathbf{M}\,\mathbf{s}\;a \tag{9}$$

R is a vector representing internal forces, a is the ground acceleration as a function of time, and **s** is a vector defining the direction of ground motion. In the case of uni-directional ground motion, e.g. in the direction x, the vector **s** consists of one unit sub-vector and of two sub-vectors equal to 0.

$$\mathbf{s}^{T} = [\mathbf{1}^{T}, \mathbf{0}^{T}, \mathbf{0}^{T}] \tag{10}$$

In the N2 method, ground motion is applied independently in two horizontal directions. Consequently, two separate analyses have to be performed with two different **s** vectors (vector (10) and a similar vector that corresponds to the ground excitation in the y-direction). A derivation, presented in (Fajfar, 2002) yields the following formulas.

The displacement and force of the equivalent SDOF system D^* and F^* are defined as

$$D^* = \frac{D_t}{\Gamma}, \quad F^* = \frac{V}{\Gamma} \qquad (11), (12)$$

where D_t is the top displacement of the MDOF system and

$$V = p\,\mathbf{\Phi}^{T}\mathbf{M}\,\mathbf{s} = pm^* \tag{13}$$

is the base shear of the MDOF model in the direction of ground motion. m^* is the equivalent mass of the SDOF system

$$m^* = \mathbf{\Phi}^T \mathbf{M} \mathbf{s} \tag{14}$$

The constant controls the transformation from the MDOF to the SDOF model and vice–versa. It is defined as

$$\Gamma = \frac{\mathbf{\Phi}^T \mathbf{M} \mathbf{s}}{\mathbf{\Phi}^T \mathbf{M} \mathbf{\Phi}} = \frac{m^*}{L^*} \tag{15}$$

Note that m^* depends on the direction of ground motion. Consequently, , D^*, and F^* also depend on the direction of ground motion. In the case of ground motion in one (x) direction (Eq. 10) and assuming a simple uncoupled displacement shape (Eq. 7), the following equations apply

$$m_x^* = \sum m_i \Phi_{x,i} \tag{16}$$

$$V_x = \sum p m_i \Phi_{x,i} = \sum P_{x,i} \tag{17}$$

$$\Gamma = \frac{\sum m_i \cdot \Phi_{x,i}}{\sum m_i \cdot \Phi_{x,i}^2} \tag{18}$$

Equation (18) is the same equation as in the case of planar structures. Consequently, the transformation from the MDOF to the SDOF system and vice versa is exactly the same as in the case of a planar structure.

$\wp$ is usually called the modal participation factor. Note that the assumed displacement shape $\mathbf{\Phi}$ is normalized – the value at the top is equal to 1. Note also that any reasonable shape can be used for $\mathbf{\Phi}$. As a special case, the elastic first mode shape can be assumed.

The same constant $\wp$ applies for the transformation of both displacements and forces (Eqs. 11 –12). As a consequence, the force –displacement relationship determined for the MDOF system (the $V - D_t$ diagram) applies also to the equivalent SDOF system (the $F^* - D^*$ diagram), provided that both force and displacement are divided by $\wp$.

In order to determine a simplified (elastic – perfectly plastic) force – displacement relationship for the equivalent SDOF system, engineering judgement has to be used. In regulatory documents some guidelines may be given. In Annex B of Eurocode 8 (CEN, 2004a) the bilinear idealization is based on the equal energy principle. Note that the displacement demand depends on the equivalent stiffness which, in the case of the equal energy approach, depends on the target displacement. In principle, an iterative approach is needed, in which a target displacement is assumed, the bilinear idealization is made and the target displacement is determined. This value is then used then as the new approximation for target displacement. According to Eurocode 8, the displacement at the formation of plastic mechanism can be used as the initial approximation for target displacement. Iteration is allowed but not required.

For the graphical procedure (visualization), used in the simple N2 method, it is convenient that the post-yield stiffness is equal to zero. This option is used in EC8, also because it yields conservative results (larger target displacement than other candidate idealizations). However, idealizations with a small to moderate post-yield stiffness can be also used. This is because the reduction factor R_μ is defined as the ratio of the required elastic strength to the yield strength and the post-yield stiffness does not influence R_μ. The influence of a moderate post-yield stiffness is incorporated in the demand spectra. It should be emphasized that a moderate post-yield stiffness does not have a significant influence on displacement demand, and that the proposed spectra approximately apply for systems with zero or moderate (up to 10%) post-yield stiffness. However, the idealization with post-yield stiffness, which is based on the equal energy rule, increases the initial (elastic) stiffness of the idealized system and yields thus a smaller target displacement than the idealization with zero post-yield stiffness.

The elastic period of the idealized bilinear system T^* can be determined as

$$T^* = 2\pi \sqrt{\frac{m^* D_y^*}{F_y^*}} \tag{19}$$

where F_y^* and D_y^* are the yield strength and displacement, respectively.

Note that, alternatively, first the bilinear idealization of the pushover curve can be made and then the transformation to the equivalent SDOF system can be made. The same equations apply.

Finally, the capacity diagram in AD format is obtained by dividing the forces in the force–deformation (F^* - D^*) diagram by the equivalent mass m^*

$$S_a = \frac{F^*}{m^*} \tag{20}$$

The procedure is applied for both horizontal directions, in each direction with + and – sign.

STEP 5: SEISMIC DEMAND FOR THE EQUIVALENT SDOF SYSTEM

The determination of the seismic demand for the equivalent SDOF system is illustrated in Fig. 1 (for medium – and long-period structures, for which the "equal displacement rule" applies; for short-period structures see e.g. (Fajfar, 2000). Both the demand spectra and the capacity diagram have been plotted in the same graph. The intersection of the radial line corresponding to the elastic period T^* of the idealized bilinear system with the elastic demand spectrum defines the acceleration demand (strength), required for elastic behaviour S_{ae}, and the corresponding elastic displacement demand S_{de}. S_{ay} represents both the acceleration demand and the capacity of the inelastic system expressed in terms of the yield acceleration. The reduction factor R_μ can be determined as the ratio between the accelerations corresponding to the elastic and inelastic systems

$$R_\mu = \frac{S_{ae}\left(T^*\right)}{S_{ay}} \tag{21}$$

Note that R_μ is not the same as the reduction (behaviour, response modification) factor R used in seismic codes. The code reduction factor R takes into account both energy dissipation and the so-called overstrength. The design acceleration S_{ad} is typically smaller than the yield acceleration S_{ay}.

If the elastic period T^* is larger than or equal to T_C, the inelastic displacement demand S_d is equal to the elastic displacement demand S_{de} (see Eqs. 3 and 5, and Fig. 1). From triangles in Fig. 1 it follows that the ductility demand, defined as $\mu = S_d / D_y^*$, is equal to R_μ

$$S_d = S_{de}(T^*) \quad T^* \geq T_C \tag{22}$$

$$\mu = R_\mu \tag{23}$$

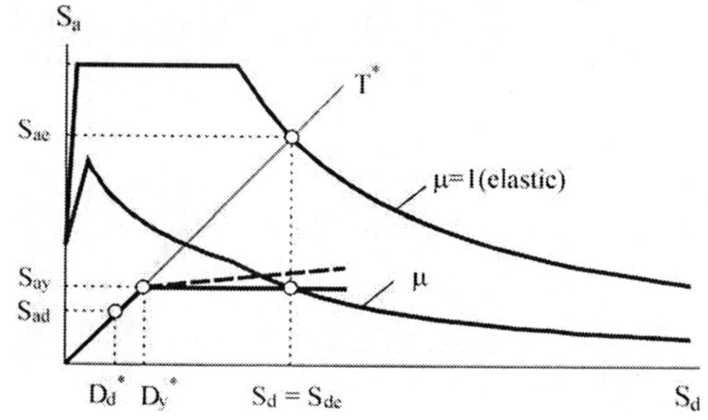

Figure 1. Elastic and inelastic demand spectra versus capacity diagram.

If the elastic period of the system is smaller than T_C, the ductility demand can be calculated from the rearranged Eq. (4)

$$\mu = \left(R_\mu - 1\right)\frac{T_C}{T^*} + 1 \qquad T^* < T_C \tag{24}$$

The displacement demand can be determined either from the definition of ductility or from Eqs. (3) and (24) as

$$S_d = \mu D_y^* = \frac{S_{de}}{R_\mu}\left(1 + \left(R_\mu - 1\right)\frac{T_C}{T^*}\right) \tag{25}$$

In both cases ($T^* < T_C$ and $T^* \geq T_C$) the inelastic demand in terms of accelerations and displacements corresponds to the intersection point of the capacity diagram with the demand spectrum corresponding to the ductility demand μ, provided that the post-yield stiffness in the capacity diagram is zero. At this point, the ductility factor determined from the capacity diagram and the ductility factor associated with the intersecting demand spectrum are equal. In the case of a post-yield stiffness different from zero, the intersection point is determined with the horizontal line through the yield acceleration rather than with the capacity diagram.

All steps in the procedure can be performed numerically without using the graph. However, visualization of the procedure may help in better understanding the relations between the basic quantities. Two additional quantities are shown in Fig. 1. S_{ad} represents a typical design strength, i.e. strength required by codes for ductile structures, and D_d^* is the corresponding displacement obtained by linear analysis.

The procedure is applied in two horizontal directions, in each direction with + and − sign. Usually, the results obtained for both signs are similar. In such a case, the larger value of two values, obtained for + and – sign, can used as the target displacement (displacement demand at CM) in each horizontal direction. Alternatively, the complete analysis can be performed for both signs and the envelopes of all relevant quantities can be taken as the end result.

STEP 6: GLOBAL SEISMIC DEMAND FOR THE MDOF MODEL

The displacement demand for the SDOF model S_d is transformed into the maximum top displacement D_t of the MDOF system (target displacement) by using Eq. (11).

STEP 7: DETERMINATION OF TORSIONAL EFFECTS

Torsional effects are determined by a linear modal analysis of the 3D mathematical model, independently for excitation in two horizontal directions and combining the results according to the SRSS rule.

STEP 8: LOCAL SEISMIC DEMAND FOR THE MDOF MODEL

Under monotonically increasing lateral loads with a fixed pattern (as in Step 3), the structure is pushed to D_t. It is assumed that the distribution of deformations throughout the height of the structure in the static (pushover) analysis approximately corresponds to that which would be obtained in the dynamic analyses. Separate 3D pushover analyses are performed in two horizontal directions.

The correction factors to be applied to the relevant results of pushover analyses are determined. The correction factor is defined as the ratio between the normalized roof displacements obtained by elastic modal analysis and by pushover analysis. The normalized roof displacement is the roof displacement at an arbitrary location divided by the roof displacement at the CM. If the normalized roof displacement obtained by elastic modal analysis is smaller than 1.0, the value 1.0 is used, i.e. no de-amplification due to torsion is taken into account. Correction factors are defined for each horizontal direction separately. Note that the correction factor depends on the location in the plan. All relevant quantities obtained by pushover analyses are multiplied with appropriate correction factors. For example, in a perimeter frame parallel to the X-axis, all quantities are multiplied with the correction factor determined with pushover results obtained for loading in the X-direction and for the location of this frame. The relevant quantities are, for example, deformations for the ductile elements, which are expected to yield, and the stresses for brittle elements, which are expected to remain in the elastic range.

STEP 9: PERFORMANCE EVALUATION (DAMAGE ANALYSIS)

Expected performance can be assessed by comparing the seismic demands, determined in Step 8, with the capacities for the relevant performance level (limit state). Global performance can be visualized by comparing displacement capacity and demand. The determination of seismic capacity for different performance levels is out of scope of this paper. Indicative values for seismic capacities of different structural elements (columns, walls, beams) are provided e.g. in FEMA 356 (FEMA, 2000). Some empirical formulas are given in EC8-3 (CEN, 2004b). A promising tool for determination of seismic capacity seems to be a non-parametric neural network-like approach which has been used by Peruš at al. (2005) for the determination of drift capacity of rectangular RC columns. For such an approach an appropriate data base of experimental results is needed.

3. Approximations and Limitations

The N2 method is, like any approximate method, subject to several limitations.

In planar analysis, there are two main sources of approximations and corresponding limitations: pushover analysis and determination of target displacement.

Non-linear static (pushover) analysis can provide an insight into the structural aspects which control performance during severe earthquakes. The analysis provides data on the strength and ductility of the structure, which cannot be obtained by elastic analysis. Furthermore, it exposes design weaknesses that may remain hidden in an elastic analysis. On the other hand, the limitations of the approach should be recognized. Pushover analysis is based on a very restrictive assumption, i.e. a time-independent displacement shape. Thus, it is in principle inaccurate for structures where higher mode effects are significant, and it may not detect the structural weaknesses, which may be generated when the structure's dynamic characteristics change after the formation of the first local plastic mechanism. A detailed discussion of pushover analysis can be found in papers by Krawinkler and Seneviratna (1998) and Elnashai (2001). Additional discussion on the relationship between MDOF and SDOF systems is presented in (Gupta and Krawinkler, 2000).

One practical possibility to partly overcome the limitations imposed by pushover analysis is to assume two different displacement shapes (load patterns), and to envelope the results. Another more complex possibility is to use lateral load distribution which changes in each step of analysis, i.e. the Adaptive pushover analysis (Elnashai, 2001). A popular approach is the Modal pushover analysis (MPA) developed by Chopra and Goel (2002). In MPA, the influence of higher modes is approximately taken into account in a similar way as in elastic analysis.

The other important source of inaccuracy is the determination of target displacement (displacement demand) for the equivalent SDOF system. Displacement demand is obtained from inelastic spectrum as a function of the initial period of the equivalent SDOF system, which is not uniquely defined. It depends on the bilinear idealization of the actual base shear – top displacement curve and is, to some extent, based on engineering judgement.

Inelastic spectra are based on statistical analyses of structural models and may not apply for structures, whose inelastic behaviour is basically different from that assumed in statistical analyses. The simplest possibility for inelastic spectra is to apply, in the medium–and long-period range, the "equal displacement rule." The equal displacement rule has been used quite successfully for almost 40 years. Many statistical studies (see, e.g., the discussion in (Fajfar, 2000)) have confirmed that the equal displacement rule is a viable approach for structures on firm sites with the fundamental period in the medium – or long-period range, with relatively stable and full hysteretic loops. In many cases a conservative estimate of the mean value of the inelastic displacement may be obtained (Cuesta el al., 2003). The equal displacement rule, however, may yield too small inelastic displacements in the case of near-fault ground motions, hysteretic loops with significant pinching or significant stiffness and/or strength deterioration, and for systems with low strength (i.e., with a yield strength to required elastic strength ratio of less than 0.2). Moreover, the equal displacement rule may be not satisfactory for soft soil conditions. In these cases, modified inelastic spectra should be used. Alternatively, correction factors for displacement demand (if available) may be applied.

In the case of short-period structures, inelastic displacements are larger than the elastic ones. The transition period, below which the inelastic to elastic displacement ratio begins to increase, is roughly equal to the characteristic period of the ground motion T_C. In the short-period range, the sensitivity of inelastic displacements to changes of structural parameters is greater than in the medium–and long period ranges. Consequently, estimates of inelastic displacement are less accurate in the short-period range. However, the absolute values of displacements in the short-period region are small and, typically, they do not control the design.

Note that for the methods, based on inelastic spectra, any realistic elastic and corresponding (compatible) inelastic spectrum can be applied. For example, for a specific acceleration time-history, the elastic acceleration spectrum as well as the inelastic spectra, which take into account specific hysteretic behavior, can be computed and used as demand spectra. Moreover, any reasonable R_μ spectrum, compatible with the elastic spectrum, can be used. (Note that elastic spectra for specific accelerograms and smooth R_μ spectra are not compatible.) Examples are presented in (Chopra and Goel, 1999; Reinhorn, 1997) .

The results of the N2 method are intended to represent mean values for the applied earthquake loading. There is a considerable scatter about the mean. Consequently, it is appropriate to investigate likely building performance under extreme load conditions that exceed the design values. This can be achieved by increasing the value of the target displacement.

For the extended version of the N2 method, which can be applied for analysis of 3D structural models, all approximations and limitations discussed above remain. Additionally, the amplification of demand due to torsion is approximated by elastic dynamic analysis, while reduction of demand due to torsion is not taken into account. Such an approach yields in most cases a conservative estimate of torsional influences. Note, however, that inelastic torsion is characterized by large inherent randomness and uncertainty.

The simplified methods usually do not take into account the effect of cumulative damage. This effect can easily be taken into account by using the so-called equivalent ductility factor (e.g. McCabe and Hall, 1989; Fajfar, 1992). The idea behind the equivalent ductility factor is to reduce the monotonic deformation capacity of an element and/or structure as a consequence of cumulative damage due to the dissipation of hysteretic energy (Fajfar and Gašperši , 1996). Alternatively, the influence of cumulative damage can be taken into account by increasing seismic demand (e.g. Cosenza and Manfredi, 1992; Chai et al., 1998).

4. Planar Test Example

As the first test example the response of a four-story reinforced concrete frame building (Fig. 2) subjected to three ground motions is analyzed. The full-scale building was tested pseudo-dynamically in the European Laboratory for Structural assessment (ELSA) of the Joint Research Centre of the European Commission in Ispra (Italy). The test results have been used for the validation of the mathematical model.

The building was designed according to European prestandard Eurocode 8 (CEN, 1994), as a high ductility structure for a peak ground acceleration of 0.3 g. The story masses from the bottom to the top amounted to 87, 86, 86, and 83 tons, and the resulting base shear coefficient amounted to 0.15. More detailed description of the structure and mathematical modeling can be found elsewhere (e.g., Fajfar and Drobni , 1998).

Our analysis will be repeated for three levels of ground motions, with the intention of checking different performance objectives. Ground motion is defined with the elastic acceleration response spectrum according to Fig. 3, which has been normalized to a peak ground acceleration a_g equal to 0.6 g, 0.3 g (the design value), and 0.15 g, respectively.

A linear displacement shape is assumed

$$\mathbf{\Phi}^T = [0.28, 0.52, 0.76, 1.00]$$

The lateral force pattern is obtained from Eq. 6 and normalized so that the force at the top is equal to 1.0

$$\mathbf{P}^T = [0.293, 0.539, 0.787, 1.000]$$

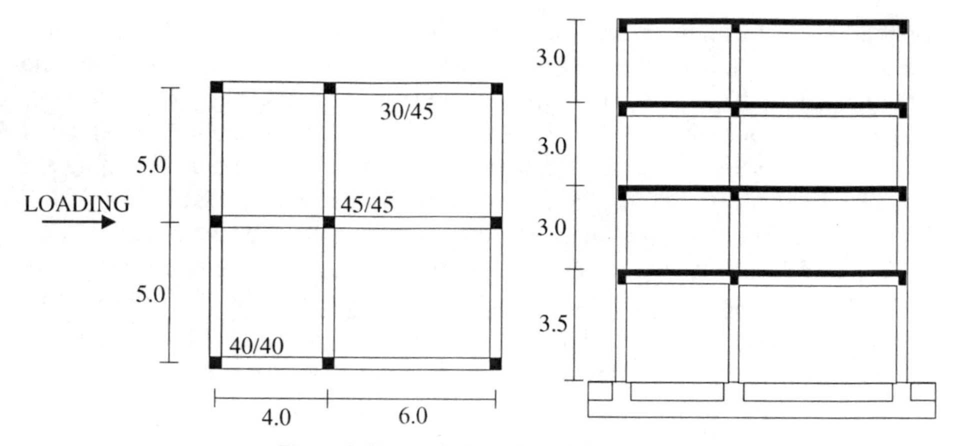

Figure 2. Plan and view of the RC building.

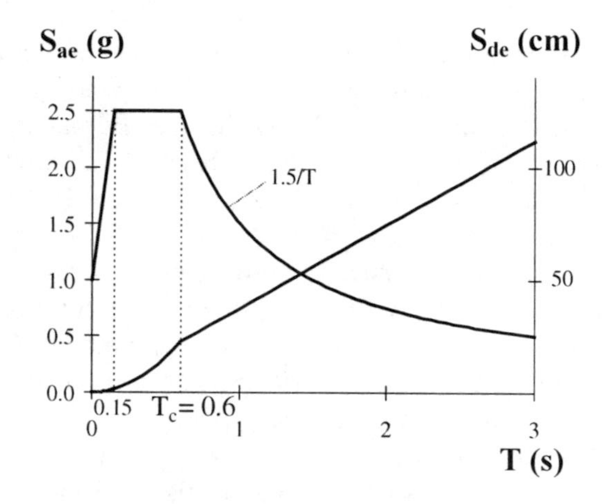

Figure 3. Elastic acceleration (S_{ae}) and displacement spectrum (S_{de}) for 5% damping normalized to 1.0 g peak ground acceleration (EC8 prestandard, soil type C).

With this force pattern, the DRAIN-2DX program (Prakash et al., 1993) yields the base shear V–top displacement D_t relationship shown in Fig. 4.

The MDOF system is transformed to an equivalent SDOF system using Eqs. (11) – (12). The equivalent mass amounts to $m^* = 217$ tons (Eqs. 14 and 16) and the transformation constant is $\wp = 1.34$ (Eqs. 15 and 18). In Fig. 4, the same curve defines both the $V - D_t$ relationship for the MDOF system, and the force F^*–displacement D^* relation for the equivalent SDOF system. The scale of the axes, however, is different for the MDOF and SDOF systems. The factor between the two scales is equal to $\wp$.

A bilinear idealization of the pushover curve is shown in Fig. 4. The yield strength and displacement amount to $F_y^* = 830$ kN and $D_y^* = 6.1$ cm. The elastic period is $T^* = 0.79$ s (Eq. 19).

The capacity diagram (Fig. 4) is obtained by dividing the forces F^* in the idealized pushover diagram by the equivalent mass (Eq. 20). The acceleration at the yield point amounts to $S_{ay} = F_y^* / m^* = 830/217 = 3.82$ m/s^2 = 0.39 g.

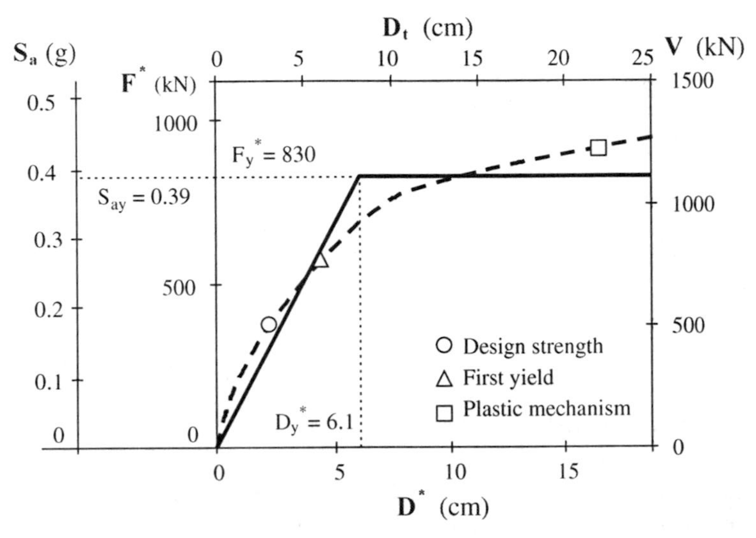

Figure 4. Pushover curve and the corresponding capacity diagram for the 4-story RC frame. Note the different scales. The top displacement D_t and the base shear V apply to MDOF system, whereas the force F^* and the displacement D^* apply to the equivalent SDOF system. The acceleration S_a belongs to the capacity diagram.

The capacity diagram and demand spectra are compared in Fig. 5. Eqs. (1) – (5) were used to obtain the inelastic demand spectra.

In the case of unlimited elastic behavior of the structure, seismic demand is represented by the intersection of the elastic demand spectrum and the line corresponding to the elastic period (T^* = 0.79 s) of the equivalent SDOF system. The values S_{ae} = 1.14 g and S_{de} = 17. 7 cm are obtained in the case of the strongest ground motion (a_g = 0.6 g). The reduction factor R_μ amounts to $R_\mu = S_{ae}/S_{ay}$ = 1.14 g/0.39 g = 2.9 (Eq. 21).

The period of the system T^* = 0.79 is larger than T_C = 0.6. Thus the equal displacement rule (Eqs. 22 and 23) applies: $\mu = R_\mu = 2.9$, $S_d = S_{de}$ = 17.7 cm.

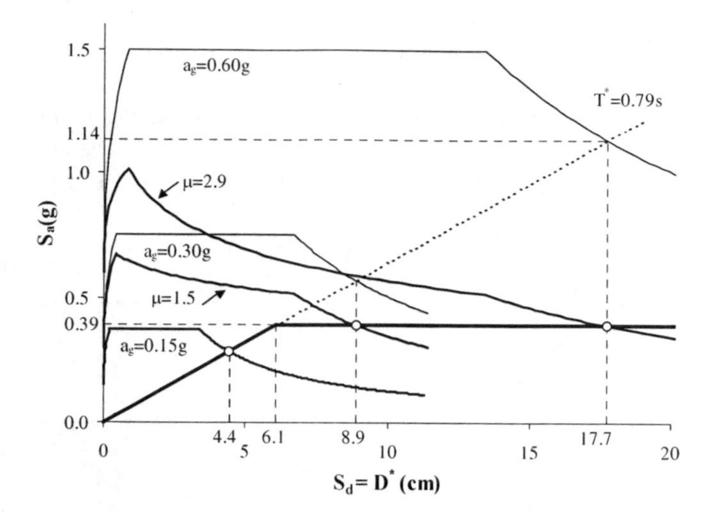

Figure 5. Demand spectra for three levels of ground motion and capacity diagram for the test example.

The seismic demand for the equivalent SDOF system is graphically represented by the intersection of the capacity curve and the demand spectrum for $\mu = 2.9$. Note, however, that the inelastic seismic demand can be determined without constructing the inelastic demand spectra.

In the next step the displacement demand of the equivalent SDOF system is transformed back to the top displacement of the MDOF system (Eq. 11): $D_t = 1.34 \cdot 17.7 = 23.7$ cm.

A pushover analysis of the MDOF model up to the top displacement D_t yields the displacement shape, local seismic demand in terms of story drifts, and joint rotations as shown in Fig. 6. Envelopes of results obtained by pushing from the left to the right and in the opposite direction are shown. The results are similar to those obtained from tests and from nonlinear dynamic analyses. A comparison for a slightly different case has been presented in (Fajfar et al., 1997). In that study, the peak ground acceleration amounted to 0.45 g and damping amounted to 1% in order to allow comparison with the results of the pseudo-dynamic tests. Nonlinear dynamic analyses were performed with eight accelerograms, which roughly corresponded to the design spectrum. Considerable sensitivity to the input ground motion was observed. The results of the N2 method were within the range of results obtained by time-history analyses, and fairly close to the test results.

The next steps include assessment of seismic capacities and performance evaluation. Discussion of these steps is out of scope of this paper.

In the case of a_g = 0.3 g, the same procedure yields $S_d = S_{de}$ = 8.9cm, μ = 1.5, and D_t = 11.9 cm. For a_g = 0.15 g, the following values are obtained: S_{de} = 4.4 cm and D_t = 5.9 cm. The idealized elasto-plastic structure remains in the elastic range. The original multi-linear pushover curve (Fig. 4) indicates that the displacement demand is approximately equal to the displacement at the first yield.

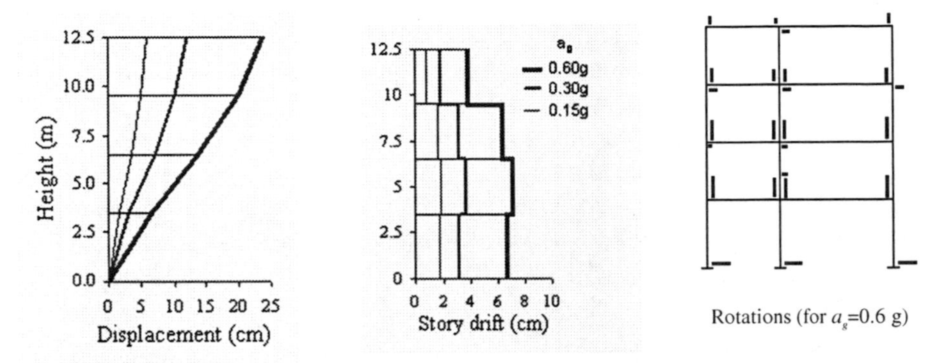

Figure 6. Displacement, story drifts, and rotations in the elements of the external frames. Rotations are proportional to the length of the mark. The maximum rotation amounts 2.2%. Only elements which yield are indicated.

5. 3D Test Example

The test structure represents a typical older three-storey reinforced concrete frame building (Fig. 7a). The storey heights amount to 3.0 meters. The structure was experimentally and numerically investigated in the SPEAR project. In the analyses, presented in this paper, a model developed before the tests (the "final pre-test model", the details of the model are presented elsewhere, e.g. at www.ikpir.com/projects/spear/), was used. The CANNY program (Li 2002) was employed. The mathematical model consists of beam elements. Flexural behaviour of beams was modelled by one-component lumped plasticity elements, composed of elastic beam and two inelastic rotational hinges. Rotational hinges were defined with the tri-linear moment-rotation envelope, which includes pre-crack, post-crack and post-yield parts, and Takeda's hysteretic rules (Cross-peak trilinear model CP3) in time-history analysis. The plastic hinge was used for the major-axis bending only. For flexural behaviour of columns also a one-component lumped plasticity model was used, with two independent plastic hinges for bending about the two principal axes. The eccentricities between the mass centres and approximate stiffness centres amount to about 10% and 14% in the X–and Y-directions, respectively. The total mass of the structure amounts to 195 tons. The three fundamental periods of vibration of the building (considering some inelastic deformations – cracks due to gravity load), amount to 0.63 s, 0.58 s, and 0.45 s. The first mode is predominantly in the X-direction, the second predominantly in the Y-direction, whereas the third mode is predominantly torsional.

In dynamic analyses, bi-directional semi-artificial ground motion records were used. The horizontal components of seven recorded ground motions were fitted to the shape of the EC8 elastic design spectrum (Type 1, soil C, Fig. 7b). The ground motions were scaled to peak ground acceleration $a_g = 0.3$. For each record eight different combinations of directions and signs of components were applied. In modal analysis, which provides results needed for the determination of the torsional influences in the N2 method, the same EC8 spectra were applied in both horizontal directions. Five percent damping was used in all analyses. In time-history analysis Reyleigh damping (with instantaneous stiffness matrix) was applied. The P- effect was not taken into account.

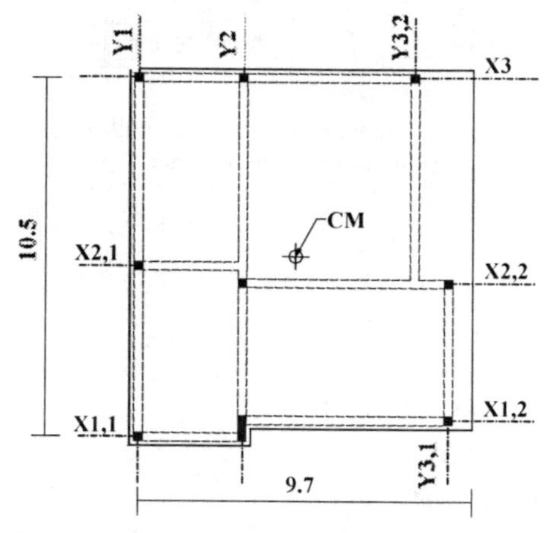

Figure 7a. Schematic plan of the SPEAR building.

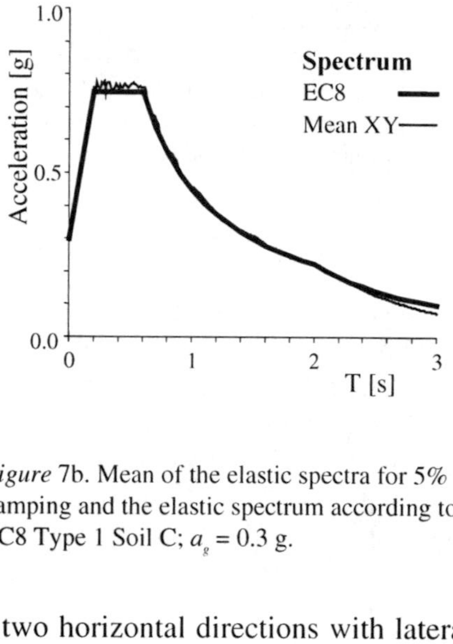

Figure 7b. Mean of the elastic spectra for 5% damping and the elastic spectrum according to EC8 Type 1 Soil C; $a_g = 0.3$ g.

Pushover analyses were performed in two horizontal directions with lateral loads based on the fundamental mode shapes in the relevant direction, i.e. x-components of the first mode shape were used in X-direction, and y-components of the second mode shape were used in Y-direction. Loading was applied with + and – sign. The P - effect was not taken into account. The results of pushover analyses are shown in Fig. 8 only for + sign, which yields in both directions larger target displacements. Two idealizations of the pushover curve were used, one with zero post-yield stiffness (denoted in figures as "bilin 1"), as required by EC8, and the other one with post-yield stiffness which closely follows the pushover curve denoted in figures as "bilin 2"). Iteration was used for determination of the idealization of pushover curves, as described in Step 4 of the procedure. (Note that in the test example the alternative with the idealization of the pushover curve for the MDOF system was used.) The idealized force – displacement relationships are plotted in Fig. 8. The effective periods amount to 1.14 s (1.2 s) and 0.99 s (0.93) in X–and Y–directions, respectively. The values in brackets correspond to the idealization with post-yield stiffness.

The capacity curves and the elastic and inelastic demand spectra are shown in Fig. 9 (only for loading with + sign). Larger displacement demands apply to the idealization with zero post-yield stiffness. For the equivalent SDOF system they amount to 12.8 and 11.1 cm in X–and Y–direction, respectively, whereas the corresponding top displacements of the MDOF system in CM amount to 15.8 and 14.2 cm. The displacement ductility demands (regarding the yield point of the idealized bilinear systems) amount to about 2.5. The corresponding values for the more natural idealization with post-yield stiffness amount to 11.5 and 10.4 cm for the SDOF system, and 14.2 and 13.3 for the MDOF system.

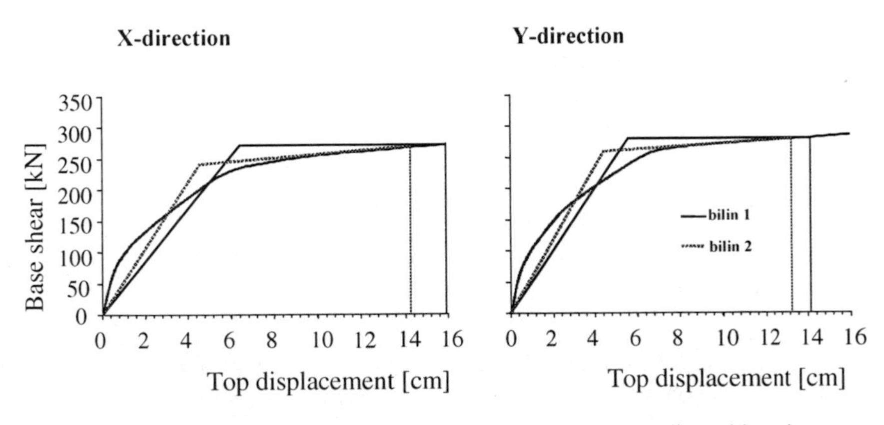

Figure 8. Pushover curves and two bilinear idealizations for loading with + sign.

Torsional effects in terms of normalized roof displacements determined by the proposed extension of the N2 method are presented in Fig. 10. The N2 results are compared with the results of elastic modal (spectral) analysis (of the essentially un-cracked structure), non-linear time-history analysis for a_g = 0.30 g, and pushover analysis.

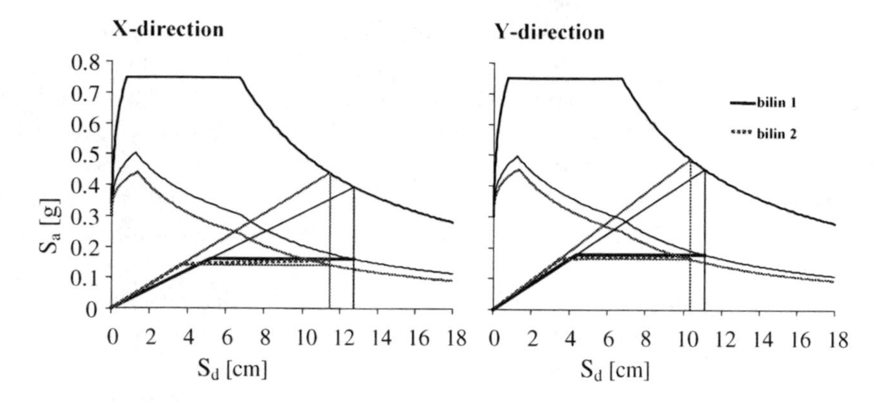

Figure 9. Elastic and inelastic demand spectra and capacity diagrams (for two idealizations and loading with + sign).

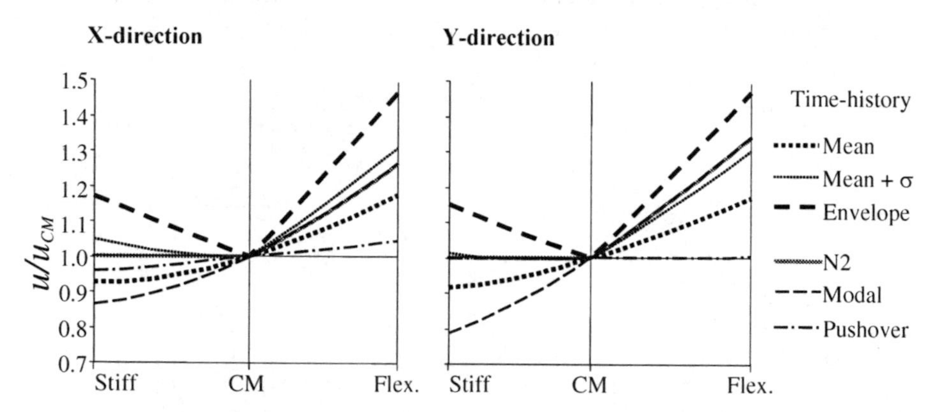

Figure 10. Torsional effects in terms of normalized top displacements obtained by the N2, modal, time-history (mean, mean + sigma values and envelope) and pushover analyses.

The static analysis suggested that some cracks (non-linear deformations) occurred already due to gravity loads. This state was assumed as the initial ("elastic") state of the building, and the modes of vibration of the building in such a condition were taken into account for the modal analysis. Modal analysis was performed independently for the loading in both horizontal directions, using the CQC rule for the combination of different modes, which is considered appropriate for structures with closely spaced modes. The results of analyses for both directions were combined by the SRSS rule.

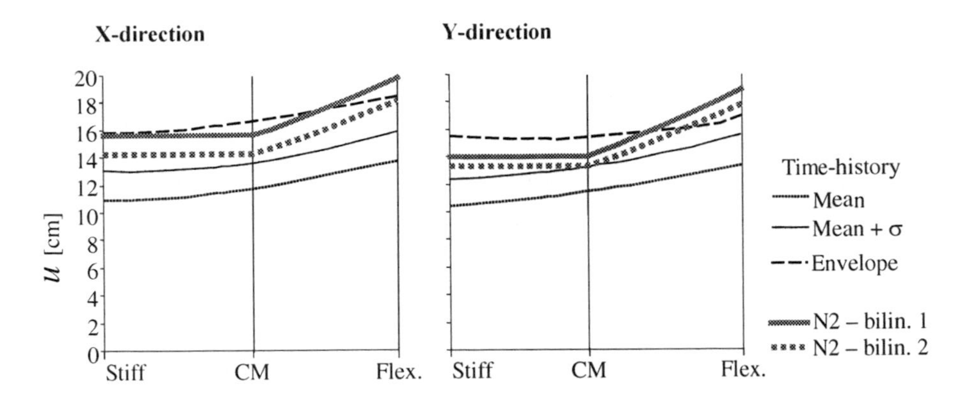

Figure 11 Displacement (in plane) at the top of the building obtained by N2 and time-history analyses.

According to the proposed extension of the N2 method, the results of elastic modal analysis are used to determine the torsional effect, provided that amplification due to torsion occurs. Consequently, the N2 results coincide with the line obtained by elastic modal analysis on the stiff side. No de-amplification due to torsion is allowed in the N2 method. So a constant value of 1.0 applies on the stiff side of the building. If compared with the mean results of nonlinear time-history analyses, the proposed N2 approach is conservative. The N2 results are close to mean + values. However, it should be noted that the torsional effects are in general higher if the ground motion intensity is lower (see (Fajfar et al., 2005). Moreover, some particular ground motions can produce very high torsional influences, as demonstrated by the envelope of results.

A pushover analysis with forces applied at the centre of masses at each floor at the same target displacement yields very small torsional rotations. According to the proposed extension of the N2 method, the results of pushover analysis are corrected by multiplying them by the ratio between the N2 normalized displacements and normalized displacements obtained by pushover analyses. The correction factors amount to 1.29, 1.22, 1.21 for columns and beams in the frames Y3,1, Y3,2, and X3 (Fig. 7a), respectively. For other frames the factors are small (from 1.00 to 1.05).

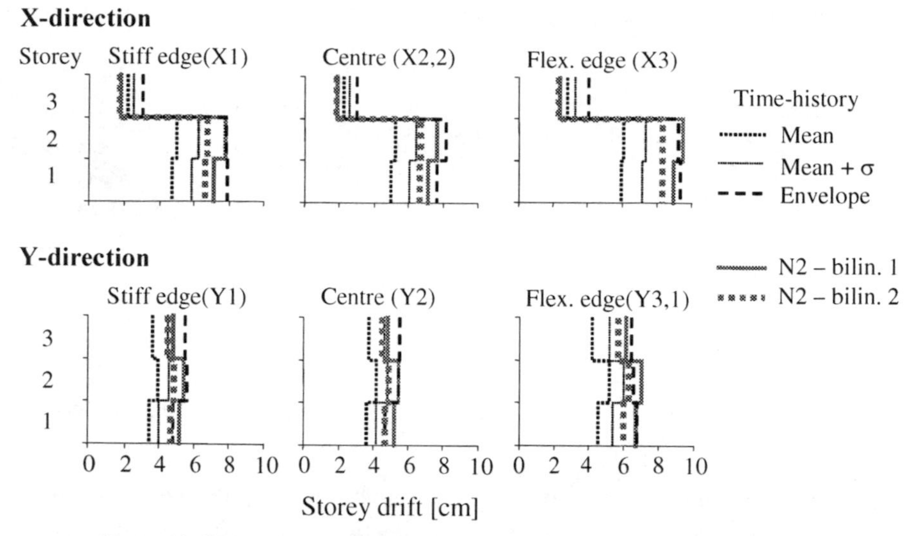

Figure 12. Storey drifts obtained by the N2 method and time-history analysis.

Absolute values of roof displacements are plotted in Fig. 11. In the case of the idealization with zero post-yield stiffness, the N2 displacements in CM are 34 % and 23 % larger than the mean values obtained by time-history analysis in X–and Y–directions, respectively, and are larger than the mean + values. Note, however, that the standard deviation of the sample of accelerograms is very small because all accelerograms are fitted to the same spectrum. In the case of recorded accelerograms, the coefficient of variation for displacements usually amounts to about 0.3. The idealization of the pushover curve with post-yield stiffness is less conservative.

Storey drifts in different frames are shown in Fig. 11. The distribution of drifts along the height of the building obtained by the N2 method is comparable with the distribution obtained by nonlinear dynamic analysis. The N2 drift estimates are conservative with the exception of the top storey in X-direction.

The seismic capacity of the structure is controlled by the capacity of columns in the bottom two stories. The ultimate capacity of columns in terms of chord rotations can be determined by empirical formulas provided in EC8-3 (CEN, 2004b). In our study an alternative approach, based on a neural network-like method (CAE method, Peruš et al., 2005), was used. In both cases, the ultimate values are intended to represent the chord rotation at a 20% drop of the strength, i.e. at the 80% of the maximum strength, what is supposed to represent the "near collapse (NC)" performance level. The most critical columns are the central column with the highest axial load and the corner column belonging to frames X3 and Y3 with the highest torsional amplifications. The ultimate chord rotations for the central column, determined with the CAE method, amount to 3.1% and 3.4% in the first and second storey, respectively, and the corresponding values for the corner column amount to 3.7% and 4.0%. Chord rotations in frames can be approximately compared with storey drifts, i.e. with values presented in Fig. 11, divided by the storey heights. In the case of the idealization with zero post-yield stiffness, the largest demand corresponds to the corner column in the bottom two storeys in X-direction. The story drift amounts to 8.9 cm or 3.0%. The corresponding demand to capacity ratio for the 1st storey amounts to 3.0/3.7 = 0.81, suggesting that this column, (and all other columns) are able to survive the ground motion defined by the spectrum in Fig. 7b.

In the case of the test structure analyzed in this example, a comparison with results of dynamic analyses suggests that the N2 results are conservative. The conservatism originates both from the determination of the target displacement at the mass centre and from the determination of torsional effects. Note that the accuracy of the estimated target displacement depends considerably on the bilinear idealization of the pushover curve, which controls the initial period of the idealized equivalent SDOF system.

6. Conclusions

Structural response to strong earthquake ground motion cannot be accurately predicted due to large uncertainties and the randomness of structural properties and ground motion parameters. Consequently, excessive sophistication in structural analysis is not warranted. The N2 method, like some other simplified non-linear methods, provides a tool for a rational yet practical evaluation procedure for building structures for multiple performance objectives. The formulation of the method in the acceleration – displacement format enables the visual interpretation of the procedure and of the relations between the basic quantities controlling the seismic response. This feature is attractive to designers.

Of course, the N2 method is, like any approximate method, subject to several limitations. In general, the results obtained using the N2 method are reasonably accurate, provided that the structure oscillates predominantly in the first mode. In the case of plan-asymmetric building structures, usually several modes substantially influence the structural response. An extended version of the N2 method, which can be used for analysis of such structures, has been developed. It combines the nonlinear static (pushover) and elastic dynamic analysis. Displacement demand (amplitude and the distribution along the height) at the mass centres is determined by the usual N2 method, which is based on pushover analysis. The amplification of demand due to torsion is determined by elastic dynamic analysis, while reduction of demand due to torsion is not taken into account.

Acknowledgement

The results presented in this paper are based on work continuously supported by the Ministry for Science and Technology of the Republic of Slovenia and, more recently, by the European Commission within the 5th and 6th Framework programs. This support is gratefully acknowledged. The author is indebted to Professor M. Fischinger for important contributions at the initial stage of development of the N2 method, to the past Ph.D. and M.Sc. students M. Dolšek, D. Drobni , P. Gašperši , V. Kilar, D. Maruši , I. Peruš, and T. Vidic, and to visiting researchers G. Magliulo (Univ.of Naples) and D. Zamfirescu (Technical Univ. of Civ.Eng., Bucharest). The results of their dedicated work are included in this paper. The editorial work was done by M. Kreslin.

References

Ayala, A. G., and Tavera, E. A., 2002, A new approach for the evaluation of the seismic performance of asymmetric buildings, *Proc. 7th Nat. Conf. on Earthquake Engineering*, EERI, Boston.

Aydinoglu, M. N., 2003, An incremental response spectrum analysis procedure based on inelastic spectral displacements for multi-mode seismic performance evaluation, *Bulletin of Earthquake Engineering* **1**(1): 3-36.

Bertero, V.V., 1995, Tri-service manual methods, in *Vision 2000*, Part 2, Appendix J, Structural Engineers Association of California, Sacramento, CA.

CEN, 1994, *Eurocode 8 – Design provisions for earthquake resistance of structures*, European prestandard ENV 1998, European Committee for Standardization, Brussels.

CEN, 2004a, *Eurocode 8 – Design of structures for earthquake resistance*, Part 1, European standard EN 1998-1, December 2004, European Committee for Standardization, Brussels.

CEN, 2004b, *Eurocode 8: Design of structures for earthquake resistance*, Part 3: Strengthening and repair of buildings (Stage 49), prEN 1998-3. May 2004, European Committee for Standardization, Brussels.

Chai, Y. H., Fajfar, P., and Romstad, K. M, 1998, Formulation of duration-dependent inelastic seismic design spectrum, *Journal of Structural Engineering, ASCE* **124**: 913-921.

Chopra, A. K., and Goel, R. K., 1999, Capacity-demand-diagram methods for estimating seismic deformation of inelastic structures: SDF systems, *Report PEER*-1999/02, Pacific Earthquake Engineering Research Center, University of California, Berkeley, CA.

Chopra, A. K., and Goel, R. K., 2002, A modal pushover analysis procedure for estimating seismic demands for buildings, *Earthquake Engineering and Structural Dynamics* **31**: 561-82.

Chopra, A. K., and Goel, R. K., 2004, A modal pushover analysis procedure to estimate seismic demands for unsymmetric-plan buildings, *Earthquake Engineering and Structural Dynamics* **33**(8): 903-927.

Cosenza, E., and Manfredi, G., 1992, Seismic analysis of degrading models by means of damage functions concept, in *Nonlinear analysis and design of reinforced concrete buildings*, P. Fajfar and H. Krawinkler, Eds., Elsevier Applied Science, London and New York, 77-93.

Cuesta, I., Aschheim, M. A., and Fajfar, P., 2003, Simplified R-factor relationships for strong ground motions. *Earthquake Spectra*, 2003, 19, 25-45.

Dolšek, M., and Fajfar, P., 2004, IN2 – A simple alternative for IDA, *Proc. 13th World Conf. Earthquake Engineering*, Paper No. 3353, Vancouver, Canada.

Dolšek, M., and Fajfar, P., 2005, Simplified non-linear seismic analysis of infilled reinforced concrete frames, *Earthquake Engineering and Structural Dynamics* **34**(1): 49-66.

Elnashai A.S., 2001, Advanced inelastic static (pushover) analysis for earthquake applications. *Structural engineering and mechanics* **12**(1): 51-69.

Fajfar, P., 1992, Equivalent ductility factors, taking into account low-cycle fatigue, *Earthquake Engineering and Structural Dynamics* **21**: 837-848.

Fajfar, P., 1999, Capacity spectrum method based on inelastic demand spectra, *Earthquake Engineering and Structural Dynamics* **28**: 979-993.

Fajfar, P., 2000, A nonlinear analysis method for performance-based seismic design, *Earthquake Spectra* **16**(3); 573-592.

Fajfar, P., 2002, Structural analysis in earthquake engineering –a breakthrough of simplified non-linear methods, *Proc. 12th European Conf. Earthquake Engineering*, London, UK, Keynote lecture.

Fajfar, P., and Drobni , D., 1998, Nonlinear seismic analysis of the ELSA buildings, *Proc. 11th European Conf. Earthquake Engineering*, Paris, CD-ROM, Balkema, Rotterdam.

Fajfar, P., and Fischinger, M., 1987, Non-linear seismic analysis of RC buildings: Implications of a case study, *European Earthquake Engineering* 1: 31-43.

Fajfar, P., and Fischinger, M., 1989, N2 – A method for non-linear seismic analysis of regular buildings, *Proc. 9th World Conf. Earthquake Engineering*, Tokyo, Kyoto, 1988, Vol.5, 111-116.

Fajfar, P., and Gašperši , P., 1996, The N2 method for the seismic damage analysis of RC buildings, *Earthquake Engineering and Structural Dynamics* **25**: 23-67.

Fajfar, P., Gašperši , P., and Drobni , D., 1997, A simplified nonlinear method for seismic damage analysis of structures, in *Seismic design methodologies for the next generation of codes*, P. Fajfar and H. Krawinkler, Eds., Balkema, Rotterdam, 183-194.

Fajfar, P., Kilar, V., Maruši , D., Peruš, I., and Magliulo, G., 2002, The extension of the N2 method to asymmetric buildings, Proc. 4th forum on Implications of recent earthquakes on seismic risk, *Technical report TIT/EERG*, 02/1, Tokyo Institute of Technology, Tokyo, 291-308.

Fajfar, P., Maruši , D., and Peruš, I., 2005, Torsional effects in the pushover-based seismic analysis of buildings, *Journal of Earthquake Engineering*, in print.

FEMA, 2000, *Prestandard and Commentary for the Seismic Rehabilitation of Buildings*, FEMA 356, Washington, D.C.: Federal Emergency Management Agency.

Freeman, S. A., Nicoletti, J. P., and Tyrell, J.V., 1975, Evaluations of existing buildings for seismic risk – A case study of Puget Sound Naval Shipyard, Bremerton, Washington, *Proc. 1st U.S. National Conf. Earthquake Engineering*, EERI, Berkeley, CA, 113-122.

Freeman, S. A., 1998, Development and use of capacity spectrum method, *Proc. 6th U.S. National Conf. Earthquake Engineering*, Seattle, CD-ROM, EERI, Oakland, CA.

Fujii, K., Nakano, Y., and Sanada, Y., 2004, Simplified nonlinear analysis procedure for asymmetric buildings, *Proc. 13th World Conf. Earthquake Engineering*, Vancouver, Canada, Paper No. 149.

Gupta, A., and Krawinkler, H., 2000, Estimation of seismic drift demands for frame structures, *Earthquake Engineering and Structural Dynamics* **29**: 1287-1305.

Isakovi , T., Fischinger, M., and Kante P., 2003, Bridges: when a single mode seismic analysis is adequate?, *Structures and Buildings* **156**: 165-173.

Kilar, V., and Fajfar, P., 1997, Simple pushover analysis of asymmetric buildings, *Earthquake Engineering and Structural Dynamics* **26**:233-49.

Kilar, V., and Fajfar, P., 2001, On the applicability of pushover analysis to the seismic performance evaluation of asymmetric buildings, *European Earthquake Engineering* **15**(1): 20-31.

Krawinkler, H. and Seneviratna, G. D. P. K., 1998, Pros and cons of a pushover analysis for seismic performance evaluation, *Engineering Structures* **20:** 452-464.

Li, K. N., 2002, *3-dimensional nonlinear static and dynamic structural analysis computer program CANNY 99*, CANNY Consultants Pte Ltd., Singapore.

McCabe, S. L., and Hall, W.J., 1989, Assessment of seismic structural damage, *Journal of Structural Engineering, ASCE* **115**: 2166-2183.

Moghadam, A. S., and Tso, W.K., 2000, 3-D pushover analysis for damage assessment of buildings, *Journal of Seismology and Earthquake Engineering* (Tehran) **2**(3): 23-31.

Peruš, I., Poljanšek, K., and Fajfar, P., 2005, Flexural deformation capacity of rectangular RC columns determined by the CAE method, Submitted to *Earthquake Engineering and Structural Dynamics.*

Prakash, V., Powell, G. H., and Campbell, S., 1993, DRAIN-2DX Base program description and user guide, Version 1.10, *Report No.UCB/SEMM*-93/17&18, University of California, Berkeley, CA.

Reinhorn, A. M., 1997, Inelastic analysis techniques in seismic evaluations, in *Seismic design methodologies for the next generation of codes*, P. Fajfar and H. Krawinkler, Eds., Balkema, Rotterdam, 277-287.

Saiidi, M., and Sozen, M. A., 1981, Simple nonlinear seismic analysis of R/C structures, *Journal of Structural Division, ASCE* **107**: 937-952.

REDUCING THE EARTHQUAKE INDUCED DAMAGE AND RISK IN MONUMENTAL STRUCTURES: EXPERIENCE AT ECOLE POLYTECHNIQUE DE MONTREAL FOR LARGE CONCRETE DAMS SUPPORTED BY HYDRO-QUEBEC AND ALCAN

P. Léger *(pierre.leger@polymtl.ca)*
Department of Civil, Geological and Mining Engineering,
École Polytechnique de Montréal, Montréal University, Quebec,
Canada

Abstract. This paper presents a summary of (a) structural analysis methodologies to assess the seismic cracking and sliding responses of concrete gravity dams including transient uplift pressures in cracks, (b) related computer programs, (c) dam strengthening using post-tension anchors, and (d) related experimental validation work. The paper refers principally to research and development work performed at École Polytechnique de Montréal (Montréal University, Canada), in recent years.

Keywords: earthquake safety; concrete dams; structural analysis techniques; concrete cracking; sliding displacements; uplift pressures, shake table and laboratory experiments

1. Introduction

Earthquakes have affected several large concrete dams in the past. Although no catastrophic failure has yet been reported unless a dam crosses a fault (Shi-Kang water supply weir, Taiwan 1999), historical events have shown that severe seismic damage could be imparted to concrete dams (Koyna India; Hsinfenkiang, China; Sefid Rud, Iran; Pacoima, USA; Rapel, Chile, Honen-Ike, Japan; Blackbrooke, UK). It has been observed that damage to concrete dams could occur for earthquake intensity that are less than the maximum value that could be expected at the site. The environmental and economic consequences associated with the release of a reservoir due to seismic failure of a large dam are catastrophic. A systematic and progressive structural analysis methodology to characterize the seismic behaviour of dam-foundation-reservoir systems, ranging from the pseudo-static method to nonlinear transient dynamic analysis, should be carried out to evaluate

A. Ibrahimbegovic and I. Kozar (eds.),Extreme Man-Made and Natural Hazards in Dynamics of Structures, 285–309.

the available safety margin as a function of the uncertainties introduced in each stage of the safety analysis. In addition to the specific conditions of the site and the dam, the consequence of failure should also be given serious consideration in the safety evaluation process. For example, the large volume of a reservoir may not contribute to increase the probability of failure, but would certainly have a significant and possibly disastrous effect on the consequence of failure.

To judge the seismic response of a dam in a damaged condition the sliding and rotational displacements that will occur are more important than stresses. To identify the seismic cracking profile, and potential displacements along the crack planes, it is required to make a number of assumptions. These relate to the crack initiation and propagation criteria, the tensile and shear strengths that could be mobilized, the seismic variations of uplift pressures, the significance of the oscillatory nature of inertia forces and vertical accelerations (Christopoulos et al. 2003), just to mention a few key parameters. Considerations should be also given to proper dynamic modelling of dam-foundation-reservoir interactions (Gogoi and Maity 2005, Léger and Bhattacharjee 1993, Léger and Boughoufalah 1989). Proper seismic safety evaluation methodology, guidance, performance criteria, and related computational tools, must be further developed and validated to optimise the resources devoted to seismic rehabilitation of existing dams or the construction cost of new dams. The validity of the safety assessment procedure can only be verified by comparison of numerical simulations predictions with observed seismic performance of actual dams, and laboratory experiments on small scale dam models that could be tested to failure, thus allowing to infer the ultimate responses of prototypes.

Utility companies responsible for the operation and maintenance of existing concrete dams are actively seeking adequately validated computational models and "user friendly" computational tools to apply in the periodic seismic safety assessment of their installations. This paper presents a summary of structural analysis methodologies to assess the seismic cracking and sliding responses of concrete gravity dams, related computer programs and experimental validation work that were developed at École Polytechnique de Montréal (Montréal University, Canada) in recent years. This work represents a contribution to enhance the state-of-the-art and the state-of-the practice related to safety assessment of concrete dams. These R&D activities inscribe themselves in the worldwide research effort performed by several research groups focusing on dam safety issues and globally summarized in the following state-of-the-art, state-of-the practice publications (Gogoi and Maity 2005, Weiland 2003, ICOLD 2001, Ghrib et al. 1997, Clough and Ghanaat 1993, Hall et al. 1992, Bhattacharjee and Léger 1992, NRC 1990, Hall 1988, Chopra 1988, Priscu et al. 1985, Tarbox et al. 1979).

2. Failure Mechanisms of Concrete Dams

Figure 1 shows the increase in seismically induced damage with the intensity of the applied ground motions for a gravity dam (DBE, design basis earthquake; MDE, maximum design earthquake; MCE maximum credible earthquake). The peak ground (or peak spectral) acceleration associated with a high frequency spike generally does not correlate well with the intensity of structural response or damage potential of significant input ground motions (Clough and Ghanaat 1993, Tarbox et al. 1979).

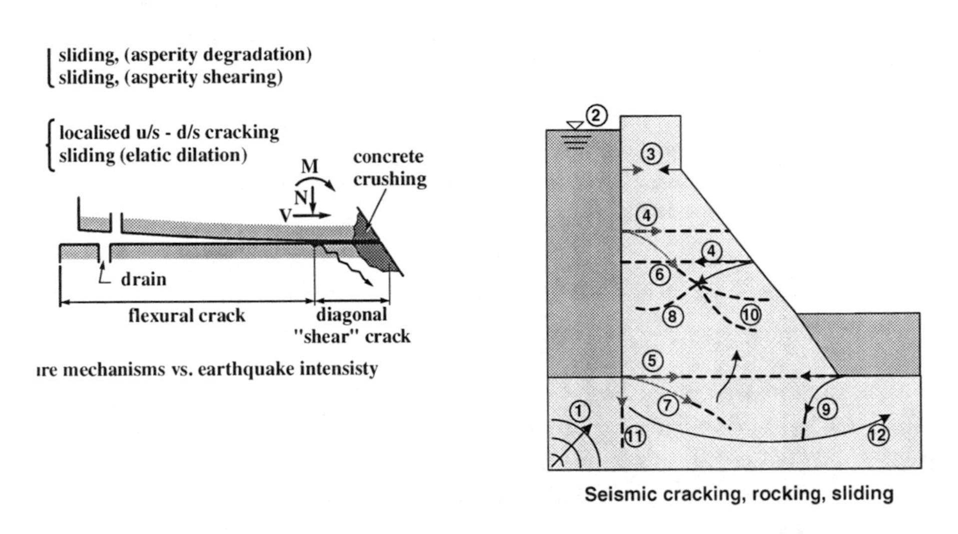

Figure 1. Seismic damage and failure mechanisms of gravity dams.

A high acceleration spike realistically applies large inertia forces (mass × accelerations) of an impulsive nature to the dam. This might be sufficient to initiate and propagate cracks in the dam body or along weak joints, but the inertia forces might not be applied in the same direction during a sufficiently long period of time to induce significant rotational or sliding displacements that are actually detrimental to the seismic and post-seismic structural stability of the cracked components (Fig. 2). It is therefore always difficult to infer potential seismic stability problem from an analysis method that does not recognize explicitly the transient oscillatory nature of the applied inertia forces.

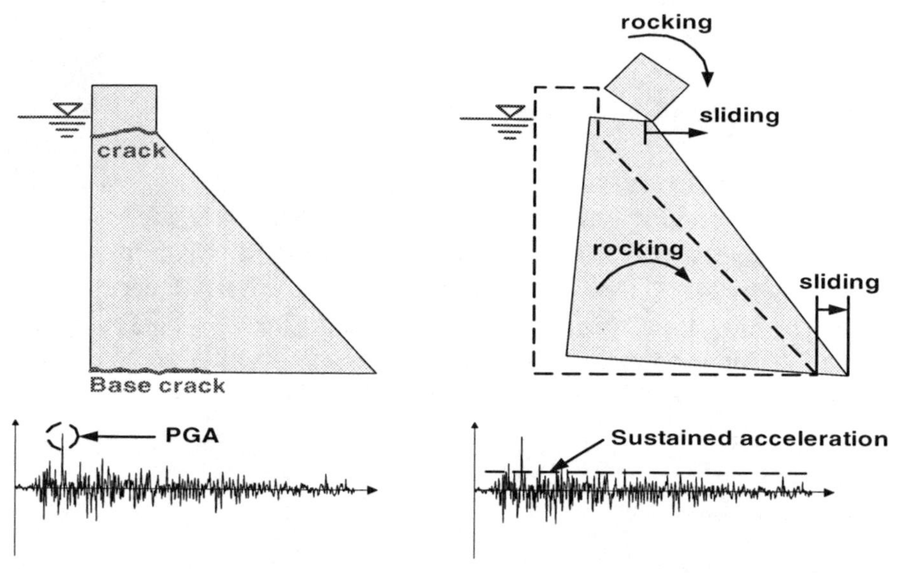

Figure 2. Seismic safety analysis of concrete dam: (a) stress analysis to compute crack length, (b) stability analysis to compute maximum rotation, sliding safety factors and residual sliding displacements.

3. Safety Assessment of Existing Concrete Dams Using Numerical Models

The prediction of damage or failure of concrete dams is made with the assistance of mathematical (numerical) models whose basic objective is to obtain results within what we could call engineering accuracy to take appropriate safety decisions with confidence. To avoid an uncontrolled release of the reservoir there are two basic requirements that are equally important. First, dynamic stability of cracked components must be maintained during the earthquake. Second, in the post-earthquake condition, static stability must be maintained considering the potential increase in uplift pressure in the dam body and the foundation as well as reduction in structural strength due to cracking and joint movements. A rapid draw down of the water level could also be detrimental if the uplift or pore pressure can not be dissipated accordingly. Due to the oscillatory nature of the induced inertia forces during an earthquake there are historical, experimental and numerical evidences that cracked dams are able to maintain dynamic stability under strong ground shaking. Therefore, predicted damage under dynamic load should always be interpreted considering the inherent limitations of the mathematical model used. For example, the applicability of small displacement theory to predict dynamic instability could be seriously questioned. The assumptions used for uplift pressure intensity and distribution during the earthquake have also a very significant impact on the earthquake safety of a dam. Yet there are very few experimental data available to model water-crack interaction with confidence in earthquake safety evaluation. In post-earthquake conditions, there are historical evidences from the 1967 Koyna earthquake (India) and the 1995 Kobe earthquake (Japan) of moderate increase in uplift pressure at the dam-foundation contact following the seismic events. A dam that survived the main shocks could thus be more susceptible to failure in the case of after shocks even if they are of smaller magnitude than the main shock. However, at Sefid Rud dam (Iran 1990) a small reduction in base uplift pressure was measured following the 7.6 magnitude Manjil earthquake that severely damaged the dam (ICOLD 2001).

4. Progressive Approach for Seismic Safety Evaluation of Gravity Dams

Concrete cracking is acceptable under severe earthquakes. The loss of dynamic stability will be associated with the development of large displacements which may not occur in a single cycle of peak earthquake response. After cracking due to excessive peak stresses, the anticipated rotational and sliding displacements should be estimated along cracked planes to evaluate the dam stability. A clear distinction must therefore be made between the (a) stress analysis seeking to define crack profiles, and (b) stability analysis seeking to estimate safety margin against detrimental sliding / rocking displacements occurring along the crack profiles. The basic question is therefore how much displacement is allowable before unacceptable performance (break of water stops and drains) or before the loss of dynamic or post-earthquake stability of cracked concrete components. To maintain an adequate safety margin, the displacement response should ideally be predominantly of the rocking type around a laterally stable equilibrium position. A clear distinction should be made between the types of dam considered. Small sliding displacements could be considered acceptable at the base of a gravity dam or along its lift joints. However, severe cracking in the body of arch dam could result from the loss of arch action due to sliding of an arch abutment leading to an unacceptable condition.

In gravity dams, cracks at the dam-foundation interface tend to reduce the stresses induced in the body of the dam. The base crack is therefore acting as a seismic isolation system (Léger and Katsouli 1988). The inertia forces could be significantly increased between the magnitude required for base crack initiation and the magnitude that will induce cracking in the upper part. For tall dams once cracking has been initiated in the upper part, the cracks tend to propagate in a very brittle manner to separate the top section from the rest of the dam (Léger and Leclerc 1996). The dynamic stability of the separated top section should be verified considering large rocking motions and potential sliding displacements that could occur under the oscillatory nature of the earthquake (Léger and Benftima 2004, Léger et al. 2003, Leclerc et al. 2002b).

5. Computational Tools to Implement a Progressive Approach for Seismic Safety Evaluation of Gravity Dams

Seismic analysis of concrete dams should be performed with a progressive methodology divided in four basic analysis levels of increasing complexity (Table 1, Ghrib et al. 1997). These are (a) the pseudo-static (seismic coefficient method), (b) the pseudo-dynamic (response spectra) method, (c) the linear or non linear dynamic finite element (FE) methods, and (d) the transient rigid body dynamic method for cracked components.

Table 1. Progressive approach for seismic stability.

Method	Excitation	Dyn. Characteristics	Response
1. Pseudo-static (seismic coefficient)	PGA (cracking) Sustained Acc. (stability)	Mass Infinite stiffness (No dyn amplification)	Non-oscillatory Equivalent static
2. Pseudo-dynamic (Chopra 1988)	Design spectra Peak (cracking) Sustained (stability)	Mass, Stiffness, Damping (Dyn. amplification)	Non-oscillatory Max. probable
3. Dynamic (FE) (Lin. / Non-Lin.)	Accelerogram	Mass, Stiffness, Damping (Dyn. amplification)	Oscillatory History (+ / -)
4. Dynamic (Rigid body)	Accelerogram	Mass, Restoring force (friction, inelastic impact)	Oscillatory

It is important to maintain consistency in modelling assumptions while comparing the results from one type of analysis to another. Obviously as the fundamental period of a dam-foundation-reservoir system tends to zero (rigid dam), the results of response spectrum or transient dynamic analysis should tend to the results obtained from the pseudo-static (seismic coefficient) method. (eg. tensile strength of concrete, use of effective (sustained) acceleration for sliding safety assessment and peak ground acceleration (PGA) for cracking assessment in modelling). A "user-friendly" computational tool, the CADAM (Computer Analysis of Concrete Gravity Dams) computer program, that has been developed to perform pseudo-static, and pseudo-dynamic (response spectra) seismic safety assessment of gravity dams (Fig. 3; Leclerc et al. 2003, Leclerc et al. 2002a). A complementary program RS-DAM (Rocking and Sliding of Dams Léger and Benftima 2004, Léger et al. 2003, Leclerc et al. 2002b) has been developed to perform the transient rigid body dynamic analysis of cracked dam components (Chopra and Zhang 1991, FERC 2002) that could be identified either using CADAM or linear or nonlinear finite element analyses using commercial computer programs or specialized program such as FRAC_DAM (Bhattacharjee 1996).

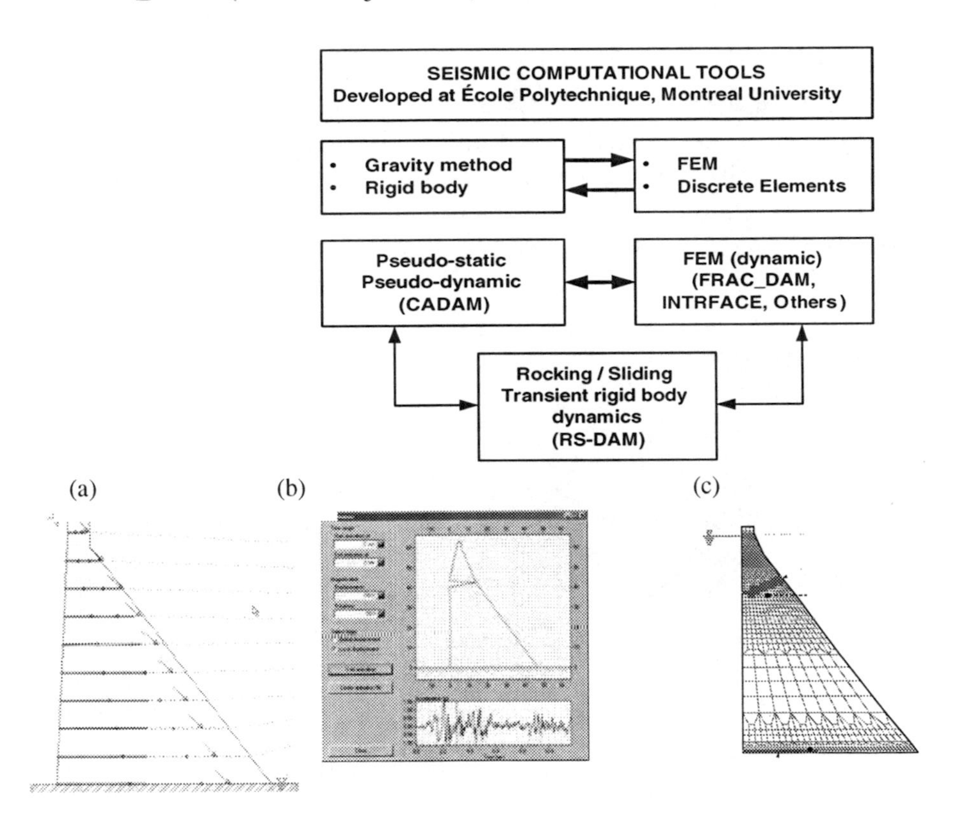

Figure 3. Computational tools to study seismic safety of gravity dams: (a) CADAM, (b) RS-DAM, (c) FRAC-DAM.

CADAM and RS-DAM are currently available from the Web. CADAM considers several features that were derived from experimental observations such as PGA for cracking that could be initiated from the upstream or the downstream faces, sustained acceleration for sliding, and modelling of seismic uplift pressures with different intensities. CADAM also present several features allowing systematic parametric analysis with ground motion of increasing intensities performed within the context of deterministic or probabilistic (Monte-Carlo) analyses. RS-DAM includes several specialised features to study the transient rigid body response of cracked dam components: (a) coupled or / uncoupled rocking, sliding large displacements, (b) translational and rotational hydrodynamic pressures, (c) energy balance computation with different upstream and downstream coefficient of impact restitution, (d) inclined failure plane, (e) graphical display of displacement response, resultant position and safety factors.

6. Seismic Cracking of Mass Concrete – Modelling and Experimental Work

A 2D FE computer program (FRAC_DAM) to perform seismic cracking analysis of concrete gravity dams using a G_f smeared crack type of non linear constitutive model for mass concrete was developed by, Bhattacharjee 1996, Léger and Bhattacharjee 1995, Bhattacharjee and Léger 1993. A somewhat equivalent approach in its end results was developed within the context of damage mechanics by Ghrib and Tinawi (1995). The fracture energy (G_f) conservation principle is used to obtain mesh objective results (Fig. 4).

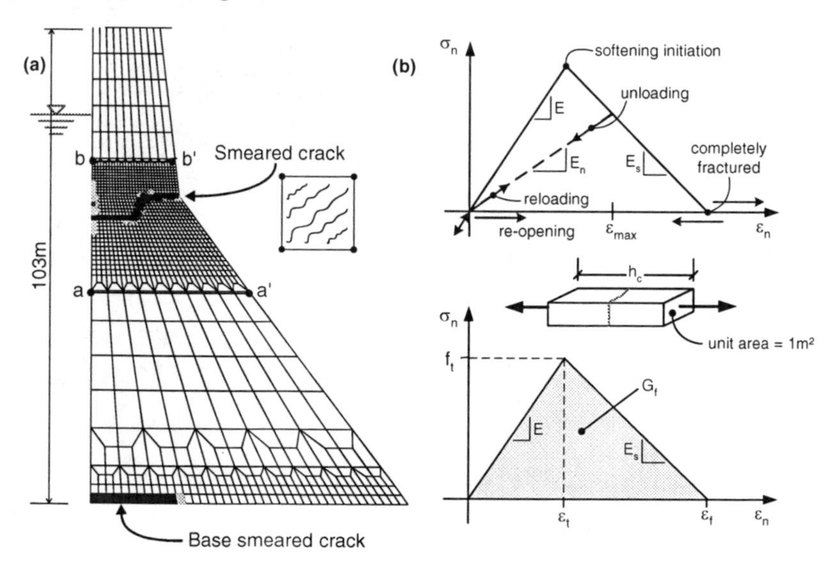

Figure 4. Seismic smeared crack analysis of Koyna dam using FRAC_DAM.

To validate FRAC_DAM shake table experiments were conducted on two 3.4 m high plain concrete gravity dam models to study their dynamic cracking response when subjected to simple triangular acceleration pulses (Tinawi et al. 2000). From the cracking tests, it was shown that triangular acceleration pulses could initiate a crack and propagate it (Fig. 5).

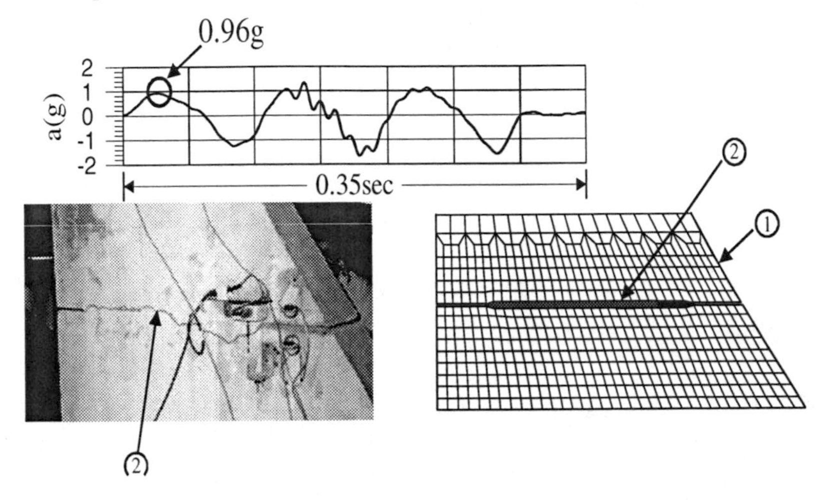

Figure 5. Shake table cracking test of a monolithic specimen: (1) FE model, (2) completely cracked section.

Viscous damping varies experimentally from 1% in an uncracked situation (Fig. 6) to approximately 10% in a fully cracked condition. For partial cracking high frictional damping (over 20%) was measured (Fig.7). The tests results correlated well with the FRAC_DAM numerical simulations if the damping values were modified as the dam switch from an uncracked to a cracked condition (Fig. 8).

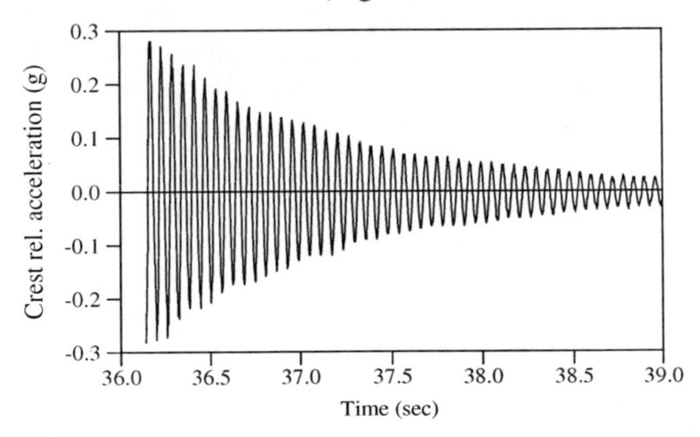

Figure 6. Impact test on uncraked specimen, exponential decay (ξ approx. 1%).

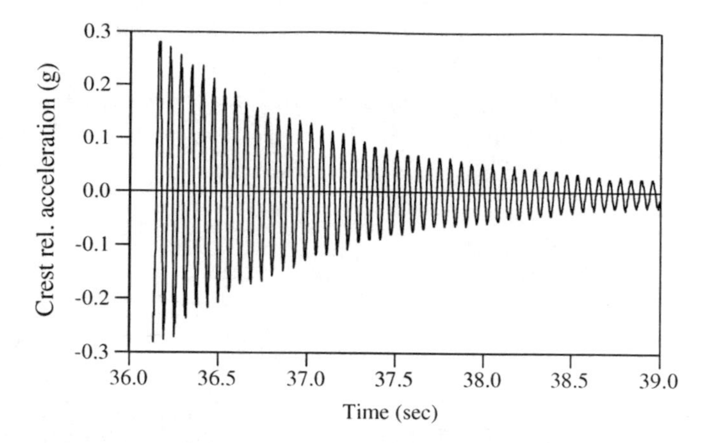

Figure 7. Impact test on partially cracked specimen (33% of the section), linear decay (ξ approx. 23%).

The static analysis suggested that some cracks (non-linear deformations) occurred already due to gravity loads. This state was assumed as the initial ("elastic") state of the building, and the modes of vibration of the building in such a condition were taken into account for the modal analysis. Modal analysis was performed independently for the loading in both horizontal directions, using the CQC rule for the combination of different modes, which is considered appropriate for structures with closely spaced modes. The results of analyses for both directions were combined by the SRSS rule.

The constitutive law for damping was based on a damping matrix for each finite element that was proportional to its non linear (cracked) stiffness thus avoiding to transfer tensile forces across a cracked element if a mass proportional term was retained in the damping matrix. A single triangular acceleration pulse can cause partial cracking in a concrete dam. Two consecutive pulses did cause complete cracking. The concrete tensile strength properties were increased by a constant dynamic magnification of 1.75 to obtain a good correlation between numerical analyses and experimental results. Even if viscous damping for the uncracked model increased significantly after partial cracking, it is not recommended, at this time, to exceed a viscous damping value of more than 10%. Research needs have been identified for proper constitutive modelling of energy dissipation effects (ex. viscous damping, frictional damping, coefficient of impact restitution) during and after the occurrence of cracks in the presence of water or not (Horii and Chen 2001, Bhattacharjee and Ghrib 1995, Léger

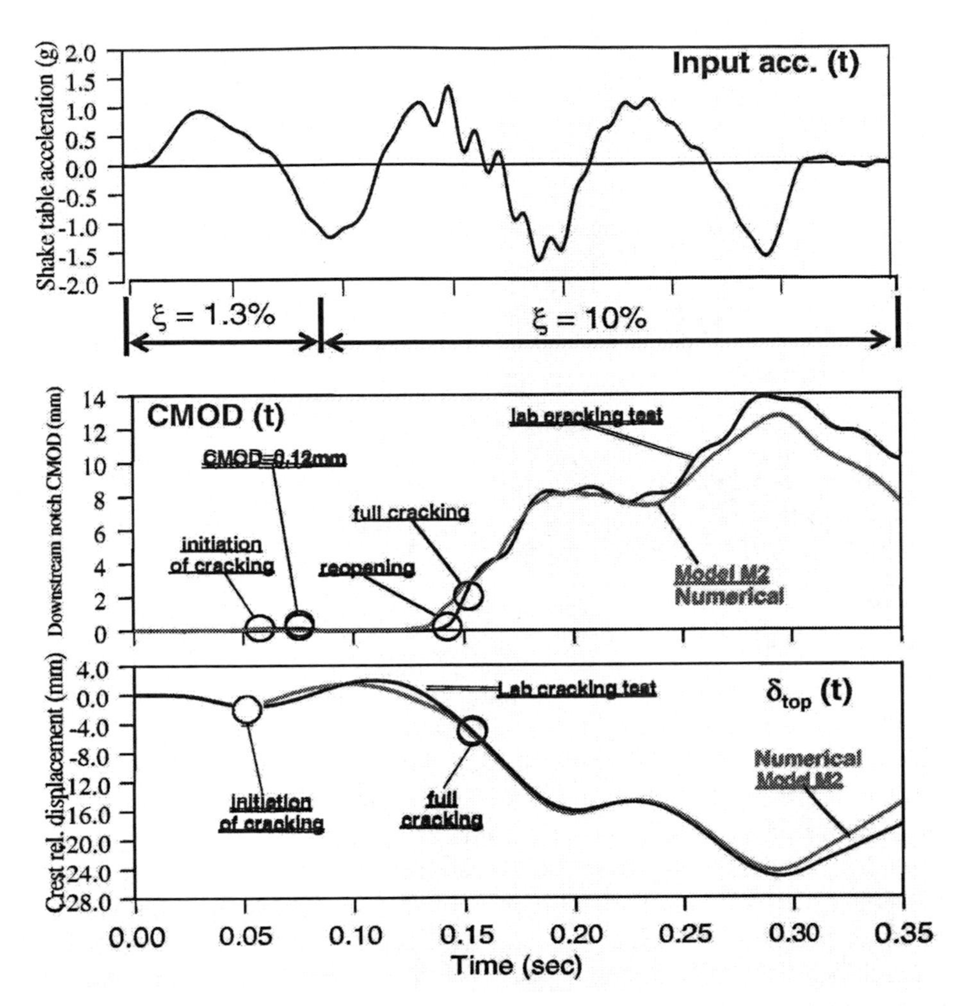

Figure 8. Cracking test on a monolithic specimen (CMOD, crack mouth opening displacement, dtop displacement at the top).

and Bhattacharjee 1994). Three-dimensional extension of the FE smeared crack approach for seismic safety analysis of concrete dams has been recently presented by Mirzabozorg and Ghaemian 2004, while 3D linear FE analyses have been most frequently used in the past (Tinawi et al. 1994).

7. Seismic Cracking, Sliding and Rocking Along Weak Lift Joints – Modelling and Experimental Work

The presence of weak concrete lift joints reduces significantly the abilities of concrete dams to resist earthquakes (Tinawi et al. 1998). Dynamic sliding shear tests on small (0.5m x 0.25m) dry lift joint specimens with different surface preparations have been performed to characterize typical cyclic load-displacement responses and related static and dynamic friction coefficients, as well as their seismic energy dissipation capabilities (Fig. 9, Fronteddu et al. 1998).

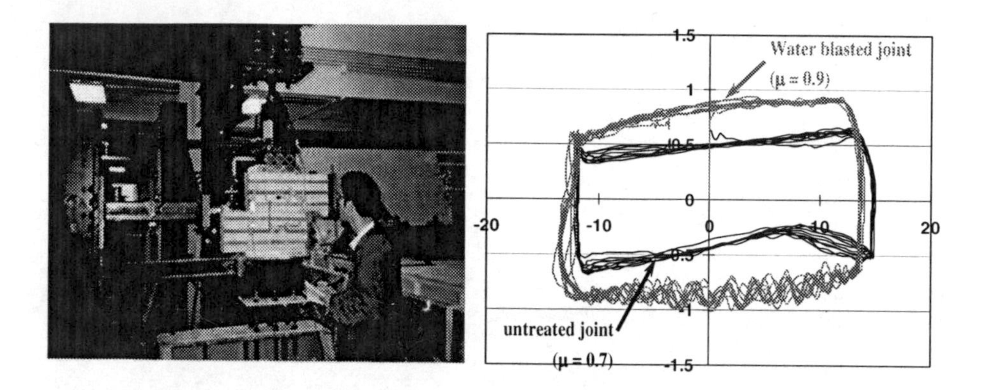

Figure 9. Transient sliding test of concrete-concrete joint specimens.

Experimental results indicate that the coefficient of friction decreases with the increase in normal load, under both static and dynamic shear. The shear strength is dependent on surface preparation. Monolithic specimens and water blasted joints show higher shear strengths than untreated joints and plane independent joint surfaces. No strength degradation is noticeable under dynamic sliding shear tests because hysteresis loops are very stable. No dependency of shear strength on frequency content of the imposed shear displacements is observed. An empirical lift joint constitutive model was developed as a function of a basic friction coefficient and a roughness friction coefficient that is dependent upon the type of surface preparation (Fronteddu et al. 1998). The proposed lift joint constitutive model was implemented in the context of a gap-friction element (Fig. 10) in an in-house FE program (INTRFACE) used to study the sliding response of a 90m concrete gravity dam. It was found that significant seismic energy could be dissipated by frictional mechanism, and that a realistic prediction of residual sliding displacements requires proper consideration of several lift joints along the dam's height.

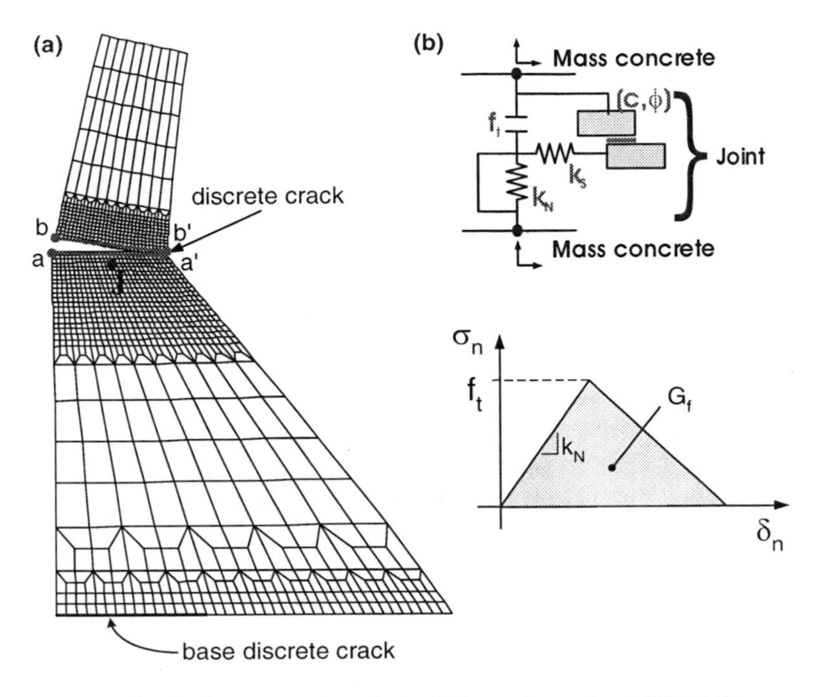

Figure 10. Seismic discrete crack analysis of Koyna dam using INTRFACE program.

Shake table tests were carried out on two 3.4m high dry concrete dam models with a weak (cold) lift joint at mid height to investigate the seismic sliding and rocking mechanisms and to compare the results with numerical simulations using rigid body dynamics (Fig. 11, Tinawi et al. 2000).

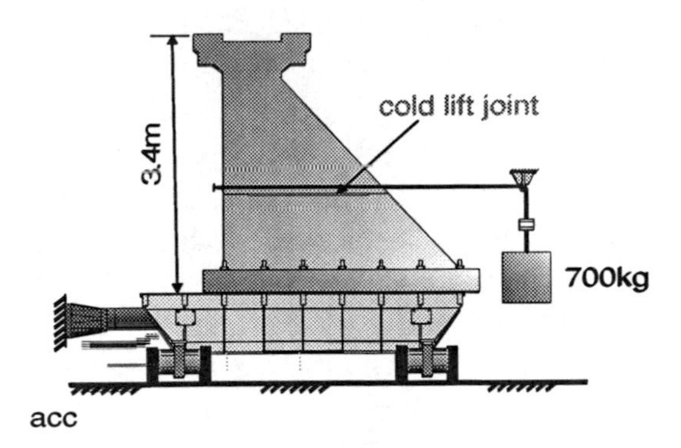

Figure 11. Shake table test set up for a specimen with a cold joint.

The influence of the frequency content of the excitation on the sliding response was investigated (Fig. 12). For the sliding mechanism, the numerical simulations used rigid body dynamics with frictional strength derived from the Mohr-Coulomb criterion. Good correlation with experimental results was obtained but it was found that the evaluation of sliding displacement is very sensitive to the assumed joint friction coefficients (μ_{stat}, μ_{dyn}). Moreover in a rigid body numerical analysis that ignores rocking to simulate only sliding, the friction coefficients have to be slightly adjusted to compensate for the transient variations of the normal forces as sliding is taking place. It was shown that for a given PGA, the sliding due to low frequency excitation (2Hz) (Western North-America) can be 3-4 times larger than the corresponding sliding due to high frequency records (10 Hz) (Eastern North-America) (Fig. 12). The experimental sliding responses and numerical analyses of the models indicate that the concept of *effective (or sustained)* acceleration, Acc_{eff}, is appropriate for pseudo-static or pseudo-dynamic analyses. The pseudo-static effective (or sustained) acceleration to evaluate the sliding safety factor using the pseudo-static (or pseudo-dynamic) method, where inertia forces are non-oscillatory, has been estimated as 0.5 PGA for high frequency excitation, and 0.67 PGA for low frequency excitation.

Figure 4 shows the predicted seismic cracking pattern of Koyna dam subjected to the 1967 earthquake using a G_f (fracture energy) type of concrete smeared crack model using the computer program FRAC_DAM. In this case, the concrete tensile constitutive model uses a secant unloading scheme passing through the origin of the stress-strain curve. The "sliding" tendency of cracked elements is represented by a reduction in the shear modulus of concrete, which is inadequate to make a reliable prediction of potential residual seismic sliding displacements. Figure 10 illustrates the seismic response computed from a FE program (INTRFACE) with a discrete representation of the seismically induced cracks at the base of the dam and at the base of the top block where there is a change in the downstream slope. In this case, the sliding response is controlled by a classical Mohr-Coulomb frictional model but small displacement theory is still used in the FE formulation. The smeared crack model and the discrete crack model are complementary to each other. First, a smeared crack model is used as a "predictor" to identify the cracking pattern likely to develop in the dam using a predefined regular mesh topology. The discrete crack model can then be used as a "corrector" while remeshing the model to follow the predicted crack geometry and compute more accurately the transient cracking response of the dam.

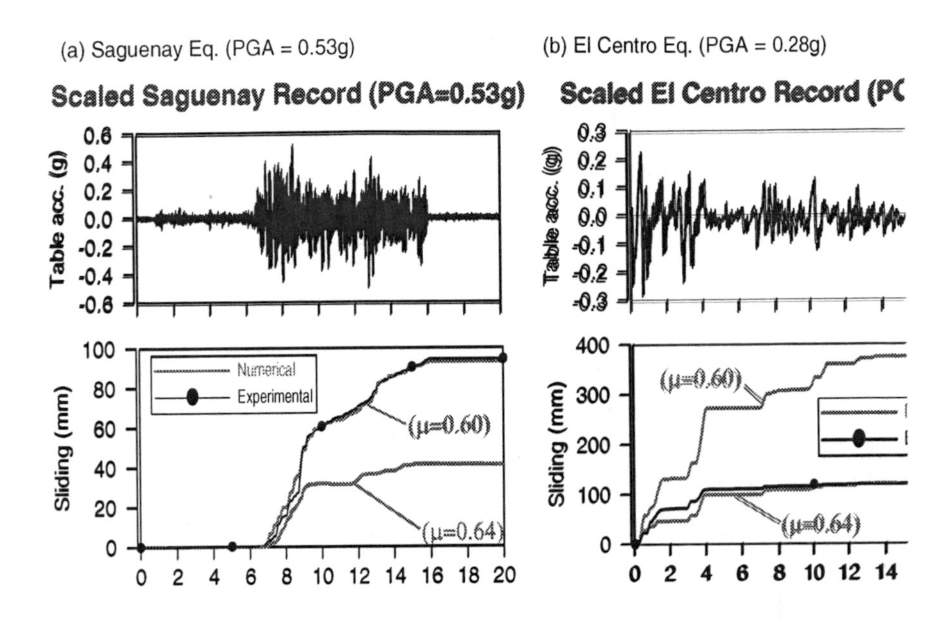

Figure 12. Sliding response of dam subjected to scaled 1988 Saguenay and 1940 El Centro earthquake records.

Figure 13 shows the corresponding RS-DAM model where the base block accelerations have been obtained from point "J" in the FE discrete crack model (Fig. 10). Figure 14 compares the rotational responses and sliding responses of the top block of Koyna dam as computed by (a) a smeared crack model (FRAC_DAM) considering the relative motions of lines a-a' and b-b' in Fig. 4, (b) a discrete crack model (INTRFACE, Fig. 10), and (c) a coupled rocking-sliding rigid body dynamic model (RS-DAM, Fig. 13). Figure 14a indicates that the rotational displacements computed by each computer program are very close to each other. The residual sliding displacement (11mm) computed by the discrete crack FE model (INTRFACE) and the rigid body model (RS-DAM) are nearly identical (Fig.14b). However the smeared crack model (FRAC_DAM) is unable to provide reliable estimate of the seismic sliding response of the top block indicating transient horizontal displacements oscillating around the initial (zero) reference position of the block. The possibility of using "floor response spectra" as seismic input to investigate the seismic stability of cracked top component has been investigated by Léger and Benftima 2004.

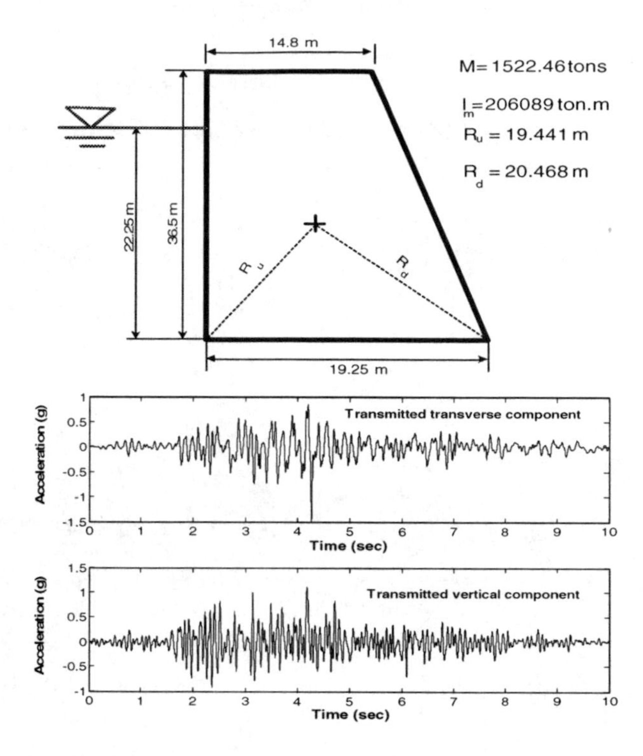

Figure 13. Transient seismic analysis of cracked top section of Koyna dam using rigid body dynamic model; RS-DAM program.

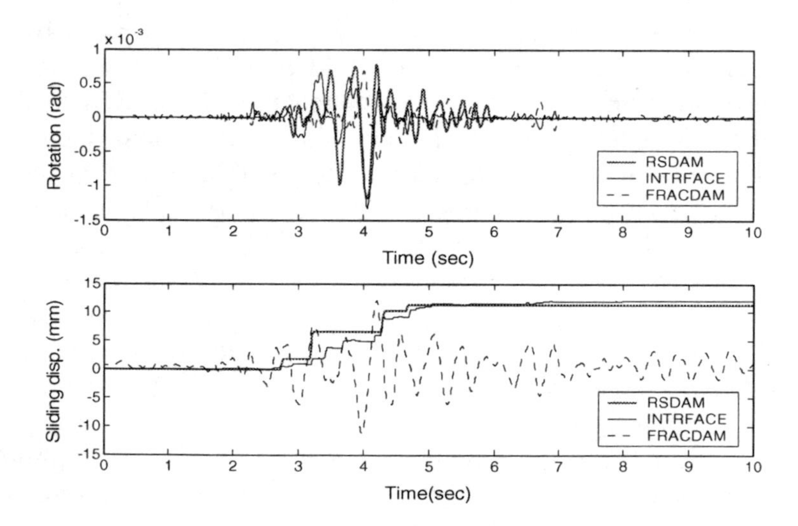

Figure 14. Transient seismic response of cracked top component of Koyna dam using various computer programs: (a) rotations, (b) sliding displacements.

8. Seismic Response of Post-Tensioned Gravity Dams – Modelling and Experimental Work

Post-tensioned anchors have often been used to strengthen concrete dams against earthquake induced damage (ICOLD 2001, Léger and Mayahari 1994, Hall et al. 1992). The presence of a post-tensioned cable near the upstream face of a dam has a tendency to promote tensile stresses at the downstream face thus favouring the initiation of a (stable) rocking motion in case of an earthquake. Displacement controlled shear tests on small concrete lift joint specimens with different surface roughness reinforced with an unbonded post-tension cable were conducted to study a related joint constitutive model, and to determine the associated failure mechanisms (Morin et al. 2002). To investigate the applicability of the joint model, shake table tests were conducted on a 3.4m high plain concrete post-tensioned gravity dam model having a cold lift joint at about mid-height (Fig. 15). Numerical simulations using the joint constitutive model were compared with the experimental results.

From the displacement controlled shear tests, it was shown that the post-tension cable provides additional strength to the joint by (a) applying an additional normal load on the joint, and by (b) the shear resistance that could be mobilized (cable dowel action). The tests performed on rough surface joints have demonstrated that the dilatancy phenomenon due to the asperities increases the post-tension force. Shake table tests on the 3.4m high dam model have shown that post-tensioning largely reduces the residual sliding displacements of the joint (Fig. 16).

This reduction is a function of the frequency content of the base excitation. However, a single post-tension cable placed near the upstream face increases the upstream rocking response of the upper block, which is also a function of the frequency content of the excitation. Strong seismic excitations can induce cable failure at a lower value than its uniaxial tensile strength when significant shear displacements are taking place. To describe this brittle failure mechanism, a proposed cable failure criterion considering the mobilized shear and tensile strengths has been found adequate (Morin et al. 2002). Moreover, the proposed post-tensioned joint model showed good agreement with the static and dynamic tests performed in evaluating the joint shear force.

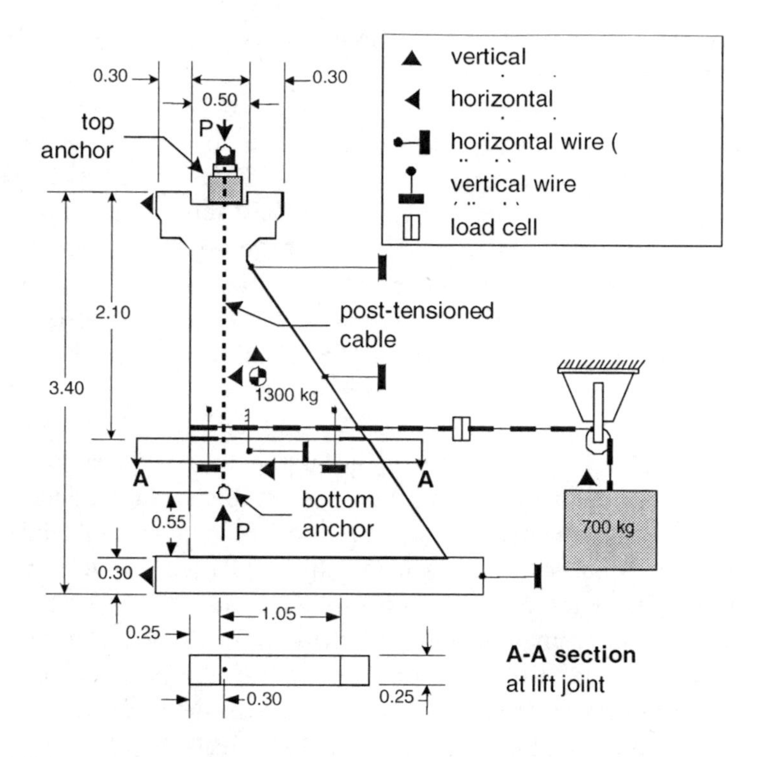

Figure 15. Seismic shake table test set up for a dam model with a post-tensioned anchor.

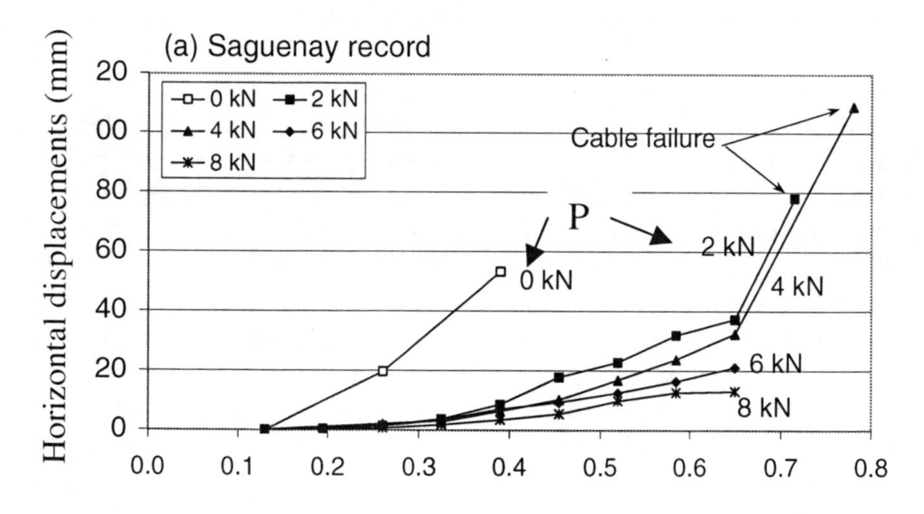

Figure 16. Shake table test residual sliding displacement for increasing values of PGA (1988 Saguenay earthquake record).

9. Transient Variations of Uplift Pressures During Earthquakes – Modelling and Experimental Work

There is not a universally accepted formulation for uplift pressure in concrete cracks during earthquakes in different dam safety guidelines. Seismic uplift pressure in cracks remains a major source of uncertainty in design and safety assessment of concrete gravity dams (NRC 1990). The importance of this problem on the seismic stability assessment of a concrete gravity dam is shown by seismic analyses using either zero, unchanged, or full uplift pressures in seismically induced cracks as recommended by different dam safety guidelines. The assumption used in modelling seismic uplift pressures has a major effect on the computed sliding safety factors and base crack lengths.

To enhance the understating of the physical phenomena associated to water-crack interactions during earthquakes, the transient evolution of uplift pressures in small (1.5 × 0.55 × 0.15m) concrete specimens (Fig. 17) with opening and closing of cracks in the 2-10 Hz frequency range as well as typical earthquake induced motions has been experimentally measured for initial static pressures up to 400 kPa (Javanmardi et al. 2005a, 2005b). Experimental results show that water pressure inside a crack increases due to closing of the crack and decreases due to opening of the crack. The range of maximum pressure increase starts from very small variations for a 2 Hz sinusoidal excitation, up to 1500 kPa for a 10 Hz excitation. It was observed that crack opening displacement and velocity, and the possibility of water extrusion and intrusion in the crack as dictated by boundary conditions, has the most significant effects on water pressure variations inside the crack. Initial static uplift pressure does not have significant effect on pressure variations, but if the pressure drop due to opening of the crack is larger than the initial static pressure, cavitations may occur.

The results of experimental tests were used to validate a constitutive model to quantify the transient uplift force induced by water-crack interaction during earthquake (Fig. 18, Javanmardi et al. 2005b, Tinawi and Guizani 1994). The model was implemented in a finite element computer program for dynamic analysis of gravity dams considering hydro-mechanical water-crack coupling. The results of analyses of a typical 90 m high concrete gravity dam subjected to typical earthquake records (Fig. 19) show that water can penetrate into part of a seismically initiated crack (Lcr) and saturates it partially (Lsat) (Fig. 18). During the crack opening mode the saturated length is small, from few centimetres to few meters,

Figure 17. Concrete specimen for transient water crack interaction test.

depending on the opening velocity and the magnitude of the opening and crack mouth pressure. Water pressure decreases along this length from crack mouth pressure to the existing void pressure at the end of the saturation length. The developed uplift forces are small and the modeling assumption of zero uplift pressure in a seismic crack in tensile opening mode appears to be justified. The magnitude of crack closing and its velocity have the most important effects on the magnitude of seismic uplift force in closing mode. The saturation length and the uplift force in crack closing mode are increasing in successive cycles. The maximum pressure can be high locally and the saturated length can be increased up to several meters, still smaller than the crack length in the system analysed. Because the pressure tends to develop in a region close to the crack mouth, detrimental effects for the global dam stability are unlikely to occur (ex. wedge effect propagating a crack filled with water as it closes).

The seismic uplift force during the heel crack opening mode is very small relative to the dam weight. Based on the limited experimental and numerical evidences gathered so far, and pending further investigations, it appears that the critical sliding safety factor (SSF) of the dam against downstream sliding could thus be computed by considering zero uplift pressure in the crack region subjected to seismic tensile opening.

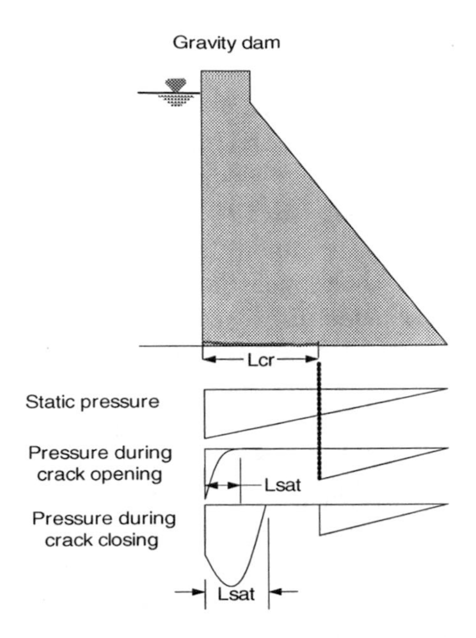

Figure 18. Transient variations of uplift pressures during earthquakes.

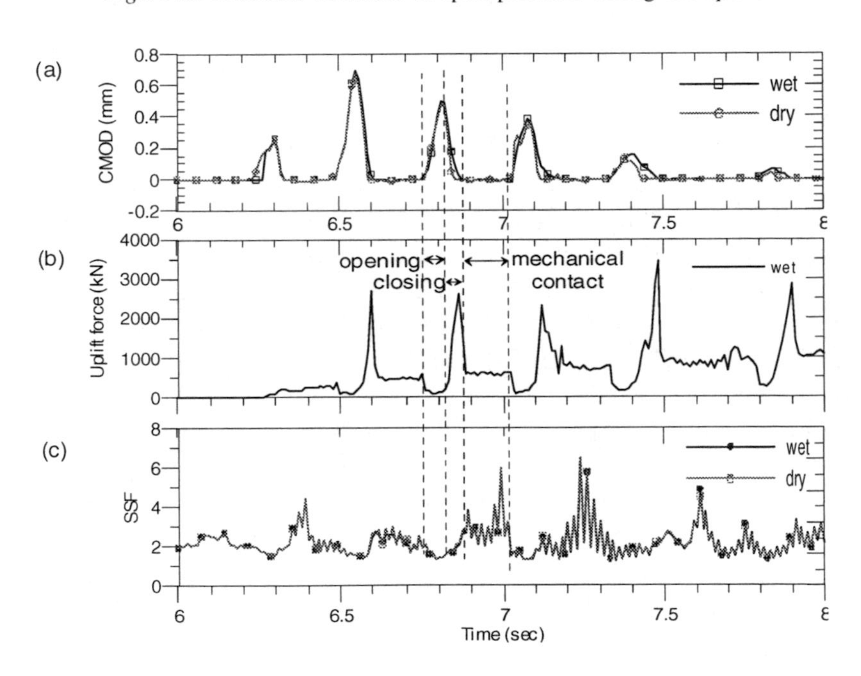

Figure 19. Details of hydro-mechanical response of a base crack for a 90m dam for a segment of the 1988 Saguenay earthquake record (scaled PGA=0.25g): (a) crack mouth opening displacement (CMOD) for wet and dry cracks, (b) uplift force, (c) sliding safety factors.

10. Conclusions

Seismic evaluation using appropriate methodologies should be considered in periodic safety re-evaluation and in rehabilitation projects of aging concrete dams. The seismic safety evaluation should be performed using a progressive approach working from the gravity method (pseudo-static and pseudo-dynamic methods), to simple linear elastic FE models, gradually introducing elaborate nonlinear constitutive material (joint) models when significant nonlinear behaviour is expected to develop. However, it must be emphasized that safety indices obtained from numerical analyses just represent indicators of potential behaviour. Numerical analyses are often sensitive to assumed initial conditions, selected material parameters, boundary conditions, and the seismic input motions used in the adopted modelling procedures. Systematic parametric analyses must therefore be undertaken to determine reasonable bounds on the potential dam-foundation damage caused by earthquakes of given intensities ranging from the design basis earthquakes (DBE), to the maximum design earthquake (MDE) or the maximum credible earthquake (MCE) (Fig.1).

The scope for research and development of accurate, efficient, reliable mathematical models and related "user friendly" computational tools to assess the seismic safety of gravity dams is very broad in scope touching, (a) seismological aspects, (b) material constitutive models (cracking, sliding, rocking) specially for 3D multiaxial loading with fast strain rates, (c) dam-reservoir-sediment-foundation interactions, (d) seismic and post-seismic water-crack interactions, (e) seismic foundation damage (grout curtain, drainage), (f) risk analysis … just to name a few aspects.

The R&D work currently undergoing at École Polytechnique de Montréal is related to (a) the development of concrete dam seismic analysis methodologies and constitutive models for safety assessment and rehabilitation, (b) related computational tools, and (c) their experimental validation. The focus is currently put on the following topics: (a) seismic water-fracture interaction using shake table experiments with small sliding specimens with a pressurised joint, (b) the use of earthbacking to strengthen dams for static and seismic loads, (c) the seismic safety of gated spillways (piers, gate lifting equipment), (d) the post-seismic safety assessment of damaged concrete hydraulic structures.

Acknowledgements

The R&D projects presented in this paper were carried out in close collaboration with Emeritus Professor R. Tinawi from École Polytechnique de Montreal to who I expressed all my thanks for our fruitful technical discussions, lively and stimulating exchanges. The collaboration of M.

Leclerc, a research engineer who contributed to several projects, is particularly acknowledged. The contributions of several graduate students through M.A.Sc and Ph.D research projects were also central to the developments presented. The author wish to gratefully acknowledge the support provided by Hydro-Quebec, Alcan as well as collaboration and discussion with their engineers that allow the realisation of the R&D work reported in this paper. The financial support provided by NSERC (National Science and Engineering Research Council of Canada) is also gratefully acknowledged.

References

Bhattacharjee, S.S. and Ghrib, F., 1995, Effects of viscous damping models in earthquake stress analysis of concrete dams, Proc. 7th Canadian Conf. Earthquake Eng., Montreal, Canada:341-348.

Bhattacharjee, S.S. and Léger, P., 1993, Seismic cracking and energy dissipation in concrete gravity dams, *Earthquake Engineering and Structural Dynamics*, **22**;991-1007.

Bhattacharjee, S.S. and Léger, P., 1992, Concrete constitutive models for seismic analysis of gravity dams - state-of-the-art, *Canadian Journal of Civil Engineering*, **19**(3):492-509.

Bhattacharjee, S.S., 1996, FRAC_DAM – A finite element analysis computer program to predict fracture and damage responses of solid concrete structures, Report No. EPM/GCS-1996-03, Department of Civil Engineering, École Polytechnique de Montréal, Quebec, Canada.

Chopra, A.K., and Zhang, L., 1991, Earthquake-induced base sliding of concrete gravity dams, *ASCE Journal of Structural Engineering*, **117**(12):3698-3719.

Chopra, A.K., 1988, Earthquake response analysis of concrete dams. In *Advanced Dam Engineering for Design, Construction, and Rehabilitation*, Edited by R.B. Jansen, Van Nostrand Reinhold, pp. 416-465

Christopoulous, C., Léger, P. and Filiatrault, A., 2003, Seismic sliding response analysis of gravity dams including vertical accelerations, *Journal of Earthquake Engineering and Engineering Vibration*, **2**(2):189-200.

Clough, R.W. and Ghanaat, Y., 1993, Concrete dams: Evaluation for seismic loading, International Workshop on Dam Safety Evaluation, Grindelwald, Switzerland, **4**:137-169.

FERC (Federal Energy Regulatory Commission), 2002, Engineering guidelines for the evaluation of hydropower projects. *Chapter III, Gravity Dams*, Department of Energy, Washington, D.C., USA.

Fronteddu, L., Léger, P. and Tinawi, R. 1998, Static and dynamic behaviour of concrete lift joints interfaces, *ASCE Journal of Structural Engineering*, **124**,(12):1418-1430.

Ghrib, F., Léger, P., Tinawi, R., Lupien, R. and Veilleux, M., 1997, Seismic safety evaluation of gravity dams, *Int. Journal Hydropower and Dams*, **4**(2):126-138.

Ghrib, F. and Tinawi, R., 1995, An application of damage mechanics for seismic analysis of concrete gravity dams, *Earthquake Engineering and Structural Dynamics*, **24**: 157-173.

Gogoi, I. and Maity, D., 2005, A review of idealisation and modelling techniques for concrete gravity dams. *Dam Engineering*, **XVI**(2):117-152.

Hall, J.F., Dowling, J.M. and EL-Aidi, B., 1992, Defensive earthquake design of concrete gravity dams, *Dam Engineering*, **III**(4):249-263.

Hall, J.F., 1988, The dynamic and earthquake behaviour of concrete dams: review of experimental behaviour and observational evidence, *Soil Dynamics and Earthquake Engineering*, **7**(2): 58-117.

Horii, H. and Chen, S.-C., 2001, Computational fracture analysis of concrete gravity dams toward engineering evaluation of seismic safety, Proc.4th Int. Conf. on Fracture Mechanics of Concrete Structures, R. de Borst et al. Ed., A.A. Balkema Pub., Cachan, France, **2**:737-747.

ICOLD (International Commission on Large Dams) 2001, *Design Features of Dams to Resist Seismic Ground Motion - Guidelines and Case Studies*, Bulletin 120, Paris France, 192pp.

Javanmardi, F., Léger, P. and Tinawi, R., 2005a, Seismic structural stability of concrete gravity dams considering transient uplift pressures in cracks, *Engineering Structures,* **27:**616-628.

Javanmardi, F., Léger, P. and Tinawi, R., 2005b, Seismic water pressure in cracked concrete gravity dams: experimental study and theoretical modeling, *ASCE Journal of Structural Engineering,* **131**(1):139-150.

Léger, P. and Benftima, M., 2004, Simplified seismic stability evaluation of gravity dams subjected to high frequency Eastern North American ground motions, Proc., Canadian Dam Association Conf., Ottawa, Ontario, Canada Sept.25-Oct.1 (CD) 10pp.

Leclerc, M., Léger, P. and Tinawi, R., 2003, Computer aided stability analysis of gravity dams - CADAM, *International Journal Advances in Engineering Software*, **34**:403-420.

Leclerc, M, Léger, P. and Tinawi, R., 2002a, Computer aided stability analysis of gravity dams – CADAM, USERS Manual, Department of Civil Engineering, Ecole Polytechnique, Montreal, Quebec, Canada (http://www.polymtl.ca/structures/telecharg/index.php)

Leclerc, M, Léger, P. and Tinawi, R., 2002b, RS-DAM Seismic rocking and sliding of concrete dams, USERS Manual, Department of Civil Engineering, École Polytechnique, Montréal, Quebec, Canada (http://www.polymtl.ca/structures/telecharg/index.php)

Léger, P., Leclerc, M. and Larivière, R., 2003, Seismic safety evaluation of concrete dams in Québec, *International Journal on Hydropower & Dams*, **10**(2):100-109.

Léger, P. and Leclerc, M., 1996, Evaluation of earthquake ground motion to predict cracking response of gravity dams, *Engineering Structures*, **18**,(3):227-239.

Léger, P. and Bhattacharjee, S.S., 1995, Seismic fracture analysis of concrete gravity dams, *Canadian Journal of Civil Engineering*, **22**(1):196-201.

Léger, P. and Mayahari, A., 1994, Finite element analysis of post-tensioned gravity dams for floods and earthquakes, *Dam Engineering*, **V**(3):5-27.

Léger, P. and Bhattacharjee, S.S., 1994, Energy concepts in seismic fracture analysis of concrete gravity dams, Proceedings of the International Workshop on Dam Fracture and Damage, Chambéry, France, March, A.A. Balkema Publishers, pp.231-240.

Léger, P. and Bhattacharjee, S.S., 1993, Reduced frequency-independent models for seismic analysis of concrete gravity dams, *Computers and Structures*, **44**(6):1381-1387.

Léger, P. and Katsouli, M., 1989, Seismic Stability of Concrete Gravity Dams, *Earthquake Engineering and Structural Dynamics*,**18**:889-902.

Léger, P. and Boughoufalah, M., 1989, "Earthquake input mechanisms for time domain analysis of dam-foundation systems, *Engineering Structures*, **11**(1):37-46.

Mirzabozorg, H. and Ghaemian, M., 2004, Non-linear behavior of mass concrete in three-dimensional problems using smeared crack approach, *Earthquake Engineering and Structural Dynamics*, **34**(3):247-269.

Morin, P., Léger, P. and Tinawi R., 2002, Seismic behaviour of post-tensioned gravity dams: shake table experiments and numerical simulations, *ASCE Journal of Structural Engineering*, **128**(2):140-152.

NRC (National Research Council), 1990, Earthquake engineering for concrete dams: design, performance, and research needs. National Academy Press, Washington, D.C.

Tarbox, G.S., Dreher, K.J. and Carpenter, L.R., 1979, Seismic analysis of concrete dams, International Commission on Large Dams 13th Congress, New-Delhi, India, Q.51-R.11, pp.963-994.

Priscu, R., Popovici, A., Stematiu, D., and Stere, C., 1985, *Earthquake Engineering for Large Dams*, Editura Academiei and John Wiley & Sons.

Tinawi, R., Léger, P., Leclerc, M. and Cipolla, G., 2000, Seismic safety of gravity dams: from shake table experiments to numerical analyses, *ASCE Journal of Structural Engineering*, **126**(4):518-529.

Tinawi, R., Léger, P., Ghrib, F., Bhattacharjee, S.S., Leclerc, M., 1998, Structural Safety of Existing Concrete Dams: Influence of Construction Joints, Canadian Electricity Association, Report CEA No.9032 G 905, Final Report; Vol.A - Review of literature and background material; Vol.B Theoretical and numerical developments and case studies.

Tinawi, R., Marchand, J.F. and Léger, P. 1994, Three-dimensional static, thermal and seismic analysis of polygonal gravity dams, *Dam Engineering*, **V**(3):29-58.

Tinawi, R. and Guizani, L., 1994, Formulation of hydrodynamic pressures in cracks due to earthquakes in concrete dams, *Earthquake Engineering and Structural Dynamics*, **23**:699-715.

Weiland, M., 2003. Seismic aspects of dams.- General Report to Question 83 – Seismic aspects of dams. Proceedings International Commission on Large Dams, 21st Congress, Montréal, Canada, 56pp.

EARTHQUAKE ENGINEERING EXPERIMENTAL RESEARCH AT JRC-ELSA

F. Javier Molina* and M. Géradin (*francisco.molina@jrc.it*)
Joint Research Centre (Ispra, I), European Commission
ELSA Laboratory, 21020 Ispra (Varese), Italy

Abstract: The ELSA laboratory is equipped with a large reaction-wall facility and has acquired its best expertise on the development and implementation of innovative experimental techniques mainly related to testing large-scale specimens by means of the pseudodynamic method. Apart from the relevant achievements within the testing techniques, such as the continuous pseudodynamic test and the development of effective techniques for the assessment of the experimental errors, the role of a reference laboratory in Europe has allowed ELSA to rely on the collaboration of many important research institutions that have contributed through the projects with and added maximum scientific value to the results of the tests. An example of pioneering tests performed for a relevant collaborative project are represented by the bi-directional tests performed on a multi-storey building, where the combination of sophisticated techniques have allowed for the first time to obtain most valuable information on the seismic experimental response of a torsionally unbalanced existing building in its original and retrofitted configurations.

Key words: Dynamic test, seismic test, pseudodynamic test, reaction wall, control errors, bidirectional test, torsional response.

1. Introduction

During the last two decades, the pseudodynamic (PsD) method has increasingly been used for seismic testing of structures. The reason for this has been the possibility of testing large-scale constructions or substructures in clear advantage with respect to shaking table testing. Testing large specimens or sub-assemblages in laboratory represents of course a way to complement the lessons obtained from

A. Ibrahimbegovic and I. Kozar (eds.),Extreme Man-Made and Natural Hazards in Dynamics of Structures, 311–351.

the behaviour of real constructions during natural earthquakes. During these years, the European Laboratory for Structural Assessment (ELSA) has substantially contributed to new developments within the PsD methodology thanks to a proper in-house design of hardware and software in which high accuracy sensors and devices are used under a flexible architecture with a fast intercommunication among the controllers. The loading capabilities of ELSA's reaction wall are shown in Figure 1 (Donea et al., 1996).

The PsD method is an hybrid technique by which the seismic response of large-size specimens can be obtained by means of the on-line combination of experimental restoring forces with analytical inertial and seismic-equivalent forces (Takanashi an Nakashima, 1986). Thanks to the use of quasistatic imposed displacements, the accuracy of the control and hence the quality of such test is normally better than for a shaking-table test, especially for heavy and tall specimens. In the classical version of the PsD method, displacements are applied stepwise allowing the specimen to stabilise at every step (subsection 2.1). The quality of the test can be further improved using a continuous version of the method as will be described in subsection 2.2. In any case, it is important to recognise that the PsD method is a sophisticated tool that may fail or produce inaccurate results in some cases depending on the systematic experimental errors (Shing and Mahin, 1987, Molina et al., 2002). Particularly, it is well known that the slight phase lags of the control system, which is used to quasi-statically deform the specimen, may considerably distort the apparent damping characteristics of the PsD response and artificially excite the higher modes (subsection 3.1). The ELSA team has undergone many relevant PsD tests, most of which have been used for the improvement of EC8 and the assessment of several categories of structures. Taking advantage of this activity, various analysis techniques have been developed and applied that try to assess the magnitude and consequences of the existing experimental errors (subsection 3.2). Some of these techniques are based on the identification of linear models using a short-time portion of the response. The identified parameters are then transformed into frequency and damping characteristics. By gaining experience in the application of these analysing techniques, it is possible to have a better knowledge of what the feasibility and the accuracy of the experiment can be depending on the type of structure and the applied PsD testing set-up.

The example of application described at section 4 of this chapter refers to the SPEAR project. There, ELSA has performed accurate bidirectional PsD tests on a real-size multi-storey building specimen by using up to nine DoFs and twelve

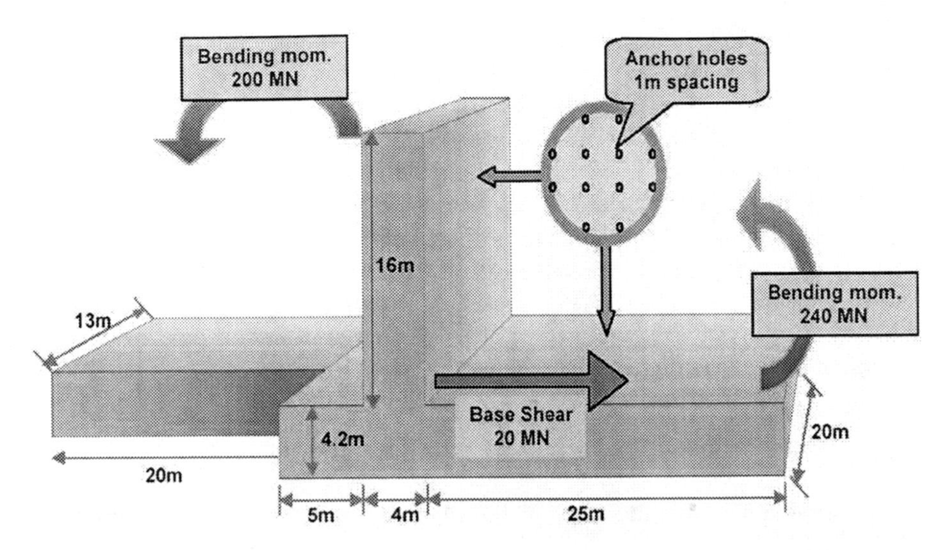

Figure 1. Dimensions and capacities of the reaction wall-strong floor system at ELSA.

actuators plus sophisticated geometric and static on-line transformations of variables between the equation of motion and the controllers. The results of these tests have widened the knowledge on the torsional response of plan-wise irregular buildings.

Out of the scope of this work, a different line of application is also worth mentioning, i.e. the use of the PsD method on structures seismically protected by passive isolators or dissipators. In such applications involving new materials and/or devices, the strain rate effect can become significant. Such effect should be reduced there by increasing the testing speed if possible and by an analytical on-line compensation of it when appropriate (De Luca et al., 2001, Molina et al., 2002b, 2004). Also important to mention are the substructuring techniques developed within the PsD method, which have proved to be very useful for obtaining the seismic response of large structures such as bridges. In that case, a testing set-up is devised in which the reduced part of the structure that has the strongest non-linear behaviour (typically some of the piers) is the actual specimen and the rest of the structure (the deck and the remaining piers) is numerically substructured (Pinto et al., 2004). In a different field of research, an important activity of ELSA is also dedicated to real dynamic tests oriented to the development of active-control systems, such as the case of attenuation of vibration on bridges (Magonette et al., 2003), vibration monitoring for damage detection or fatigue testing on large cable specimens. The common denominator of all these tests has been the use of innovative testing techniques or the development of advanced structural systems.

Finally, during the last years ELSA has also put an important effort in the development of techniques for telepresence, teleoperation and, in general, distributed laboratory environment, which have been successful and now are becoming compatible with the Network for Earthquake Engineering Simulation (NEES) (Pinto et al., 2006). NEES currently integrates the major US laboratories, making it possible for researchers to collaborate remotely on experiments, computational modelling, and education.

2. Implementation of the Pseudodynamic Test Method

PsD testing consists of the step-by-step integration of the discrete-DoF equation of motion

$$\mathbf{Ma} + \mathbf{r(d)} = \mathbf{f}(T) \tag{1}$$

where $\mathbf{M}$ is the theoretical matrix of mass, $\mathbf{a}$ and $\mathbf{d}$ represent the unknown vectors of acceleration and displacement and $\mathbf{f}(T)$ are known external forces that, in the case of a seismic excitation, are obtained by multiplying the specified ground acceleration by the theoretical masses. The unknown restoring forces $\mathbf{r(d)}$ are experimentally obtained at every integration step by quasistatically imposing, generally by means of a hydraulic control system, the computed displacements.

For every PsD test, we will define the prototype time T (Figure 2), which corresponds to the one of the original problem with an earthquake excitation that may last for a few tens of seconds, and the experimental time t, which may extend to several hours for the execution of the test in the laboratory. We may call λ the time scale factor, defined as the proportion between experimental and prototype times, i.e.,

$$\lambda = \frac{t}{T} \tag{2}$$

that usually ranges from one hundred to a few thousands.

2.1. CLASSICAL STEPWISE PSD METHOD

In the classical PsD method, the time increment ΔT for the integration of the equation of motion is chosen small enough to satisfy the stability and accuracy criteria of the integration scheme. The ground accelerogram must be discretised with the prototype record increment (Figure 2). The smaller this time increment is chosen, the larger the number of integration steps will be to cover the duration of the earthquake. The execution of every step will take place in the corresponding experimental time lapse Δt, which uses to be in the order of several seconds. In fact, the experimental time Δt is split in four phases (Figure 2):

1. A stabilising hold period Δt_{h1} of the system motion after the ramp of the reference signal at the controller. In practise, this period allows the specimen to reach the computed displacement. If the computed displacement has not been accurately achieved, the measured force will not correspond to the computed displacement.
2. A measuring hold period Δt_{h2} that allows reducing the signal noise by averaging a number of measures. This could be important to reduce some random errors in the solution.
3. A computation hold period Δt_{h3} for solving the next step at the integration algorithm. This period includes also the transmission time if the system equations are solved in a different CPU than the controller itself.
4. A period of ramp Δt_{ramp} at the reference signal in order to smoothly change to the new computed displacement d_{n+1} -- during the three previous hold periods, the reference was maintained constant at the previous value of the computed displacement d_n.

The accuracy in the imposed displacement and the measured force depends, apart from the characteristics of the experimental set-up, on the selected periods for stabilising, measuring and ramp, while the computation period is determined by the system equations and processor characteristics.

2.2. CONTINUOUS PSD METHOD

In the continuous PsD method (Figure 3), as a difference with respect to the classical one, the execution of every integration step takes just one sampling period (S.P.) of the digital controller of the control system, e.g. 2 ms in the ELSA implementation (Magonette et al., 1998). The ramp and stabilising periods are reduced to zero duration and the measuring plus the computation periods must be feasible within those few milliseconds of experimental time. The surprising fact is that, since the hydraulic control system is unable to respond significantly at frequencies in the range of the sampling frequency of its controller, e.g. 500 Hz, the missing periods of ramp and stabilising are not needed at all.

Under these conditions, the accuracy in the imposed displacement depends basically on the testing speed, which is characterised by the time scale factor λ (Eq. 2). In order to reduce the testing speed while the experimental time step is kept fixed to one S.P., every original time increment in the prototype domain is subdivided into a number of internal steps N_{int} (Figure 3), so that, by increasing this number, the time scale factor is enlarged as

$$\lambda = \frac{\Delta t}{\Delta T} = \frac{N_{\text{int}}\,\text{S.P.}}{\Delta T} \tag{3}$$

As shown in Figure 3, at every ΔT the required internal values of the input ground accelerogram are linearly interpolated from the original record values. The total number of integration steps in a continuous PsD test can be of several millions, in comparison with several thousands as typically required for a classical PsD test for the same specified earthquake. This fact implies that, on the one hand, an explicit time integration algorithm can always be used without concern about stability or integration error and, on the other hand, there is no longer need for an averaging period at the measuring of the force since any high-frequency noise at the load cells will automatically be filtered out in the solution. Such filtering effect is due to the equation response characteristics for the frequency associated to such small time increment.

Additionally, working with the continuous PsD method, it is usually possible to perform the test in a shorter experimental time, but with a better accuracy than with the classical method for the same experimental hardware. This is because, as a consequence of the mentioned characteristics of the continuous version, the absence of alternation between ramp and hold periods in the controller reference signal notably improves the control quality.

3. Assessment of Experimental Errors in Pseudodynamic Tests

The experimental errors within the PsD method can be of different types (Shing and Mahin, 1987) and may come from a lack of reliability of the used instruments as well as from the limitations in the adopted discrete model, which obviously includes a discrete mass representation, and the strain-rate effects coming from the slow application of the load. However, the kind of errors whose effects are the most difficult, in principle, to predict are the ones coming from the control system. In this section we will concentrate on this type of error that can be critical for the success of any PsD test.

3.1. LINEAR MODEL OF THE TESTING SET-UP

The development of a linear model of the testing set-up including the specimen and the hydraulic system can be a very helpful tool for understanding the effects of the experimental errors in a PsD test. For an ideal case without control errors, running the test with a time scale factor of λ, the history of every variable in the experimental response should equal the one of the prototype except for the time scale. For example, that means that, in the time t of the laboratory, the experimental response eigenmodes and damping ratios would be the same as for the prototype structure but the eigenfrequencies will be λ times smaller than in the prototype (Molina et al., 2002a):

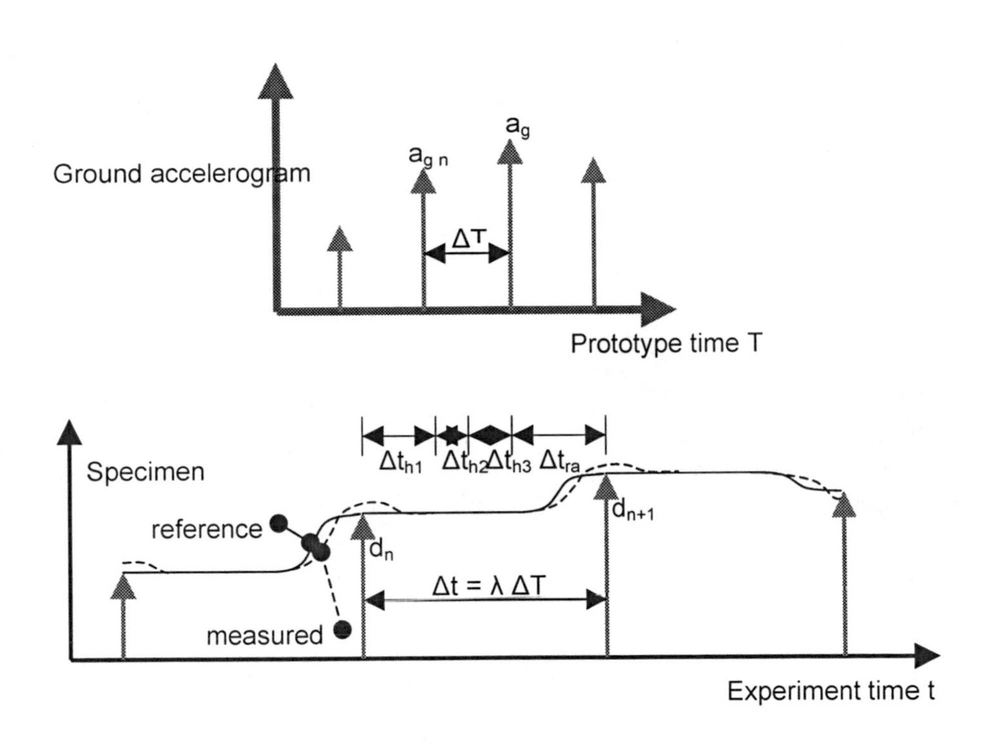

Figure 2. Classical stepwise PsD method.

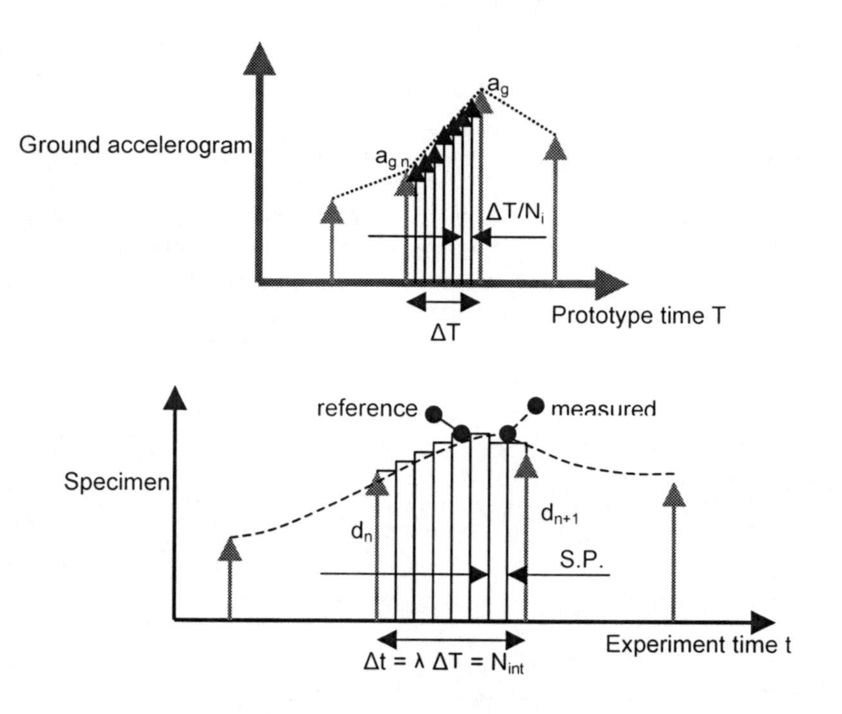

Figure 3. Continuous PsD method.

$$\left(\omega_n\right)_t^{id} = \frac{\omega_n}{\lambda} \qquad ; \qquad \left(\zeta_n\right)_t^{id} = \zeta_n \tag{4}$$

In the presence of errors introduced by the hydraulic system, the obtained experimental frequencies and damping ratios will differ from the ideal ones so that respective associated errors can be defined as

$$\frac{\left(\omega_n\right)_t - \left(\omega_n\right)_t^{id}}{\left(\omega_n\right)_t^{id}} \qquad ; \qquad \left(\zeta_n\right)_t - \left(\zeta_n\right)_t^{id} \tag{5}$$

Using the proposed linear model, the behaviour of these errors can be analysed as a function of the characteristic of the set-up components, such as the pistons and servovalves dimensions, and the testing parameters. The testing speed and the controller proportional gain are among the testing parameters that most strongly may influence the accuracy of the test. In the case of the classical version of the method, the selected hold and ramp periods (Figure 2) may also be determinant, but here we will concentrate our study on the continuous version of the method. An example of this type of analysis will be shown here. In this example (Figure 4), a piston deforms a single-DoF specimen with prototype mass of 4150 kg, eigenfrequency

$$\omega_n = 2.5\,Hz \tag{6}$$

and damping ratio

$$\zeta_n = 0.10 \tag{7}$$

In practice, the tested specimens generally exhibit hysteretic damping, in which case this parameter must be understood as a "viscous-equivalent" damping ratio. However, within the model, the damping of the specimen is of viscous type with a damping constant inversely proportional to the testing speed. This scheme allows having a damping force that does not change with the time scale as one can expect from hysteretic damping behaviour.

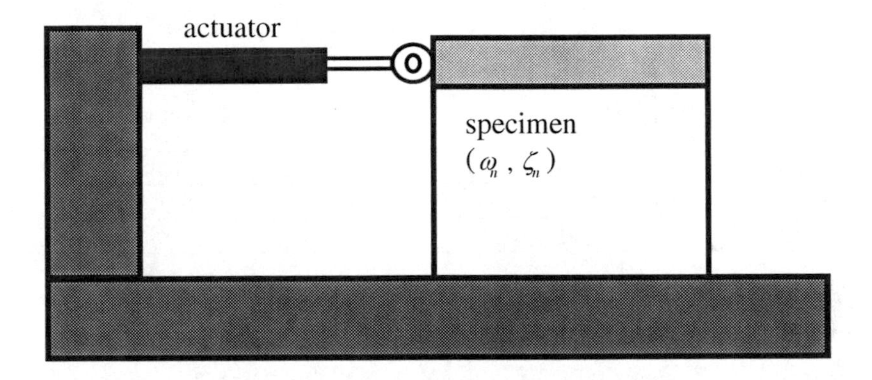

Figure 4. Example of PsD set-up with a single DoF specimen.

The model will include a proportional controller and an ideal linear servovalve. The compliance of the oil column inside the cylinder will be considered by means of a linear spring of $5{\cdot}10^7$ N/m and the internal friction in the cylinder by a linear damper of $6{\cdot}10^4$ Ns/m. The mass of the piston moving with the specimen mass will be 67.5 kg.

Initially, the proportional gain of the controller-servovalve-piston set is given a value of

$$P_{con} = 2.5\ s^{-1} \tag{8}$$

defined within this model in terms of piston theoretical velocity (for uncompressible oil) divided by displacement error.

The results of the application of the model to a continuous PsD test at different testing speeds are shown in Table and Figure 5. If the test is performed at a speed λ times slower than the reality (see the first column at the table), the ideal value of the response frequency and damping in the laboratory time are given by Eq. (4). The ideal frequency is given at the second column of the table while the ideal damping ratio is 0.10 for all the cases. The response eigenfrequency and damping ratio as obtained by solving the described PsD linear model during the test are shown at columns 4 and 5. The respective error measures on the eigenfrequency and damping are also shown at columns 6 and 7. It is clearly seen that their values grow with the testing speed.

Table 1. PsD test example at several testing speeds.

λ	$(\omega_n)_t^{id}$	(HP_{con})	$(\omega_n)_t\ (Hz)$	$(\zeta_n)_t$	$\frac{(\omega_n)_t - (\omega_n)_t^{id}}{(\omega_n)_t^{id}}$	$(\zeta_n)_t - (\zeta_n)_t^{id}$	$\frac{1}{2}\lfloor H(\omega_{id})\rfloor_r$
1000	0.00250	2.5	0.002503	0.0969	0.0010	-0.0031	-0.0032
100	0.02500	2.5	0.025121	0.0680	0.0048	-0.0320	-0.0320
10	0.25000	2.5	0.229815	0.1572	-0.0807	-0.2572	-0.2855

The damping errors are generally much higher than the frequency errors. Thus, in this example we see that,

- for the test at a speed 1000 times slower than reality, the response frequency is very good (0.1% error) while the damping is just acceptable (3% error) and
- for the test at a speed 100 times slower than reality, the response frequency is still quite good (0.5% error) while the damping is questionable (32% error), but,
- for the test at a speed 10 times slower than reality, the response frequency is still reasonably accurate (8% error) while the damping is completely aberrant (257% error!).

In fact, the fastest case would show a negative apparent damping ratio of −0.1572 with which the structure, even without excitation, would exponentially increase the amplitude of oscillation at its eigenfrequency. That test would be useless and dangerous.

A graphical representation the PsD frequency response function (FRF) obtained from the model (displacement response divided by external force) is shown for all three cases at Figure 5. The amplitude is represented in the upper graph while the phase (in degrees) is represented in the lower graph. For each one of the three values of λ, the experimental FRF (solid line) is represented as well as the ideal one (dashed line). One can see again how the damping error increases with testing speed and, in particular for the fastest case, the presence of negative damping ratio is detected by the positive phase of the displacement with respect to the force.

In order to understand the origin of these errors within this simple model, the representation of the closed-loop FRF of the control system for this set-up can be helpful (Figure 6). Such FRF, now defined as the performed displacement divided by reference displacement, depends on the characteristics of the set-up and on the controller parameters such as the proportional gain, but does not depend on the PsD testing speed or on the type of test to perform with this set-up. Looking at the first curve, corresponding to the current case with a proportional controller gain of 2.5, one can, effectively see that:

- For the test at a speed 1000 times slower than reality, the dominant frequencies are around $(\omega_n)_t^{id} = 0.0025\,Hz$ and in that area the closed-loop FRF is almost perfect in amplitude (upper graph) and shows just a slight negative phase (lower graph). As a consequence, no important error is introduced by the control system in the PsD FRF (Figure 5).

- For the test at a speed 100 times slower than reality, the dominant frequencies are around $(\omega_n)_t^{id} = 0.025\,Hz$ and in that area the closed-loop FRF is still quite accurate in amplitude but a significant negative phase is present. There, considerable error is now introduced by the control system in the PsD FRF.

- Finally, for the test at a speed 10 times slower than reality, the dominant frequencies are around $(\omega_n)_t^{id} = 0.25\,Hz$ and in that area the closed-loop FRF is no longer accurate in amplitude and shows an exaggerated negative phase. There, unacceptable error is introduced by the control system in the PsD FRF as was already observed.

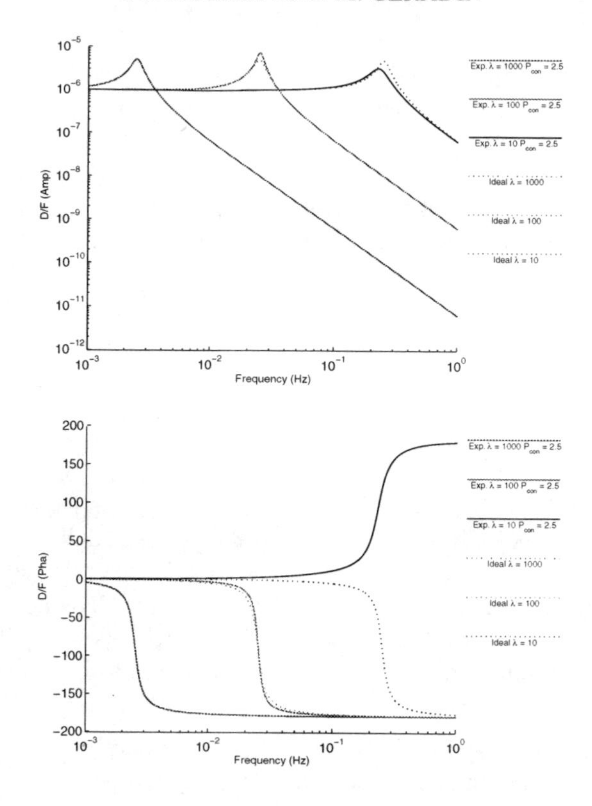

Figure 5. PsD test example at several testing speeds. FRF amplitude (top) and phase (bottom).

In general, the errors introduced by the control system in a PsD test are critical in terms of phase of the closed-loop FRF. Quantitatively, for a SDoF system, a good approximation of the damping error can be given by the formula (Molina et al, 2002a):

$$(\zeta_n)_t - (\zeta_n)_t^{id} \approx \frac{1}{2} \underline{|H(\omega_{id})}_{rad} \tag{9}$$

where the right-hand side is one half of the phase of the closed-loop FRF of the control system at the ideal eigenfrequency of the test. The value of this approximation is also included at the last column of the table so that it can be compared with the damping error included in the previous column.

It is interesting to notice that, according to Eq. (9), an out-of-phase on the control system of only one degree in the area of the PsD-response eigenfrequency, will imply an error in the order of

$$(\zeta_n)_t - (\zeta_n)_t^{id} \approx \frac{1}{2}(-1.0\frac{\pi}{180}) = -0.0087 \tag{10}$$

which means almost a decrease of 0.01 in damping ratio in absolute terms. For example, for a specimen in elastic range with a damping ratio of, say 0.02, that decrease would divide the performed damping by two, which is clearly unacceptable.

Let us now apply the linear model to study the effects of the variation of the controller proportional gain. To that purpose, the testing time scale will be kept now constant at 100 times slower than reality, but the proportional gain will be given the values of 2.5, 1.0 and 0.5 as shown at Table 1, Figure 6 and Figure 7. The meaning of the columns of the table and the curves of the graphs is similar to the ones of the previous study. Notice that in this study the ideal response frequency is constantly 0.025 Hz since the testing speed is also invariant, however the experimental response does change depending on the selected value of proportional gain.

Table 1. PsD test example at several values of the controller proportional gain.

λ	$(\omega_n)_t^{id}\ (Hz)$	P_{con}	$(\omega_n)_t\ (Hz)$	$(\zeta_n)_t$	$\frac{(\omega_n)_t - (\omega_n)}{(\omega_n)_t^{id}}$	$(\zeta_n)_t - (\zeta_n)_t^{i}$	$\frac{1}{2}\lfloor H(\omega_{id})$
100	0.02500	2.5	0.025121	0.068	0.0048	-0.0320	-0.0320
100	0.02500	1.0	0.025104	0.019	0.0042	-0.0807	-0.0796
100	0.02500	0.5	0.024628	-0.054	-0.0149	-0.1547	-0.1557

It is again observed from this study that the damping error is generally much higher than the frequency error. In the worst case, the experimental damping ratio can be again negative, but in this case due to a poor selection of the proportional gain (last row in the table).

Again, the origin of the errors can be understood by looking at the closed-loop FRF of the control system in Figure 6. There, for the ideal frequency of 0.025 Hz, the different errors introduced by the control in terms of amplitude and phase are clearly seen.

As a conclusion, it is observed that higher values of the proportional gain of the controller generally imply lower errors in the control and in the response parameters of the PsD test. Nevertheless, from the control theory (Phillips and Harbor, 1999), it is known that, by excessively increasing the proportional gain, the control system may become unstable. This fact can be assessed by looking at the open-loop FRF of the control system for the same values of that gain (Figure

8). For this kind of system, the gain margin can be defined as the inverse of the open-loop FRF amplitude at the frequency at which the FRF phase is equal to −180 degrees. The system will become unstable when the gain margin is less than one.

In practical cases, a gain margin considerably larger than one must be used in order to guaranty the stability even when some of the conditions of the system change. At the figure we see that for

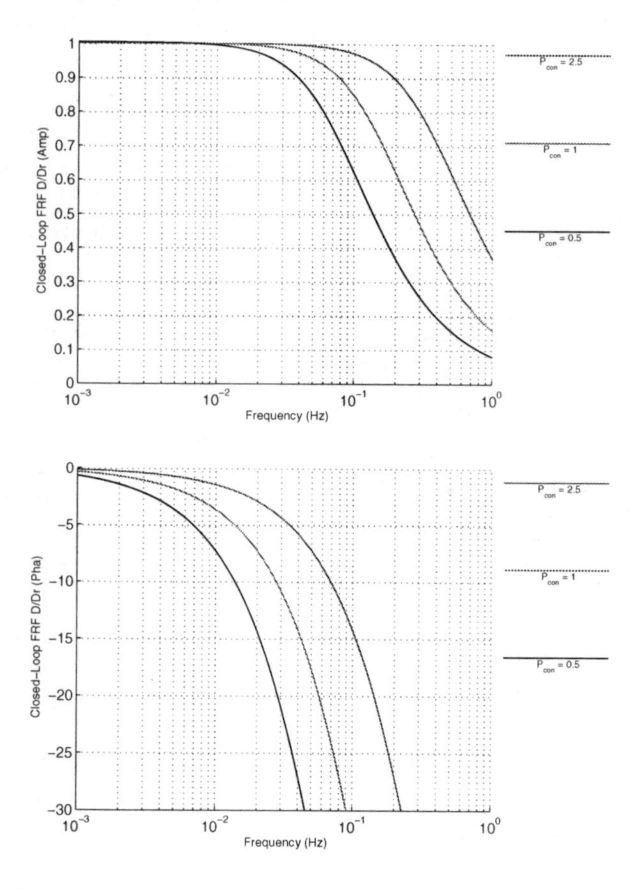

Figure 6. Example control system. Closed-loop FRF amplitude (top) and phase (bottom).

$P_{con} = 2.5$ the gain margin is in the order of 10, which probably means that the control system is safely stable. However, from this value down, the gain margin may start to be considered in a dangerous range and the system may show unstable due unpredictable changes coming from factors such as the oil temperature or the nonlinearities of the components. In practice, the alternative way to improve the quality of the control without compromising its stability consists of the introduction of the integral gain. Therefore, structural tests in displacement control such as the PsD tests are generally performed using a PI controller.

3.2. ESTIMATION OF PSD RESPONSE PARAMETERS

As already shown in subsection 3.1, the experimental quality of the results from a PsD test is strongly

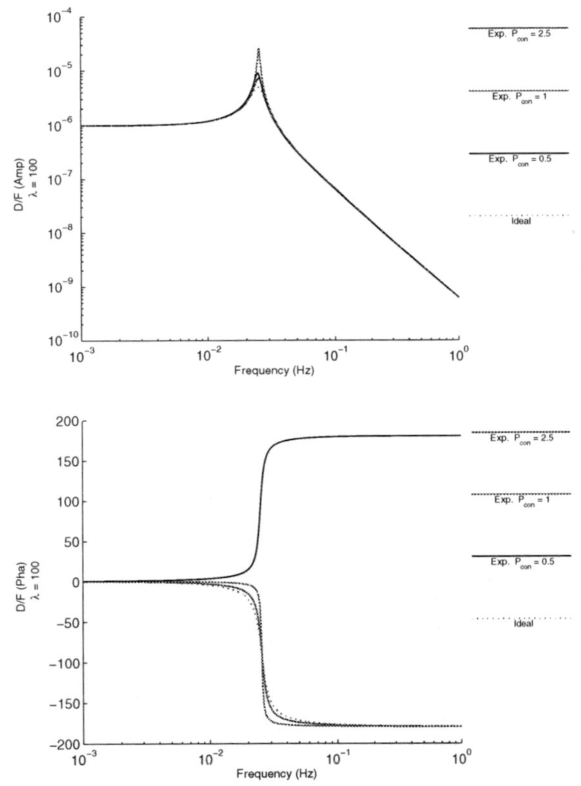

Figure 7. PsD test example at several values of the controller proportional gain. FRF amplitude (top) and phase (bottom).

determined by the magnitude and the characteristics of the errors in the control system during the test. Since it is not always feasible to develop a calibrated model such as the one developed in subsection 3.1, different techniques will be proposed here for the on-line estimation of parameters related to the response errors during a PsD test (Molina et al., 2002a). The first technique is simple to apply but provides only global information while the second one is more complex to apply but provides information for every mode in the response.

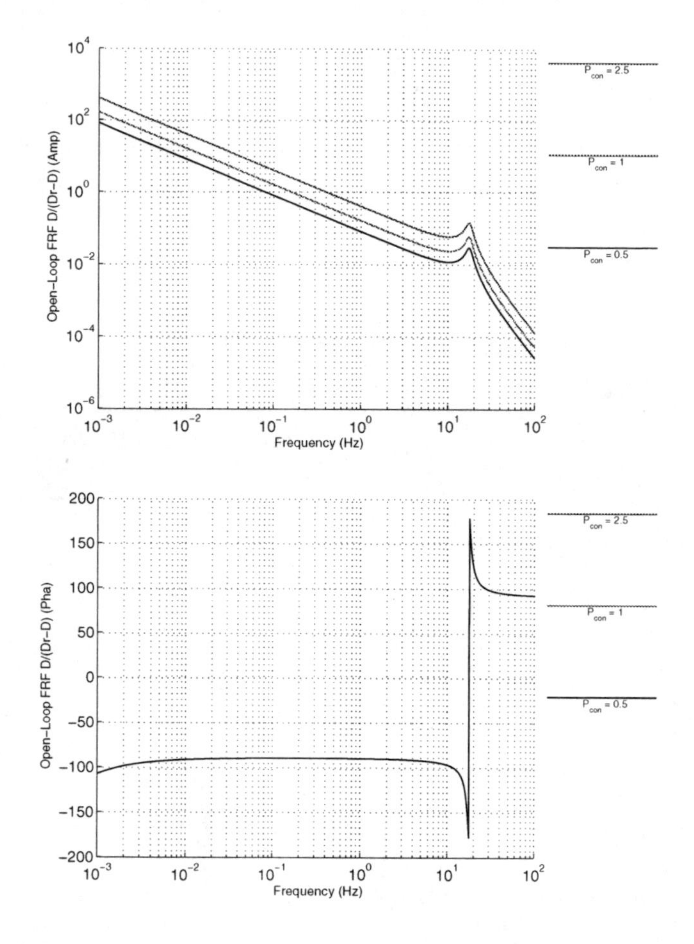

Figure 8. Example control system. Open-loop FRF amplitude (top) and phase (bottom).

3.2.1. *Energy Error*

The energy absorbed by the structure during the test may be computed from the measured restoring forces and displacements as

$$E = \int \mathbf{r}^T d\mathbf{d} \tag{11}$$

In can also be estimated in terms of the variables seen by the integration algorithm, i.e. the measured restoring forces and reference displacements,

$$E_r = \int \mathbf{r}^T d\mathbf{d}_\mathbf{r} \tag{12}$$

Now, let us define the energy error as the difference between (11) and (12), which is also the work done by the restoring forces on the displacement error:

$$\Delta E = E_r - E = \int \mathbf{r}^T d\mathbf{d}_\mathbf{r} - \int \mathbf{r}^T d\mathbf{d} = \int \mathbf{r}^T d(\mathbf{d}_\mathbf{r} - \mathbf{d}) = \int \mathbf{r}^T d\varepsilon \tag{13}$$

By comparing this energy error with the total energy of Eq (11), a global upper estimate of the damping error effect due to the control error is obtained (Thewalt and Roman, 1994, Molina et al., 2002). When applying this variable as a global indicator, our experience is that in general, for an acceptable quality in the test results, the energy error should not exceed, say, a 5% of the total absorbed energy at any moment. Once the testing parameters such as the controller gains and the testing speed have been settled, a short-duration very-low-intensity test should be done to check this criterion. If the criterion is satisfied for that preliminary test in the linear range of the structure, the quality of the results should in general improve for larger amplitudes because, with the degradation of the structure, the damping ratio should grow while the eigenfrequency decreases and, as a consequence, the damping error introduced by the control should also decrease (look for example at Eq. 9 and Figure 6).

3.2.2. *Spatial Model*

This technique is based on the identification of a linear model of the structure using a narrow time window of the experimental response. The identified parameters are then transformed into frequency and damping characteristics that give very useful information regarding the quality of the performed test. By gaining experience in the application of this identification, it is possible to have better knowledge of what the feasibility and the accuracy of the experiment can be depending on the type of structure and the applied PsD testing set-up.

For the identification of the eigenmodes of the structure from the results of a PsD test, in fact, two different time-invariant linear models have been proposed that can be used working in the time domain (Molina et al., 1999a). The advantage of working in the time domain is in the easiness to automate the identification process. The simplest and most robust model is a spatial one that is going to be described here. Within this model, it is assumed that the measured restoring forces and the corresponding displacements and velocities are linked, for every discrete time n, in the form

$$\mathbf{r}(n) = \mathbf{K}\mathbf{d}(n) + \mathbf{C}\mathbf{v}(n) \tag{14}$$

where $\mathbf{r}(n)$, $\mathbf{d}(n)$ and $\mathbf{v}(n)$ result from the test. Of course, the linear stiffness and viscous damping of this model must be interpreted as an approximation to the real non-linear hysteretic behaviour. For identification purpose, the model can be rewritten in the form

$$\begin{bmatrix} \mathbf{d}^T(n) & \mathbf{v}^T(n) & \mathbf{1} \end{bmatrix} \begin{bmatrix} \mathbf{K}^T \\ \mathbf{C}^T \\ \mathbf{o}^T \end{bmatrix} = \mathbf{r}^T(n) \tag{15}$$

where a constant force offset term $\mathbf{o}$ has also been added and $\mathbf{1}$ is a column of ones. Here, if n_{DoF} is the number of DoFs in the structure, $\mathbf{K}$, $\mathbf{C}$ and $\mathbf{o}$ contain $2n_{DoF}^{\ 2} + n_{DoF}$ unknowns and the number of available equations is $N\, n_{DoF}$, so that, the minimum required number of discrete-time data sets is

$$N \geq 2n_{DoF} + 1 \tag{16}$$

Once $\mathbf{K}$, $\mathbf{C}$ and $\mathbf{o}$ have been estimated by a least squares solution, the complex eigenfrequencies and mode shapes can be obtained by solving the generalised eigenvalue problem (Maia and Silva, 1997)

$$s\begin{bmatrix} \mathbf{C} & \mathbf{M} \\ \mathbf{M} & \mathbf{0} \end{bmatrix}\boldsymbol{\varphi} + \begin{bmatrix} \mathbf{K} & \mathbf{0} \\ \mathbf{0} & -\mathbf{M} \end{bmatrix}\boldsymbol{\varphi} = \mathbf{0} \tag{17}$$

where $\mathbf{M}$ is the theoretical mass matrix. The complex conjugate eigenvalues can be expressed in the form

$$s_n, s_n^{\ *} = \omega_n(-\zeta_n \pm j\sqrt{1-\zeta_n^2}) \tag{18}$$

where ω_n is the natural frequency and ζ_n the damping ratio. The corresponding mode shape is also given by the first n_{DoF} rows of the associated eigenvector $\boldsymbol{\varphi}_n$.

Since this model assumes an invariant system, at any selected time instant, an identification may be done based on a data time window, of duration roughly equal to the period of the first mode, centred around that instant. The adopted time window has to be narrow enough so that the system does not change too much inside of it, but, at the same time, it has to contain enough data to allow the compensation of different existing data noises and nonlinearities. The selection of the most appropriate window length is done by trial and error. It is usually possible to estimate the eigenfrequencies and damping ratios of all the modes at any time instant.

In order to use the method for the assessment of the error on the PsD response, the identification process may be repeated for two sets of variables entering in the model of Eq (14)), that is,

- First set: measured forces, measured displacements and derived velocities,
- Second set: measured forces, reference displacements and derived velocities.

The first set of variables takes into account the synchronous measured forces and displacements on the specimen without any influence from the control errors so that the identified frequencies and damping from this set are considered the real ones. For the second set, the considered forces and displacements are the ones entering in the PsD equation and the eigenfrequencies and damping linking them are the ones that explain the test response. For every frequency or damping ratio of the structure, the difference between the identified value with both sets of data is considered as the error introduced by the control system. As for the energy error criterion application, it is recommended to set the testing parameters so that, for the small-amplitude checking test, the damping error for every mode is kept under a 5% of the total damping of that mode. Looking at Eq. (9) and Figure 6, it is easy to understand that the higher modes with low damping use to be the most conditioning ones for the PsD test quality (Molina et al., 2000). Some results of the application of this estimation technique will be shown in the following section for a particular performed test within the SPEAR project.

As a different application, during the large-intensity tests, the identified response parameters may also be used to predict the response for a short time lapse. This is useful for piloting the test and may help taking the decision of stopping it before a dangerous configuration is attained by the structure.

4. Bidirectional PsD Testing (SPEAR Project)

For PsD testing of building specimens in one direction without torsion, normally two hydraulic actuators acting in that direction are used per floor. A first bidirectional PsD test was successfully performed on a real-size frame in 1997 at

ELSA by using two actuators acting on each direction for everyone of its three floors (Molina et al., 1999b). At every integration step of the equation of motion, the developed procedure included, firstly, the transformation of measured restoring forces from the four actuator coordinates to the three DoFs of every floor, secondly, the finite-difference prediction of the new displacement and, thirdly, its transformation into actuator displacements. Those new target displacements were then imposed to the structure.

The testing technique described in this section has been applied for obtaining the experimental bidirectional response of a reinforced concrete (RC) three-storey building within the SPEAR project. The SPEAR project (Seismic Performance Assessment and Rehabilitation) focused on existing buildings because of the current economic and social relevance of their seismic performance. The experimental campaign on this structure consisted of rounds of seismic tests on the original structure and on two retrofitted configurations. This section describes essentially the testing technique and the corresponding experimental set-up; more information about the project itself, the structural characteristics of the specimen and the test results has been given by Mola et al. (2004) and Negro and Mola (2005).

4.1. TESTING TECHNIQUE

This subsection covers many of the aspects of the applied PsD testing technique, starting with the formulated general equation of motion (including three DoFs per floor) and time integration algorithm, and following with the specific force and displacements coordinate transformations needed for the adopted control system and strategy. Details on the implemented step-by-step testing procedure and on testing hardware and software characteristics are also given.

4.1.1. *Equation of Motion*

The mass of the building is assumed to be concentrated at the rigid floors, so that the equation of motion of every floor in the horizontal plane takes the form

$$\mathbf{ma} + \mathbf{r} = -\mathbf{mja}_{\mathbf{g}} \tag{19}$$

where the floor mass matrix

$$\mathbf{m} = \begin{bmatrix} \mathrm{m} & 0 & 0 \\ 0 & \mathrm{m} & 0 \\ 0 & 0 & \mathrm{I} \end{bmatrix} \tag{20}$$

is expressed in terms of the floor mass m and moment of inertia I at the Centre of Mass (CM), and

$$\mathbf{a} = \begin{bmatrix} a_x & a_y & a_\theta \end{bmatrix}^{\mathrm{T}} = \frac{d^2}{dt^2}\mathbf{d} = \frac{d^2}{dt^2}\begin{bmatrix} d_x & d_y & d_\theta \end{bmatrix}^{\mathrm{T}} \tag{21}$$

is the vector of relative accelerations, which is the second derivative with respect to time of the floor relative displacements d_x, d_y and rotation d_θ at the CM (Figure 9). Since the laboratory time will not be used at all within this section, the classical symbol t will refer to prototype time.

The vector of conjugated generalised restoring forces

$$\mathbf{r} = \begin{bmatrix} r_x & r_y & r_\theta \end{bmatrix}^{\mathrm{T}} \tag{22}$$

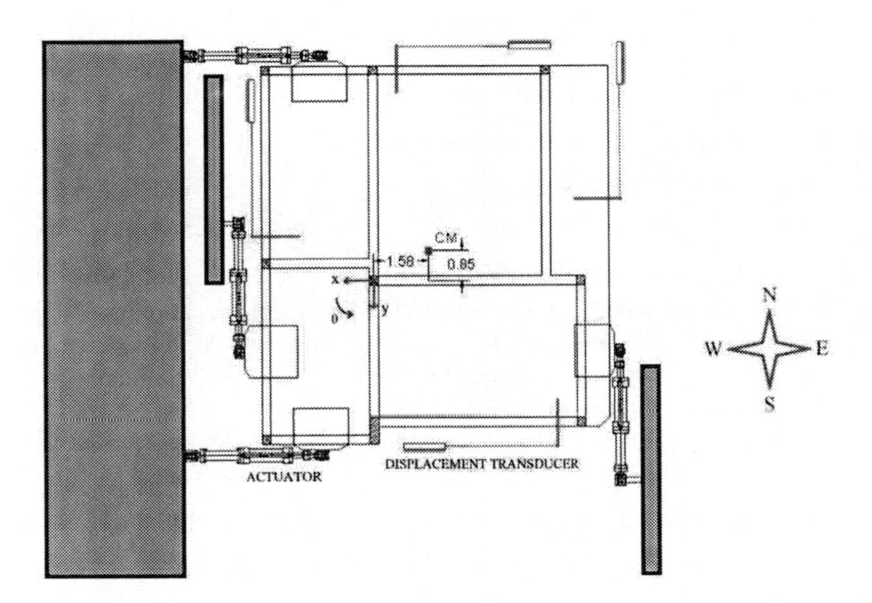

Figure 9. Floor axis, actuators and displacement transducers.

contains the resultant forces and moment at the CM which are a function of the deformations of the building. The right-hand side of Eq. (19) is expressed in terms of the ground accelerations in directions x and y and the influence matrix:

$$\mathbf{a_g} = \begin{bmatrix} a_{gx} & a_{gy} \end{bmatrix}^{\mathrm{T}} \quad , \quad \mathbf{j} = \begin{bmatrix} 1 & 0 \\ 0 & 1 \\ 0 & 0 \end{bmatrix} \tag{23,24}$$

The system of equations of motion for the multi-storey structure can thus be written as

$$\mathbf{MA} + \mathbf{R} = -\mathbf{MJa_g} \tag{25}$$

with the matrices

$$\mathbf{M} = \begin{bmatrix} \ddots & & 0 \\ & \mathbf{m} & \\ 0 & & \ddots \end{bmatrix} ; \; \mathbf{A} = \begin{bmatrix} \vdots \\ \mathbf{a} \\ \vdots \end{bmatrix} ; \; \mathbf{R} = \begin{bmatrix} \vdots \\ \mathbf{r} \\ \vdots \end{bmatrix} ; \; \mathbf{J} = \begin{bmatrix} \vdots \\ \mathbf{j} \\ \vdots \end{bmatrix} \tag{26}$$

collecting the contributions from the different floors of the building.

4.1.2. *Time Integration Algorithm*

Equation (25) is now generalised as

$$\mathbf{MA} + \mathbf{CV} + \mathbf{R} = \mathbf{F} \tag{27}$$

where $\mathbf{C}$ is a viscous damping matrix, $\mathbf{V}$ is the vector of relative velocities and $\mathbf{F}$ is a general external force vector including either seismic equivalent forces or directly applied forces. For classic materials such as steel or concrete, most of the damping is hysteretic, which is included in the velocity-independent restoring forces $\mathbf{R}$, so that $\mathbf{C}$ matrix is taken null for the PsD model.

Two different time integration algorithms are classically used at ELSA for PsD testing (Molina, 1999b), i.e.,

- the Explicit Newmark algorithm (equivalent to the central difference scheme) and
- the α Operator Splitting algorithm.

The second method requires and estimation of the stiffness matrix of the structure which may affect the results if not properly chosen and updated. For this reason, the Explicit Newmark algorithm is preferred whenever a time increment Δt may be chosen such that the algorithm is stable and the test duration is feasible. The stability is guaranteed when

$$\Delta t \leq T_{\min} / \pi \tag{28}$$

where $T_{\min}$ is the minimum period of the structure.

The Explicit Newmark algorithm was adopted for the SPEAR test. It may be written in the form

$$\begin{aligned} &\mathbf{MA}_{n+1} + \mathbf{CV}_{n+1} + \mathbf{R}_{n+1} = \mathbf{F}_{n+1} \\ &\mathbf{D}_{n+1} = \mathbf{D}_n + \Delta t \mathbf{V}_n + \Delta t^2 \left[\left(\frac{1}{2} - \beta \right) \mathbf{A}_n + \beta \mathbf{A}_{n+1} \right] \\ &\mathbf{V}_{n+1} = \mathbf{V}_n + \Delta t \left[(1 - \gamma) \mathbf{A}_n + \gamma \mathbf{A}_{n+1} \right] \end{aligned} \tag{29}$$

with the specific parameter values

$$\beta = 0; \ \gamma = \frac{1}{2} \tag{30}$$

Contrarily to an integration based on a finite element model of the structure, in a PsD test, the restoring forces at every time step are not computed. They are directly measured on the physical model after imposition of the corresponding displacements:

$$\mathbf{R}_{n+1} = \mathbf{R}(\mathbf{D}_{n+1}) \tag{31}$$

Starting from known initial values $\mathbf{D}_1, \mathbf{V}_1, \mathbf{A}_1$, at time step n=1 (t=0), the successive computation stages of the method at every time step are:

- Stage 1) Compute at instant n+1 the values of displacements $\mathbf{D}_{n+1}$ using Eq. (29).
- Stage 2) Impose to the structure the new displacement $\mathbf{D}_{n+1}$ and measure the associated restoring force of Eq. (31).

- Stage 3) Compute the acceleration at new time $n+1$ from the equilibrium equation

$$\mathbf{A}_{n+1} = \left[\mathbf{M} + \Delta t/2\,\mathbf{C}\right]^{-1}\left[\mathbf{F}_{n+1} - \mathbf{R}_{n+1} - \mathbf{C}\left(\mathbf{V}_n + \Delta t/2\,\mathbf{A}_n\right)\right] \tag{32}$$

- Stage 4) Increment the step counter $n \leftarrow n+1$ and go back to stage 1 until the final time is reached.

4.1.3. *Kinematic Transformation from Generalised to Transducer Displacements*

At every step of the PsD test, the computed generalised displacement of the floor is imposed through the actuators with feedback on the displacement transducers. Thus, in order to determine the target displacements at transducer level, a geometric transformation is needed. A minimum number of 3 high-resolution linear displacement transducers are attached to each floor for control purpose.

As shown in Figure 10, each control displacement transducer consists of a slider G1-G2 on which a body M translates and gives a measure of its relative position on the slider. The slider is attached to a fixed reference frame while the mobile body M is connected to the measuring point D on the structure through a pin-jointed rod.

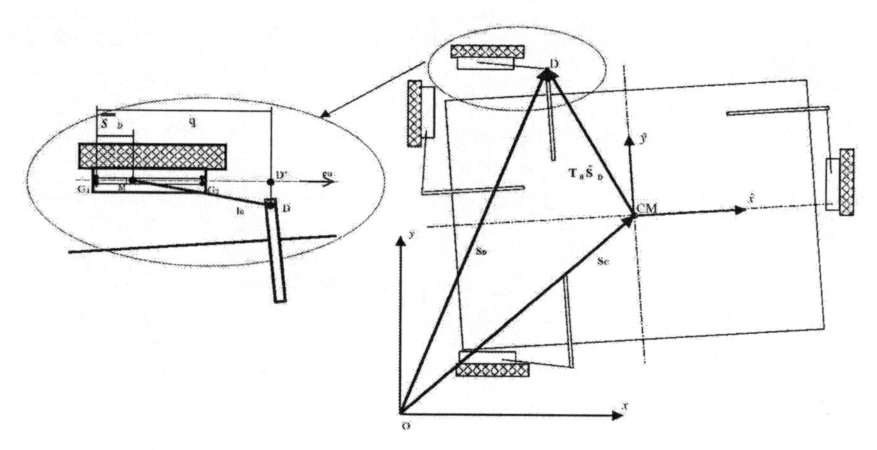

Figure 10. Control displacement transducer.

For every floor, starting from a known set of generalised displacements **d** (21), the position of the floor CM is updated as

$$\mathbf{S}_\mathrm{C} = \mathbf{S}_\mathrm{C}^0 + \mathbf{d} \tag{33}$$

where

$$\mathbf{S_C} = \begin{bmatrix} x_C & y_C & \theta_C \end{bmatrix}^{\mathrm{T}} \tag{34}$$

are its global coordinates and $\mathbf{S_C^0}$ are their reference value for zero displacement. Then, the position of the measuring point D is likewise updated as

$$\mathbf{S_D} = \begin{bmatrix} x_D \\ y_D \end{bmatrix} = \begin{bmatrix} x_C \\ y_C \end{bmatrix} + \mathbf{T_{\theta_C}} \mathbf{\hat{S}_D} \tag{35}$$

where

$$\mathbf{\hat{S}_D} = \begin{bmatrix} \hat{x}_D \\ \hat{y}_D \end{bmatrix} \tag{36}$$

are the coordinates of point D in a reference system local to the floor, centred at its CM, and

$$\mathbf{T_{\theta_C}} = \begin{bmatrix} \cos\theta_C & -\sin\theta_C \\ \sin\theta_C & \cos\theta_C \end{bmatrix} \tag{37}$$

is a rotation matrix. Assuming an infinitely rigid floor, the local coordinates of Eq. (36) are constant.

In order to express the position of body M (Figure 10) of the transducer along its slider $G_1 - G_2$, the relative position of the measuring point D with respect to the slider origin G_1 is computed as

$$\mathbf{\tilde{S}_D} = \overrightarrow{G_1 D} = \mathbf{S_D} - \mathbf{S_{G1}} \tag{38}$$

and the projection of vector of Eq. (38) along the slider is obtained as

$$q = \overline{G_1 D'} = \mathbf{\tilde{S}_D}^T \mathbf{e_G} \tag{39}$$

where $\mathbf{e_G}$ is the constant unit vector defined in the direction of positive measurement along the slider.

The position of body M along the slider is then given by

$$\overline{\overline{S}}_D = \overline{G_1 M} = \overline{G_1 D'} - \overline{MD'} = q - \mathrm{sign}(q)\sqrt{l_D^2 - \overline{DD'}^2} = q \tag{40}$$

being l_D the constant length of the rod. Finally, the corresponding measure at the transducer is

$$d_D = \overline{\overline{S}}_D - \overline{\overline{S}}_D^{\,0} \tag{41}$$

where $\overline{\overline{S}}_D^{\,0}$ is a prescribed reference position.

4.1.4. *Kinematic Transformation from Measured to Generalised Displacements*

The actual displacements and rotation at the CM will differ from the desired ones and this is due to geometry and control errors as well as to the flexibility of the floor. In order to get an estimate of the actual generalised displacements, the measures given by all the control displacement transducers may be exploited. Since the obtained relations between generalised and transducer displacements are non-linear and cannot easily be inverted in closed form, the solution is achieved through a Newton-Raphson iteration procedure starting from a first estimate of the generalised displacements, e.g.,

$$\mathbf{d}^{EST} = \mathbf{0} \tag{42}$$

The non-linear equations giving the associated transducer displacements $\mathbf{d}_\mathbf{D}$ are applied by substituting $\mathbf{d}$ by $\mathbf{d}^{EST}$ into Eq. (33) and applying Eqs. (35) to (41)

$$\mathbf{d}_\mathbf{D}^{EST} = \mathbf{d}_\mathbf{D}\left(\mathbf{d}^{EST}\right) \tag{43}$$

The difference between this estimate and the measured transducer displacements $\mathbf{d}_\mathbf{D}$ is used to provide an new estimate of the generalised displacements

$$\mathbf{d}^{EST} \leftarrow \mathbf{J}^{-1}\left(\mathbf{d}_\mathbf{D} - \mathbf{d}_\mathbf{D}^{EST}\right) + \mathbf{d}^{EST} \tag{44}$$

where $\mathbf{J}$ is the Jacobian matrix

$$\mathbf{J} = \left.\frac{\partial \mathbf{d}_\mathbf{D}}{\partial \mathbf{d}}\right|_{\mathbf{d}^{EST}} \tag{45}$$

computed at $\mathbf{d}^{EST}$.

Taking into account Eq. (40), every single row of J can be obtained as

$$\frac{\partial d_D}{\partial \mathbf{d}} = \left[\frac{\partial d_D}{\partial \mathbf{S_D}}\right]\left[\frac{\partial \mathbf{S_D}}{\partial \mathbf{d}}\right] = \left[\frac{\partial \overline{GD'}}{\partial \mathbf{S_D}} - \frac{\partial \overline{MD'}}{\partial \mathbf{S_D}}\right]\left[\frac{\partial \mathbf{S_D}}{\partial \mathbf{d}}\right] = \left[\mathbf{e}_\mathbf{G}^\mathrm{T} - \frac{q\mathbf{e}_\mathbf{G}^\mathbf{T} - \mathbf{S_D}^\mathrm{T}}{\overline{MD'}}\right] \quad (46)$$

If the number of control transducers on the floor exceeds 3, Eq. (25) must be solved in a least squares sense, in which case the inverse of $\mathbf{J}$ in Eq. (44) is replaced by the pseudo-inverse

$$\mathrm{psinv}(\mathbf{J}) = \left[\mathbf{J}^\mathrm{T}\mathbf{J}\right]^{-1}\mathbf{J}^\mathrm{T} \quad (47)$$

The steps of Eqs. (43)-(44) are iteratively repeated until a specified tolerance is reached.

4.1.5. *Static Transformation from Actuator to Generalised Forces*

In a PsD test, the restoring forces are experimentally obtained from the specimen. Once the prescribed displacements are imposed, the acting axial force at every actuator is measured by its load cell. However, in order to express these forces as resultant generalised forces at the CM of the floor, a static transformation is needed. Since the ends of the actuator are pin-jointed, it is assumed that it acts with a purely linear force along the line PR (Figure 11) connecting the ends of the actuator. Starting from the load-cell measure r_P of this force and assuming that the current position of the floor CM is known, the global position of the loading point P is obtained as

$$\mathbf{S_P} = \begin{bmatrix} x_P \\ y_P \end{bmatrix} = \begin{bmatrix} x_C \\ y_C \end{bmatrix} + \mathbf{T}_{\theta_C}\hat{\mathbf{S}}_\mathbf{P} \quad (48)$$

where $\mathbf{T}_{\theta_C}$ is given by Eq. (37). Similarly to Eq. (36), $\hat{\mathbf{S}}_\mathbf{P}$ contains the coordinates of point P in the local reference system to the floor.

Then, the global components of the piston force are computed as

$$\mathbf{p_P} = r_P\mathbf{e_P} \quad (49)$$

where $\mathbf{e_P}$ is a unit vector in the direction of $\overrightarrow{PR}$. The floor generalised restoring forces of Eq. (22) are obtained by summing up the effects of all pistons acting on the floor

$$\mathbf{r} = \sum_{\mathbf{P}} \mathbf{T_P p_P} \tag{50}$$

where

$$\mathbf{T_P} = \begin{bmatrix} 1 & 0 \\ 0 & 1 \\ -y_P & x_P \end{bmatrix} \tag{51}$$

4.1.6. *Optimal Distribution of Piston Loads*

When using a number of actuators larger than the number of DoFs at a floor, it is convenient to maintain an acceptable distribution of loads among all the pistons. This can be done by implementing an algorithm capable of optimising the distribution of piston loads for a known set of generalised floor loads. Even distribution of forces is also desirable because it leads to a better approximation of the distributed inertial forces of a real dynamic event.

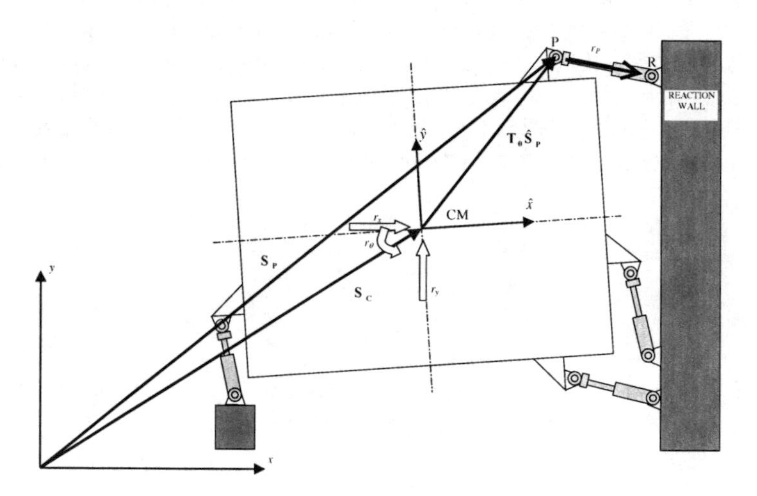

Figure 11. Force applied by the actuator.

To this purpose, the piston forces will be defined as 'optimal' if statically equivalent to the specified generalised loads and while minimising a penalty function that becomes infinite when any piston force reaches its working limit. The adopted expression for the penalty function is

$$f(\mathbf{r_P}) = \sum_P \frac{1}{M_P^{\ 2} - r_P^{\ 2}} \tag{52}$$

where M_P is the working limit of the absolute value of the piston load r_P. Clearly, minimising the function of Eq. (52) will guarantee that all piston loads are kept far from their limit. Using Eqs. (49) - (50), the conditions of static equivalence of the piston loads with the known set of generalised loads give a constraint on the minimisation problem which can be written in the vector form

$$\mathbf{g}(\mathbf{r_P}) = \sum_P \mathbf{T}_P \mathbf{e}_P r_P - \mathbf{r} = \mathbf{0} \tag{53}$$

where $\mathbf{r}$ is the known set of generalised forces of Eq. (22). The constrained minimisation problem is then

$$\begin{cases} f(\mathbf{r_P}) \quad minimum \\ \quad \mathbf{g}(\mathbf{r_P}) = \mathbf{0} \end{cases} \tag{54}$$

By introducing a set of Lagrange multipliers associated to the constraint equations, it can be restated in unconstrained form, leading now to the stationarity condition

$$\frac{\partial h}{\partial \mathbf{x}} = \mathbf{0} \tag{55}$$

of the augmented functional

$$h(\mathbf{x}) = f(\mathbf{r_P}) + \mathbf{g}^{\mathrm{T}}(\mathbf{r_P})\boldsymbol{\lambda} \tag{56}$$

The solution vector

$$\mathbf{x} = \begin{bmatrix} \mathbf{r_P} \\ \boldsymbol{\lambda} \end{bmatrix} \tag{57}$$

contains as unknowns the piston loads and Lagrange multipliers

$$\boldsymbol{\lambda} = [\lambda_1 \quad \lambda_2 \quad \lambda_3]^{\mathrm{T}} \tag{58}$$

The system of Eq. (55) may also be written as

$$\mathbf{A}(\mathbf{x}) = \mathbf{0} \tag{59}$$

where

$$\mathbf{A}(\mathbf{x}) = \begin{bmatrix} \begin{bmatrix} \vdots \\ \dfrac{2r_P}{\left(M_P{}^2 - r_P{}^2\right)^2} \\ \vdots \end{bmatrix} + \mathbf{R}^{\mathrm{T}}\boldsymbol{\lambda} \\ \mathbf{R}\mathbf{r}_{\mathbf{P}} - \mathbf{r} \end{bmatrix} \tag{60}$$

and, from Eq. (53),

$$\mathbf{R} = \begin{bmatrix} \ldots & \mathbf{T}_{\mathbf{P}}\mathbf{e}_{\mathbf{P}} & \ldots \end{bmatrix} \tag{61}$$

Since Eqs. (59) are nonlinear, their solution may be obtained through Newton-Raphson iteration in a similar way as done for obtaining the generalised displacements. Thus, starting from an initial estimate $\mathbf{x}^{\mathbf{EST}}$ of the unknowns, each iteration would consist in computing

$$\mathbf{B}^{\mathbf{EST}} = \mathbf{A}\left(\mathbf{x}^{\mathbf{EST}}\right) \tag{62}$$

from Eq. (60) and updating the estimated optimum parameters by doing

$$\mathbf{x}^{\mathbf{EST}} \leftarrow \mathbf{J}^{-1}\left(\mathbf{0} - \mathbf{B}^{\mathbf{EST}}\right) + \mathbf{x}^{\mathbf{EST}} \tag{63}$$

where the Jacobian matrix is now

$$\mathbf{J} = \left.\frac{\partial \mathbf{A}}{\partial \mathbf{x}}\right|_{\mathbf{x}^{\mathbf{EST}}} = \begin{bmatrix} \ddots & & 0 & \\ & \dfrac{2M_P{}^2 + 6r_P{}^2}{\left(M_P{}^2 - r_P{}^2\right)^3} & & \mathbf{R}^{\mathrm{T}} \\ 0 & & \ddots & \\ & \mathbf{R} & & \mathbf{0} \end{bmatrix} \tag{64}$$

4.1.7. *Piston Internal Displacements*

A further step for indirectly imposing a desired load at the redundant pistons, mentioned at the preceding subsection, is the calculation of the displacement of their internal transducer. This is because using the internal displacement as feedback variable leads to a stable control of the redundant actuators.

Every actuator has an internal displacement transducer measuring the displacement d_P which represents the excursion of the piston. This displacement is a function of the floor generalised displacements $\mathbf{d}$. Once the position of the CM in Eq. (34) is known and the position $\mathbf{S_P}$ of the point P (Figure 11) of attachment of the piston to the floor calculated by Eq. (48), the current length of the piston is computed as

$$\overline{\overline{S}}_P = \sqrt{(\mathbf{S_P} - \mathbf{S_{PR}})^T (\mathbf{S_P} - \mathbf{S_{PR}})} \tag{65}$$

and the internal displacement is

$$d_P = \overline{\overline{S}}_P^{\,0} - \overline{\overline{S}}_P \tag{66}$$

where $\overline{\overline{S}}_P^{\,0}$ is the reference length. A positive value of d_P means a reduction of the actuator length.

4.1.8. *Marching Procedure*

The time integration of the 3-DoF-per-floor model has then be achieved following an experimental step-by-step procedure. The algorithm starts from known initial values $\mathbf{D}_1, \mathbf{V}_1, \mathbf{A}_1$ and the operations are organised in the following stages:

- Stage 0) Let $n = 0$.
- Stage 1) Transform the generalised displacements into target displacements $\mathbf{d_D}$ at the control transducers computed from Eq. (41):

$$(\mathbf{d_D})_{n+1}^{TARGET} = \mathbf{d_D}(\mathbf{D}_{n+1}) \tag{67}$$

- Stage 2) Send these target displacements of Eq. (67) to the controllers which impose them to the specimen.
- Stage 4) Measure the attained displacements $(\mathbf{d_D})_{n+1}^{MEAS}$ and restoring loads $(\mathbf{r_P})_{n+1}$ at the controllers, $\mathbf{r_P}$ being the forces applied by the pistons.

- Stage 5) From the measured control displacements, by using the iterative least-squares solution of Eq. (44), estimate the generalised displacements on each floor

$$\mathbf{D}_{n+1}^{MEAS} = \mathbf{D}\left(\left(\mathbf{d}_{\mathbf{D}}\right)_{n+1}^{MEAS}\right) \tag{68}$$

- Stage 6) From the measured piston forces, by using the transformation Eq. (50), compute the new generalised restoring forces

$$\mathbf{R}_{n+1} = \mathbf{R}\left(\left(\mathbf{r}_{\mathbf{P}}\right)_{n+1}, \mathbf{D}_{n+1}^{MEAS}\right) \tag{69}$$

- Stage 7) If $n > 1$, by using Eq. (32), compute the new accelerations

$$\mathbf{A}_{n+1} = \mathbf{A}_{n+1}\left(\mathbf{F}_{n+1}, \mathbf{R}_{n+1}, \mathbf{V}_n, \mathbf{A}_n\right) \tag{70}$$

- Stage 3) Let $n \leftarrow n+1$.
- Stage 8) Predict the generalised displacement at the next time increment by using the finite-difference approximation Eq. (29) of the integration algorithm:

$$\mathbf{D}_{n+1} = \mathbf{D}_{n+1}(\mathbf{D}_n, \mathbf{V}_n, \mathbf{A}_n) \tag{71}$$

- Stage 9) Go back to stage 1) until reaching the final time.

4.1.9. *Control Strategy*

Let us assume that PID controllers are used and that, at each floor, three actuators are controlled using as feedback the corresponding displacement transducer on the structure. In that case, there are, in principle, three options in the choice of the feedback transducer to control the remaining redundant actuators:

- a) the aligned displacement transducer on the structure (as for the other three actuators),
- b) the actuator load cell force or
- c) the actuator internal displacement transducer.

The first option is not practicable in general because the control system becomes unstable even for very low controller gains due to the high stiffness of the floor slab relatively to the stiffness of the pistons. The second option may lead to a stable control system, but, usually, force control strategies result in poor accuracy. It could be acceptable for a cyclic test, but not for a PsD test in which relatively small control errors may result in a large distortion of the integrated response as commented in the previous sections.

The third option may give an accurate and stable control system since it is a displacement control strategy, as in the first option, but associated to a transducer which sees a more flexible subsystem. In fact, the displacements measured by the internal transducer of the actuator are considerably larger than those coming from the flexibility of the floor slabs since they comprise also the deformation of the reaction wall and actuator attachments as well as a part of the floor slab deformation. Thus, the actuator internal transducers have been adopted as feedbacks for the redundant pistons. In practice, the computed target for the redundant pistons is slightly adjusted at every integration step in order to control the distribution of loads on the floor at the same time. So, instead of using Eq. (67), their target is computed as

$$\left(d_P\right)_{n+1}^{TARGET} = d_P\left(\mathbf{D}_{n+1}\right) + \left(\overline{d}_P\right)_{n+1} \tag{72}$$

where the first term on the right-hand side is the theoretical elongation of the piston in Eq. (66) and the second one is a correction introduced in order to modify the force. The latter is updated at every time step in the cumulative form

$$\left(\overline{d}_P\right)_{n+1} = \left(\overline{d}_P\right)_n + \frac{\left(r_P^{OPTIMUM} - r_P^{MEASURED}\right)_n}{K_P} \tag{73}$$

where the term added to the previous correction is the difference between the computed optimum force of Eq. (63) of the piston and the measured one at the former step, divided by a stiffness parameter K_P empirically selected. In general terms, the smaller the parameter, the faster is the convergence of the force to the optimum value. However, using a too small value may result in instability, which would make the force to oscillate out of control in very few steps.

4.2. HARDWARE SET-UP

The servo-control units used for this experiment were MOOG actuators with ±0.5 m stroke and load capacity of 0.5 MN. The control displacement transducers on the structure were optical HEIDENHAIN sensors with a stroke of ±0.5 m and 2 μm of resolution. Every actuator was equipped with a load cell, a TEMPOSONICS internal displacement transducer and a PID controller based on a MSM486DX CPU executing the digital control loop at a sampling period of 2 ms. The four controllers for the four pistons of each floor were governed by a powerful master CPU which is able to transmit and receive the controllers signals at real time (every 2 ms) through a high-speed communication channel based on dual port RAM boards. The three master units for the three floors had a common clock signal for the 2 ms interrupt. Those master controller units are programmed in C and have been designed to be able to apply complex algorithms in which the signals of several actuators are combined in real time. The sampling period of 2 ms has shown to be fast enough for stable digital control of structural specimens by hydraulic jacks but could be too short for cases in which the computations for those algorithms may become cumbersome. This hardware may permit the implementation of a PsD test with three DoFs per floor on a real time continuous basis only if the algorithm computation for each floor at every step can be performed in less than 2 ms.

For the classical implementation of the PsD method (subsection 2.1) adopted for this test, all master units were connected to a Windows workstation by means of ETHERNET. On such workstation, an application developed at ELSA and called STEPTEST executed the PsD test. This application is written in interpreted language (MATLAB) and implements the described model and marching procedure with interactive capabilities for monitoring and controlling the test (Figure 12).

4.3. TESTS PERFORMED IN THE FRAMEWORK OF SPEAR PROJECT

Within the SPEAR project, a full-size RC building model has been tested at ELSA. The mock-up is representative of non-seismic construction of the 60-70's in southern European countries, irregularities in plan and detailing included (Negro and Mola, 2005). The structure is a simplification of an actual Greek three-storey building and was designed for gravity loads alone using the Greek design code applied from 1954 to 1995. As seen in Figure 13, it has a doubly unsymmetric plan configuration, but it is regular in elevation. It is made of three 2-bay frames spanning from 3 to 6 m in each direction with a storey height of 3 m.

For practical reasons, the specimen was built outside of the laboratory, starting from a thick RC base slab on which the ground columns were anchored. Afterwards, the base was rolled over PVC rolls (Figure 15) in order to transport the built specimen up to its final position on the strong floor of the laboratory, where it was clamped with post-tensioning bars.

As shown in Figure 9, four actuators with four associated control displacement transducers were used for the bidirectional testing of the structure. In principle, these actuators were connected to the floor in positions not too close to the structural joints. Additional RC stiffeners were created in the floor slabs

(Figure 15) in order to properly distribute the local force applied by the pistons. The clamping of the piston attachments included post-tensioning bars. The piston bodies were supported by the reaction wall or by a supplementary reaction structure while the control displacement transducers were fixed on unloaded reference frames (Figure 9).

Apart from the load cells and control displacement transducers, additional instrumentation was installed on the specimen. Damage was expected mostly on top and bottom of the columns of the first two stories. Accordingly, clinometers were installed on up to three levels of each column. Additionally, some potentiometer extensometers were also installed on some localised areas. For the first time at ELSA, photogrammetry techniques were applied to a bidirectional test and, at two locations, a couple of cameras were used in order to record stereo images and estimate the bidimensional displacements of marked targets.

The CM, total mass and moment of inertia for the PsD equations were obtained from the theoretical mass distribution for the seismic assessment of the building according to Eurocode 8. Since the corresponding weight of the model was lower, additional water containers (Figure 16) were set on the floors of the specimen in the laboratory in order to have realistic gravity loads at the members. The input signals for the test were semi-artificial records modulated consistently with Eurocode 8. After some preliminary small tests to verify the testing system and tune the control parameters, two big PsD tests were performed for peak ground acceleration of up to 0.20 g, on the original specimen, and 0.30 g on the two retrofitted configurations. For each test, the intensity of the excitation was equal in the horizontal x and y components but the history of ground acceleration was different.

Details on the obtained response and damage at the different configurations are given by Negro and Mola (2005). During the different tests, the obtained maximum displacements at the third floor were in the order of 200 mm in the "x" (weak) direction, 150 mm in the "y" direction and 23 mrad of torsion. This level of torsion was higher than the predicted one using regulated methods.

The tests were performed with very low experimental error. For example, the errors produced at the controller loops were always lower than 0.1 mm and those estimated in the generalised displacement at the CM were lower than 0.3 mm. The latter are mostly due to the existing flexibility of the floors since three structural displacements were used for the control while all four of them entered in the estimation of the generalised displacement. An overall manner to assess the quality of a PsD test in relation to the existing control errors is through the computation of the energy of error (as we have commented in subsection 3.2.1), which amounted some 50 J, in comparison to the total absorbed energy in the structure, which was around 130 kJ for the second test for example. A way to look in more detail at the effects of the control errors on each mode consists of the identification of the eigenfrequency and damping from the measured forces and from the measured and reference displacements (see subsection 3.2.2). For example, for the 0.20 g test on the original specimen, the identified values for the first three modes at different time instants are shown in Figure 17. The results for the other estimated six modes are not much worse. The conclusion is that the control errors verified during the test have not affected at all these clue parameters for the response of the structure.

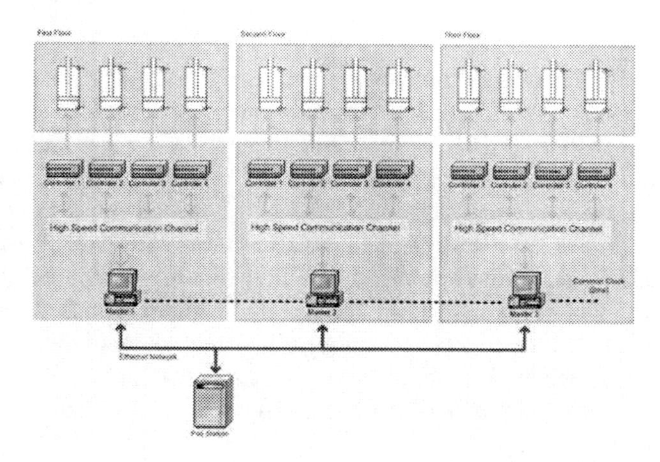

Figure 12. Controller set-up for the PsD tests.

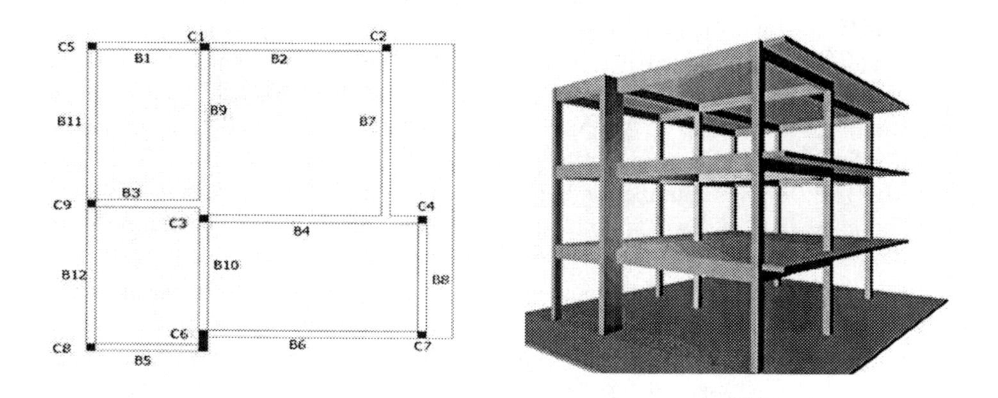

Figure 13. Plan and 3D views of the SPEAR RC building specimen.

After the testing campaign on the original model, a second round of tests was run on the first retrofitted configuration which tempted to increase the poor ductility capacity of the original columns and joints by means of the application of uniaxial and multiaxial glass-fibre wraps (Figure 18). For this configuration, the introduced ground acceleration level arrived up to 0.30g which produced 200mm displacement in the X direction and 120mm in the Y direction. After that, the glass fibre was removed and an alternative retrofitting intervention was introduced by means of RC-jacketing of two of the columns at all the levels. This second kind of intervention was focused to reduce the eccentricity between the CM and the centre of strength in the X and Y direction, without regarding the ductility capacity of the remaining members. For this second retrofitted configuration, the 0.30g earthquake produced displacements of 160mm in the X direction and 130mm in the Y direction.

Figure 14. Transportation of the building specimen inside of ELSA.

Figure 15. View of the structure ready for the PsD tests.

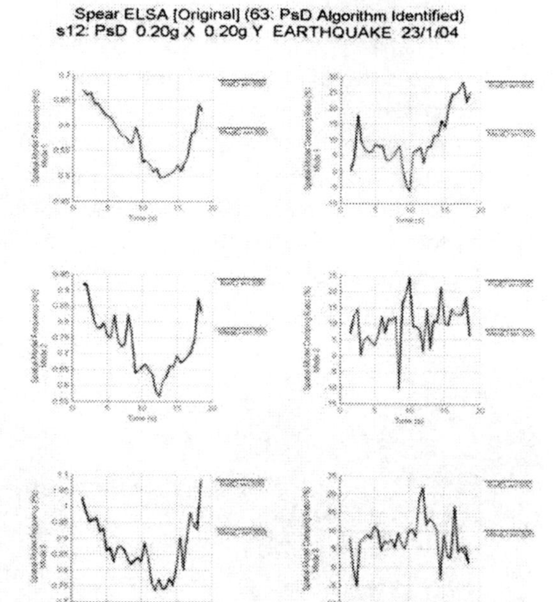

Figure 16. Identified frequencies and damping ratios from PsD results.

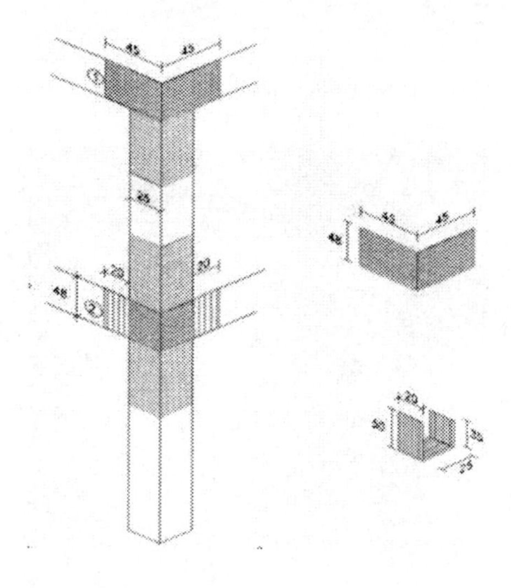

Figure 17. Adopted FRP retrofitting design on the SPEAR model.

5. Conclusions

Some of the main capabilities and achievements in structural testing at the ELSA laboratory have been summarised in this chapter. As a complement to other laboratories in Europe, ELSA has specialised itself in tests on large-size models and with sophisticated computer-controlled load-application conditions. Internationally recognized pioneering steps have been achieved for the development of the PsD testing method and its full-scale implementation. The ELSA contribution includes also some cases of real dynamic tests of active and semi-active control systems as well as vibration monitoring.

As a difference with respect to a shaking-table test, the control and measuring errors in a PsD test can be very low thanks to the low speed of execution. However, as has been shown in this chapter, even small time delays at the control may still significantly reduce the apparent response damping of the structure. Assessment methods that have been proposed give an on-line quantitative measure of the effect of the existing errors. They represent useful tools for controlling those effects and taking decisions in order to improve the quality of a PsD test. In the case of unacceptable distortion of the response, some parameters, such as the testing speed or the controller gains, may be changed during the test. Such parameter tuning is mainly performed during the preliminary small-intensity tests in order to assure the required accuracy.

In the framework of the activity of the SPEAR research project, three rounds of bidirectional PsD tests were carried out on a full-scale three-storey plan-wise irregular RC frame structure. The applied PsD technique considers three DoF per floor and includes rigorous geometrically non-linear transformation of coordinates for the actuator forces and displacements transducers attached to the floors. It also allows using more than three pistons and displacements transducers per floor, in which case it guarantees and even distribution of piston forces. The performed tests within the SPEAR project have been quite successful from the point of view of verified experimental error. The applied methodology showed to be quite appropriate for large-displacement bidirectional testing of full scale specimens. The results of these tests are of high relevance for understanding the torsional response of unsymmetric buildings.

Acknowledgements

Many members of the ELSA staff as well as project partners have provided significant contribution to the success of the activities described in this chapter. The authors gratefully acknowledge their contribution. Project SPEAR was funded by the EC under the "Competi tive and Sustainable Growth" Programme,

Contract N. G6RD-2001-00525 and the access to the experimental facility took place by means of the EC contract ECOLEADER N. HPRI-1999-00059.

REFERENCES

De Luca, A., Mele, E., Molina, F. J., Verzeletti, G., Pinto, A. V., 2001, Base isolation for retrofitting historic buildings: evaluation of seismic performance through experimental investigation, *Earthquake Engineering & Structural Dynamics*, **30**: 1125-1145.

Donea J., Magonette G., Negro P., Pegon P., Pinto A., 1996, Verzeletti, G. Pseudodynamic capabilities of the ELSA laboratory for earthquake testing of large structures. *Earthquake Spectra,* **12**:163-180.

Magonette, G., Pegon, P., Molina, F. J., Buchet, Ph., 1998, Development of fast continuous substructuring tests, *Proceedings of the 2nd World Conference on Structural Control.*

Magonette, G., F. Marazzi, H. Försterling, 2003, Active control of cable-stayed bridges: large scale mock-up experimental analysis, *ISEC-02 - Second International Structural Engineering and Construction Conference*, Rome.

Maia, N. M. M. and Silva, J. M. M. (editors), 1997, *Theoretical and Experimental Modal Analysis*, Research Studies Press, John Wiley.

Mola, E., Negro, P., Pinto, A. V., 2004, Evaluation of current approaches to the analysis and design of multi-storey torsionally unbalanced frames, *Proceedings of the 13th World Conference on Earthquake Engineering*, Vancouver, Paper N. 3304.

Molina, F. J., Pegon, P. and Verzeletti, G. 1999a, Time-domain identification from seismic pseudodynamic test results on civil engineering specimens, *2nd International Conference on Identification in Engineering Systems*, University of Wales Swansea.

Molina, F. J., Verzeletti, G., Magonette, G., Buchet, Ph., Geradin, M., 1999b, Bi-directional pseudodynamic test of a full-size three-storey building, *Earthquake Engineering & Structural Dynamics*, **28**: 1541-1566.

Molina, F. J., Gonzalez, M. P., Pegon, P., Varum H. and Pinto, A., 2000. Frequency and Damping Evolution During Experimental Seismic Response of Civil Engineering Structures. *COST F3 Conference on System Identification & Structural Health Monitoring.*

Molina, F. J., Magonette, G., Pegon, P., 2002a, Assessment of systematic experimental errors in pseudodynamic tests, *Proc. of 12th European Conference on Earthquake Engineering*, Elsevier Science , Paper 525.

Molina F. J., Verzeletti, G., Magonette, G., Buchet, Ph., Renda, V., Geradin, M., Parducci, A, Mezzi, M, Pacchiarotti, A, Federici, L., Mascelloni, S., 2002b, Pseudodynamic tests on rubber base isolators with numerical substructuring of the superstructure and strain-rate effect compensation, *Earthquake Engineering & Structural Dynamics*, **31**: 1563-1582.

Molina, F. J., S. Sorace, G. Terenzi, G. Magonette, B. Viaccoz, 2004, Seismic tests on reinforced concrete and steel frames retrofitted with dissipative braces, *Earthquake Engineering & Structural Dynamics*, **33**: 1373-1394.

Negro, P., Mola, E., 2005, Full scale PsD testing of the torsionally unbalanced SPEAR structure in the 'as-built' and retrofitted configurations, *SPEAR Workshop*, EC-JRC Ispra, Italy.

Network for Earthquake Engineering Simulation; http://www.nees.org.

Phillips, C. L. and Harbor, R. D., 1999, *Feedback Control Systems*, Prentice Hall.

Pinto, A. V., Pegon, P., Magonette, G., Tsionis, G., 2004, Pseudo-dynamic testing of bridges using non-linear substructuring, *Earthquake Engineering & Structural Dynamics*, **33**: 1125-1146.

Pinto, A. V., Pegon, P., Taucer, F., 2006, Shaking table facilities and testing for advancement of earthquake engineering: international cooperation, experiences, values, chances, *First European Conference on Earthquake Engineering and Seismology.*

Shing, P. B., Mahin, S. A., 1987, Cumulative experimental errors in pseudo-dynamic tests, *Earthquake Engineering & Structural Dynamics*, **15**: 409-424.

Takanashi, K., Nakashima, M., 1986, A State of the Art: Japanese Activities on on-Line Computer Test Control Method, *Report of the Institute of Industrial Science*, **32**, 3, University of Tokyo.

Thewalt, C. and Roman, M., 1994, Performance Parameters for Pseudodynamic Tests, *ASCE Journal of Struct. Eng.*, **120**, 9.

THE HISTORY OF EARTHQUAKE ENGINEERING AT THE UNIVERSITY OF CALIFORNIA AT BERKELEY AND RECENT DEVELOPMENTS OF NUMERICAL METHODS AND COMPUTER PROGRAMS AT CSI BERKELEY

E.L. Wilson
Professor Emeritus of Structural Engineering
University of California at Berkeley
1050 Leneve Place, El Cerrito, CA 94530 USA

Abstract. The purpose of this paper is to summarize, from a personal viewpoint, some research in structural dynamics within the Department of Civil Engineering at the University of California at Berkeley during the period of 1950 to 1990. The second part of the paper is to present a few recently developed numerical algorithms for dynamic analysis that are required in the design of wind, wave and earthquake resistant structures. These algorithms have been incorporated into the SAP 2000 programs (developed by Computers and Structures, Inc. in Berkeley) and have been used in the analysis of hundreds of large structural systems. This most recent research and development work was conducted since the author's retirement in 1991 from teaching at the University.

Key words: Earthquakes; Eigenvalues; Error Analysis; Static Vectors; Termination of LDR Vectors, Fast Nonlinear Analysis; Periodic Loading; Graduate Study at Berkeley

1. Structural Dynamics Research at UC Berkeley 1950 to 1990

1.1. INTRODUCTION

The University of California was established in 1868 and was authorized to teach courses in several different disciplines, including Military Science and Civil Engineering, CE. At that time the other engineering disciplines did not exist. By the time this author started his undergraduate work in early nineteen

A. Ibrahimbegovic and I. Kozar (eds.),Extreme Man-Made and Natural Hazards in Dynamics of Structures, 353–379.

fifty the undergraduate courses in dynamics were conducted by faculty members from the department of Mechanical Engineering, with emphasis on rigid body dynamics.

The Building Codes that were used in structural engineering design courses for buildings and bridges did not contain the words earthquake or seismic forces. Also, the majority of the structural engineering faculty at Berkeley did not seem to be concerned about the effects of dynamic loading.

In 1949 Professor Ray Clough joined the CE faculty and began teaching a graduate structural dynamics course for students in CE and the Naval Architecture departments. Clough had also been a weather officer in the US Army Air Corps during WWII and had developed an appreciation of natural disasters.

In 1953 Professor Joseph Penzien joined the CE faculty[3]. Both Ray and Joe had received their undergraduate degrees from the University of Washington and their Doctors' degrees from Massachusetts Institute of technology. In addition, both had extensive practical experience with the analysis of aircraft structures and blast analysis of all types of structures. They appeared to have a similar educational and practical background; however, their approach to the solution of dynamic problems was often very different. Their collaboration during the next 25 years resulted in the publication of the classical textbook *Dynamics of Structures* in 1975. In 2003, the latest edition of the book was reprinted and can be obtained from CSI in Berkeley[1].

During the early nineteen fifties very little research funding was available for research in the area of structural dynamics. Therefore, both Clough and Penzien were forced to take summer employment at companies, such as Boeing, and to work for local engineering firms as consultants.

However, during the late nineteen fifties and early nineteen sixties funded research projects increased significantly for the following reasons:

First; the CE faculty members hired during the twenty years after World War II were young and energetic with excellent practical experience and impressive educational backgrounds. Also, the majority of the new faculty was capable of conducting both experimental and theoretical research.

Second; the introduction of the digital computer required that new numerical analytical methods be developed. The early work of Professor Clough in the creation of the Finite Element Method is an example of a new method that revolutionized many different fields in engineering and applied mathematics.

Third; the Federal Government and the California Department of Transportation were rapidly expanding the freeway system in the state and were sponsoring research at Berkeley, led by Professors Scordelis and Monismith, concerning the behavior of bridges and overpass structures.

Fourth; it was the height of the Cold War and the Defense Department was studying the cost and ability to reinforce buildings and underground structures to withstand nuclear blasts.

Fifth; the manned space program was a national priority. Professors Pister, Penzien, Popov, Sackman, Taylor and Wilson were very active conducting research related to these activities.

Sixth; the offshore drilling for oil in deep water and the construction of the Alaska pipeline required new technology for steel structures, which was developed by Professors Popov, Bouwkamp and Powell.

Finally; after the 1964 Alaska earthquake (and small tsunami that hit northern California) the National Science Foundation initiated a very significant program on Earthquake Engineering Research.

To support this research and development a new Structural Engineering Building, Davis Hall, was completed in 1968 on the Berkeley campus and a large shaking table, to simulate earthquake motions, was constructed at the Richmond Field Station (1971).

1.2. THE EARTHQUAKE ENGINEERING RESEARCH CENTER

Professors Penzien and Clough wrote a formal proposal to the University requesting the creation of the Earthquake Engineering Research Center, EERC. The proposal was approved by the University Administration and Professor Penzien was named as the first director of the Center in December 1967.

EERC was approved as an Organized Research Unit, ORU; however, the center was not given a budget. Therefore, Professors Penzien, Clough and many other members of the UC faculty spent a significant amount of time, in addition to their normal teaching responsibilities, to obtain more permanent funding for the Center. During the first several years of EERC the National Science Foundation provided a significant amount of the funding for the Center and the construction of the world's largest shake table.

1.3. IMPACT ON THE STRUCTURAL ENGINEERING ACADEMIC PROGRAM

The existence of EERC and the pioneering Finite Element Research Programs in the Division of Structural Engineering and Structural Mechanics, SESM, within the Department of Civil Engineering, made the Graduate Program at Berkeley very popular. The finite element research was complimented by the unique and modern dynamic experimental facilities at EERC.

The SESM program attracted a very large number of bright national and international students. In addition, many post doctoral research scholars and faculty members from other universities spent one or more years at Berkeley.

1.4. WORKING UNDER THE DIRECTION OF PROFESSOR CLOUGH

In 1960-63 the author, as a doctoral student working under the supervision of Professor Ray Clough, developed the first fully automated finite element program for two-dimensional plane structures. Also, several other students working with Professor Clough extended the program to include the effects of nonlinear material crack closing and creep. In addition to completing a Doctor's degree with Professor Clough, the author had the opportunity to assist him with consulting activities involving the earthquake and blast response analysis of many different buildings and bridges.

1.5. EXPERIENCE WORKING IN THE AEROSPACE PROGRAM

The author worked as a Senior Development Engineer at Aerojet General Corporation in Sacramento, CA during the period 1963 to 1965. He was responsible for the development of numerical methods and computer programs for the structural analysis of the Apollo spacecraft and other rocket components. The experience was very frustrating since engineers were not allowed to conduct dynamic analysis if they were assigned to a static analysis group. However, the author did learn that it was possible to design elastic structures to withstand forces greater that ten times the force of gravity. In August 1965 the author returned to UC Berkeley as an Assistant Professor of Structural Engineering.

1.6. THE SAP SERIES OF COMPUTER PROGRAMS

In the mid and late nineteen sixties, many different Universities and companies were working on general purpose programs which assumed that each joint has six displacement degrees of freedom (three displacements and three rotations). In general, most of these programs were developed by groups of ten or more engineers and programmers.

In 1969 the Author realized that it was possible to develop a general purpose **S**tructural **A**nalysis **P**rogram[4] (SAP) that was simple and efficient. The key to the simplicity of SAP was to create, within the program, an equation ID array dimensioned 6 by the number of joints (nodes). The boundary conditions and the type of elements connected to each joint determined if an equilibrium equation existed. Therefore, the stiffness matrix could be directly formed in a compact form. The majority of the program was written by the Author. Five graduate students, working part-time, incorporated several different elements from previously written special purpose programs. The total time required for the program development was less than a man-year. The development was funded with small grants from local structural engineering companies. The three-dimensional solid element was sponsored by the Walla Walla District of the Corps of Engineers. The first version of SAP[3] was released within a year (1970). The dynamic response was based on the automatic generation of one set of Ritz vectors. At the time of release it was the fastest and largest capacity structural analysis program of its time.

Development of SAP continued for the next few years with sponsorship from users of the program. Dr. Klaus Jürgen Bathe was responsible for the final version of the program[6], SAP IV, in 1972. Dr. Bathe incorporated his 'Subspace Iteration Algorithms" for the evaluation of the exact eigenvalues and vectors of very large structural systems[5]. In addition, he was responsible for documentation and international distribution of the program.

In addition, Dr. Bathe developed a completely new static and dynamic computer analysis program, NONSAP[7]. It was designed to solve general structures with nonlinear material, large strains and large displacements. Later at MIT Professor Bathe continued this research and developed the ADINA program.

All computer programs developed during this early period at Berkeley were freely distributed worldwide, allowing practicing engineers to solve many new problems in structural dynamics. Hence, the research was rapidly transferred to the engineering profession. In many cases the research was used professionally prior to the publication of a formal paper.

Regarding the distribution of the FORTRAN programs from the University, it was made clear that the programs were not to be resold. However, this was not the case. Many companies and other Universities created, distributed and sold modified versions of the program. In addition, some obtained government funding to modify the UC programs. At the same time, proposals by this author to NSF for research on numerical methods for dynamic analysis were not funded

1.7. RESEARCH AT UC DURING THE NINETEEN EIGHTIES

The author had several excellant doctoral students during the nineteen eighties that worked on many important problems in computational mechanics, incremental construction, dam-reservoir interaction, parallel processing and structural dynamics. One of the most significant accomplishments during the period was proving that the exact dynamic eigenvalues and vectors were not the best basis for performing dynamic analysis by the mode superposition method [8,9,10,11]. It was shown that Load Dependent Ritz vectors could be generated by a simple algorithm that was faster than the subspace iteration algorithm and always produced more accurate results than the use of the exact eigenvalues and vectors.

1.8. THE INTRODUCTION OF THE INEXPENSIVE PERSONAL COMPUTER

On a leave from the University, in 1980 the author purchased an inexpensive ($6,000) personal computer with the CPM operating system. It had a FORTRAN compiler; however, it had only 64k of 8 bit high-speed memory and limited low-speed disk storage. Therefore, it was impossible to move large mainframe programs to these small personal computers. However, it was an opportunity to write a completely new general purpose structural analysis program that would incorporate the latest research in finite element technology and numerical methods that had been developed since the release of SAP IV in 1973. The most significant changes was in the dynamic analysis method where **L**oad **D**ependant **R**itz (LDR) modal vectors[8] were introduced as and option.

The small amount of high-speed memory on the first personal computers required that a structural analysis program be subdivided into separate program modules which were executed in sequence via batch files. For example, the preprocessing, formation of each different type of element stiffness matrix, the assembly of the global stiffness matrix, solution of equation, solution for eigenvalues, integration of modal equations, calculation of element forces plotting and post processing were all separate programs. Therefore, the first version of SAP 80 was very easy to develop and maintain. Also, it automatically had a restart option. In order to avoid a conflict of interest with his research at the University, the author did not use one line of code from SAP IV and personally did all the SAP 80 development work at his home office on his personal computer.

After the release of the 16 bit IBM personal computer in 1983 with 64 bit floating point hardware, large capacity hard disks, and a standard colored graphics terminal, it was possible to solve large practical structural systems. At this point in time it was apparent that the development of SAP80 required additional staff. The author made an agreement with a former UC student, Ashraf Habibullah president of Computers and Structures, Inc. (CSI), to add design options and graphical pre and post processors to SAP 80 in order that it would better serve the needs of the structural engineering profession and to provide professional support to the users of the program. This association with CSI has continued for nearly 25 years and has been very successful in the transfer of the latest research to the structural engineering profession. In addition, CSI has created a special purpose program of SAP2000 for multistory buildings, ETABS This program has special pre and post processing capabilities and automatic design options for earthquake engineering.

2. Research and Development at CSI to Present

2.1. SAP 90 AND SAP 2000

In 1990 the Author had a mild heart attack and decided to retire from teaching at the University and work at a more leisurely pace with CSI on the development of the next version of the program, SAP 90. A new improved SOLID element, three degrees of freedom per node, and a new thin or thick shell element, six degrees of freedom per node, were implemented.

By 1990 the earthquake engineering profession was beginning to use nonlinear base isolation and energy dissipation devices to reduce the response, and damage, of structures to earthquake motions. Therefore, there was a strong motivation to add this option to SAP 90.

The approach used to solve this class of nonlinear problems was to move the nonlinear force associated with these elements to the right hand side of the node point equilibrium equations; then, solve the uncoupled modal equations by iteration as the mode equations are integrated[11]. Since the nonlinear forces are treated as external loads, **LDR** vectors must be generated for each of these nonlinear degrees of freedom. This new approach is named the **F**ast **N**onlinear **A**nalysis, **FNA**, method.

2.2. EXAMPLES OF NONLINEAR ELEMENTS

Local buckling of diagonals, uplifting at the foundation, contact between different parts of the structures and yielding of a few elements are examples of structures with local nonlinear behavior. For dynamic loads, it is becoming common practice to add concentrated damping, base isolation and other energy dissipation elements. Figure 1 illustrates typical nonlinear problems. In many cases, these nonlinear elements are easily identified. For other structures, an initial elastic analysis is required to identify the nonlinear areas

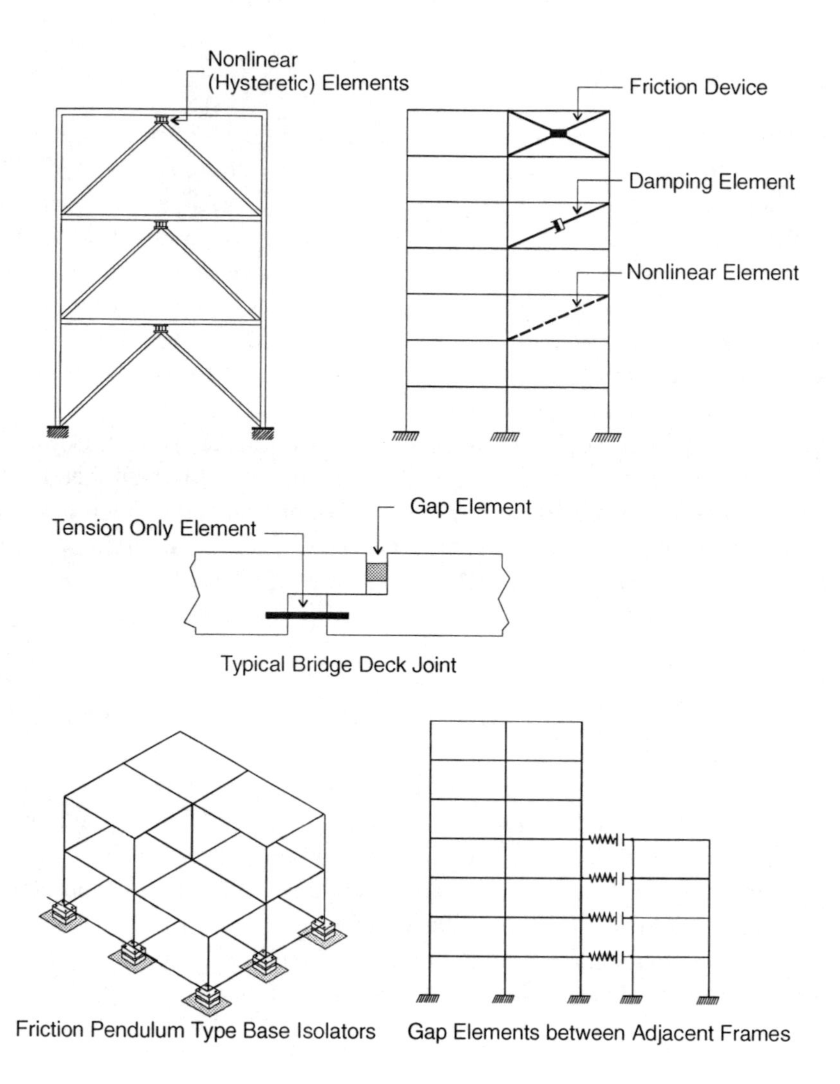

Figure 1. Examples of Nonlinear Elements.

After several years experience with the application of the FNA method on hundreds of large structural systems, it was found that the method could fail if there was very small or no mass associated with the nonlinear force. Therefore, it was necessary to modify the LDR vector algorithm in order that certain static vectors are included in the FNA method.

3. New Load Dependent Ritz Vector Algorithm and Error Analysis

3.1. THE COMPLETE EIGENVALUE SUBSPACE

In the analysis of structures subjected to three base accelerations there is a requirement that one must include enough modes to account for 90 percent of the mass in the three global directions. However, for other types of loading, such as base displacement loads and point loads, there are no guidelines as to how many modes are to be used in the analysis. In many cases it has been necessary to add static correction vectors to the truncated modal solution in order to obtain accurate results. One of the reasons for these problems is that number of eigenvectors required to obtain an accurate solution is a function of the type of loading that is applied to the structure. However, the major reason for the existence of these numerical problems is that all the LDR vectors of the structural system are not included in the analysis.

In order to illustrate the physical significance of the complete set of LDR vectors for a structure consider the unsupported beam shown in Figure 1a. The two-dimensional structure has six displacement DOF, three rotations (each with no rotational mass) and three vertical displacements (each with a vertical lumped mass).

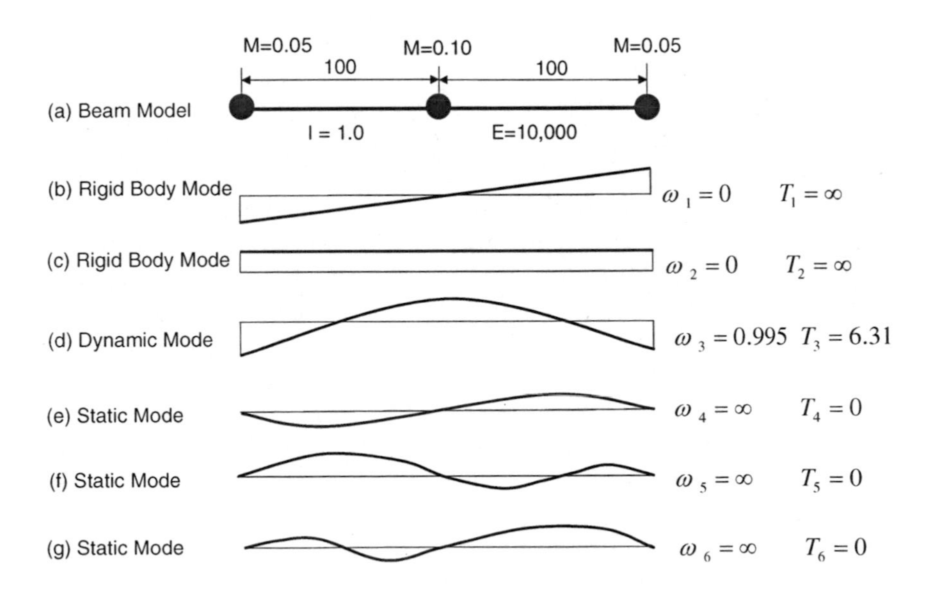

Figure 2. Rigid-Body, Dynamic and Static Modes for Simple Beam .

The six stiffness and mass orthogonal eigenvectors with frequency ω_n, in radians per second, and period T_n, in seconds, are shown in Figure 1b to 1g. The maximum number of natural eigenvectors that are possible is always equal to the number of displacement DOF. The static vectors (modes) have infinite frequencies; therefore, it is not possible to use the classical definition that the eigenvalues are equal to ω_n^2 if the eigenvalues are to be numerically evaluated. A new definition of the natural eigenvalues and the new algorithm used to numerically evaluate the complete set of natural eigenvectors will be presented in detail later in the paper.

Note that the rigid-body modes only have kinetic energy and the static modes only have strain energy. Whereas, the free vibration dynamic modes contain both kinetic and strain energy; and, the sum of kinetic and strain energy at any time is a constant. Also, the eigenvectors with identical frequencies are not unique vectors. Any linear combination of eigenvectors, with the same frequency, will satisfy the orthogonality requirements.

3.2. STRUCTURAL EQUILIBRIUM EQUATIONS

The static and dynamic node-point equilibrium equations for any structural system, with N_d displacement degrees-of-freedom (DOF), can be written in the following general form:

$$\mathbf{M}\ddot{\mathbf{u}}(t) + \mathbf{K}\mathbf{u}(t) = \mathbf{R}(t) + \mathbf{R}_D(\mathbf{u}, \dot{\mathbf{u}}, t) = \overline{\mathbf{F}}\mathbf{g}(t) \tag{1}$$

At time t the node acceleration, velocity, displacement and external applied load vectors are defined by $\ddot{\mathbf{u}}(t)$, $\dot{\mathbf{u}}(t)$, $\mathbf{u}(t)$ and $\mathbf{R}(t)$, respectively.

The unknown force vectors, $\mathbf{R}_D(\mathbf{u}, \dot{\mathbf{u}}, t)$, are the forces associated with internal energy dissipation such as damping and nonlinear forces. In most cases, these forces are self-equilibrating and do not contribute to the global equilibrium of the total structure.

The sum of $\mathbf{R}$ and $\mathbf{R}_D$ can always be represented by the product $\overline{\mathbf{F}}\mathbf{g}(t)$, where $\mathbf{F}$ is an N_d by L matrix of L linearly independent spatial load vectors associated with both linear and nonlinear behavior, and $\mathbf{g}(t)$ is a vector of L time functions. These time functions are directly specified for linear analysis, and are evaluated by iteration for nonlinear elements. For many problems, nonlinear forces may be restricted to a subset of all DOF, so that $L < N_d$, although this is not required in what follows.

The node-point lumped mass matrix, $\mathbf{M}$, need not have mass associated with all degrees-of-freedom; therefore, it may be singular and mathematically positive semi-definite. Also, external loads may be applied to displacement DOF that do not have mass and produce only static displacements.

The linear elastic stiffness matrix $\mathbf{K}$ may contain rigid-body displacements, as is the case for ship and aerospace structures; therefore, it need not be positive-definite. In order to overcome this potential singularity the term $\rho\mathbf{Mu}(t)$ may be added to both sides of the equilibrium equations, where ρ is an arbitrary positive number. Or, Equation (1) can be written as

$$\mathbf{M}\ddot{\mathbf{u}}(t) + \overline{\mathbf{K}}\mathbf{u}(t) = \mathbf{Fg}(t) + \rho\mathbf{Mu}(t) = \overline{\mathbf{R}}(\mathrm{t}) \tag{2}$$

While $\mathbf{K}$ and $\mathbf{M}$ may be singular, it is assumed here that the effective-stiffness matrix, $\overline{\mathbf{K}} = \mathbf{K} + \rho\mathbf{M}$, is nonsingular. Therefore, the effective-stiffness matrix represents a real structure with the addition of external springs to all mass DOF; these springs have stiffness proportional to the mass matrix.

The purpose of this paper is to present a general solution method for the numerical calculation of displacement and member forces. The proposed method can be used for both static and dynamic loads and has the ability to include arbitrary damping and nonlinear energy dissipation. The derivation of the vector-generation algorithm presented in this paper is self-contained and only uses the fundamental laws of physics and mathematics. Near the end of the paper, it will be pointed out that each step in the solution algorithm is nothing more than the application of well-known numerical techniques that have existed for over fifty years. It is an extension of Load Dependent Ritz vectors that have been previously described Wilson, 2003.

3.3. CHANGE OF VARIABLE

Equation (2) is an exact equilibrium statement for the structure at all points in time. The first step in the static or dynamic solution of this fundamental equilibrium equation is to introduce the following change of variable:

$$\mathbf{u}(t) = \mathbf{\Phi Y}(t) \quad \text{and} \quad \ddot{\mathbf{u}}(t) = \mathbf{\Phi}\ddot{\mathbf{Y}}(t) \tag{3}$$

The N_d by N matrix $\mathbf{\Phi}$ of spatial vectors are calculated and normalized to satisfy the following orthogonality equations:

$$\mathbf{\Phi}^T\mathbf{M\Phi} = \mathbf{\Psi} \tag{4}$$

$$\mathbf{\Phi}^T\overline{\mathbf{K}}\mathbf{\Phi} = \mathbf{I} \quad \text{or,} \quad \mathbf{\Phi}^T\mathbf{K\Phi} = \mathbf{I} - \rho\mathbf{\Psi} \tag{5}$$

The N by N diagonal matrices are $\mathbf{I}$ for the unit matrix and $\mathbf{\Psi}$ for the generalized mass matrix associated with each vector. Therefore, Equation (2) can be written as a set of uncoupled equations of the following form:

$$\mathbf{\Psi}\ddot{\mathbf{Y}}(t) + \mathbf{IY}(t) = \mathbf{\Phi}^{\mathrm{T}}\overline{\mathbf{R}}(t) \tag{6}$$

If N equals N_d, the introduction of this simple change of variables into Equation (2) does not introduce any additional approximations. The number of nonzero terms in the diagonal matrix $\boldsymbol{\Psi}$ indicates the maximum number of dynamic vectors and is equal to the number of lumped masses in the system (or, mathematically, the rank of the mass matrix). If a vector has zero generalized mass it indicates that it is a static response vector.

It is not practical to calculate all N_d static and dynamic shape functions for a large structure. First, it would require a large amount of computer time and storage. Second, a large number of vectors that are not excited by the loading may be calculated. Therefore, a truncated set of N natural eigenvectors will be calculated that will produce an accurate solution for an ***optimum number of LDR vectors***.

In order to minimize the number of shape functions required to obtain an accurate solution the static displacement vectors produced by L linearly independent spatial functions $\overline{\mathbf{F}}^{(1)}$ associated with the loading $\overline{\mathbf{R}}(t)$ will be used to generate the first set of vectors. The linear independent spatial load functions $\overline{\mathbf{F}}^{(1)}$ can be automatically extracted from $\overline{\mathbf{R}}(t)$ based on the type of external global loading and the location of the nonlinear elements.

3.4. CALCULATION OF STIFFNESS ORTHOGONAL VECTORS

The first step in the calculation of the orthogonal vectors defined by Equation (4) and (5) is to calculate a set of stiffness orthogonal vectors $\mathbf{V}$ where each vector satisfies the following equation:

$$\mathbf{v}_{\mathrm{m}}^{\mathrm{T}}\overline{\mathbf{K}}\mathbf{v}_{n} = \mathbf{v}_{\mathrm{m}}^{\mathrm{T}}\mathbf{f}_{n} = \begin{Bmatrix} 1 \text{ for } m = n \\ 0 \text{ for } m \neq n \end{Bmatrix} \tag{7}$$

These stiffness orthogonal displacement and load vectors are calculated and stored in the following arrays:

$$\begin{aligned} \mathbf{V} &= \begin{bmatrix} \mathbf{v}_1 & \mathbf{v}_2 & \mathbf{v}_3 & - & - & \mathbf{v}_N & - & - & - & - & \mathbf{v}_{N_d} \end{bmatrix} \\ \mathbf{F} &= \begin{bmatrix} \mathbf{f}_1 & \mathbf{f}_2 & \mathbf{f}_3 & - & - & \mathbf{f}_N & - & - & - & - & \mathbf{f}_{N_d} \end{bmatrix} \end{aligned} \tag{8}$$

All vectors are generated in sequence $n = 1, 2, ----N$. After each vector is made stiffness orthogonal and normalized it is inserted into position N. For example, consider a new displacement candidate vector $\overline{\mathbf{v}}$ (produced by the load vector $\overline{\mathbf{f}}$) that is not stiffness orthogonal as defined by Equation (7). This vector can be modified to be stiffness orthogonal by conducting the following numerical operations:

Normalization vector by the application of the following equations:

$$\hat{\mathbf{v}} = \hat{\beta}\,\overline{\mathbf{v}} \text{ and } \hat{\mathbf{f}} = \hat{\beta}\,\overline{\mathbf{f}} \text{ where } \hat{\beta} = \sqrt{\overline{\mathbf{v}}^{\mathrm{T}}\overline{\mathbf{f}}} \text{ ; therefore } \hat{\mathbf{v}}^{\mathrm{T}}\hat{\mathbf{f}} = \mathbf{1} \qquad (9)$$

Remove from $\hat{\mathbf{v}}$ all previously calculated stiffness orthogonal vectors by the application of the following equations:

$$\tilde{\mathbf{v}} = \hat{\mathbf{v}} - \sum_{n=1}^{N-1} \alpha_n \, \mathbf{v}_n \text{ and } \tilde{\mathbf{f}} = \hat{\mathbf{f}} - \sum_{n=1}^{N-1} \alpha_n \, \mathbf{f}_n \qquad \text{(10a and b)}$$

Multiplication of Equation (10a) by $\mathbf{v}_n^T\mathbf{K}$ yields the following equation:

$$\mathbf{v}_n^T\mathbf{K}\tilde{\mathbf{v}} = \mathbf{v}_n^T\mathbf{K}\hat{\mathbf{v}} - \alpha_n \mathbf{v}_n^T\mathbf{K}\mathbf{v}_n \qquad n = 1 \text{ to } N \qquad (11)$$

If the new vector $\tilde{\mathbf{v}}$ is to be stiffness orthogonal $\mathbf{v}_n^T\mathbf{K}\tilde{\mathbf{v}}$ must equal zero. Therefore,

$$\alpha_n = \mathbf{v}_n^T\mathbf{K}\hat{\mathbf{v}} \qquad (12)$$

After Equations (10a) and (10b) are evaluated they must be normalized by the application of the equations:

$$\mathbf{v}_N = \beta\,\tilde{\mathbf{v}} \text{ and } \mathbf{f}_N = \beta\,\tilde{\mathbf{f}} \text{ where } \beta = \sqrt{\tilde{\mathbf{v}}^{\mathrm{T}}\tilde{\mathbf{f}}} \text{ ;} \qquad (13)$$

therefore $\mathbf{v}_N^T\mathbf{f}_N = \mathbf{1}.$

It is now possible to check if the candidate vector $\overline{\mathbf{v}}$ was linearly independent of the previously calculated vectors by checking if the proposed new vector $\mathbf{v}_N$ is nothing more than numerical round-off. Therefore,

$$\text{If } \beta < tol \text{ reject } \mathbf{v}_{\mathbf{N}} \text{ as a new stiffness orthogonal vector} \qquad (14)$$

The value for *tol* is selected to be approximately 10^{-7}.

The first block of candidate vectors is obtained by solving the following set of equations, where the static loads $\overline{\mathbf{F}}^{(1)}$ and displacements $\overline{\mathbf{V}}^{(1)}$ are N_d by L matrices:

$$\overline{\mathbf{K}}\overline{\mathbf{V}}^{(1)} = \mathbf{LDL}^{\mathrm{T}}\overline{\mathbf{V}}^{(1)} = \overline{\mathbf{F}}^{(1)} \qquad (15)$$

Note that the effective stiffness matrix need be triangularized, $\overline{\mathbf{K}} = \mathbf{LDL}^{\mathrm{T}}$, only once. Additional blocks of candidate vectors can be generated from the solution of the following recursive equation:

$$\overline{\mathbf{K}}\overline{\mathbf{V}}^{(i)} = \mathbf{M}\overline{\mathbf{V}}^{(i-1)} = \overline{\mathbf{F}}^{(i)} \qquad (16)$$

If, during the orthogonality calculation, a new displacement or load vector in the block is identified as the same (parallel) as a previously calculated vector it can be discarded from the block and the algorithm is continued with a reduced block size. If the block size is reduced to zero, prior to the production of N_d vectors, it indicates that all of the static and dynamic vectors, excited by the initial load patterns, have been found.

3.5. MASS ORTHOGONAL VECTORS

After all blocks of the stiffness orthogonal vectors are calculated they can be made orthogonal to the mass matrix by the introduction of the following transformation:

$$\Phi = \mathbf{VZ} \tag{17}$$

Substitution of Equation (17) into Equation (4) produces the following N by N eigenvalue problem:

$$\overline{\mathbf{M}}\mathbf{Z} = \Psi \tag{18}$$

where $\overline{\mathbf{M}} = \mathbf{V}^{\mathrm{T}}\mathbf{MV}$. The stiffness and mass orthogonal vectors are then calculated from Equation (17). The static modes have zero periods, or $\Psi_n = 0$. Therefore, in order to avoid all potential numerical problems, it is recommended that the classical Jacobi rotation method[2] be used to extract the eigenvalues and vectors of this relatively small eigenvalue problem.

Equation (2) can now be rewritten as

$$\mathbf{M\ddot{u}}(t) + \overline{\mathbf{K}}\mathbf{u}(t) - \rho\mathbf{Mu}(t) = \mathbf{Fg}(t) \tag{19}$$

The transformation to modal coordinates produces the following uncoupled model equations:

$$\mathbf{\Psi\ddot{Y}}(t) + [\mathbf{I} - \rho\mathbf{\Psi}]\mathbf{Y}(t) = \mathbf{\Phi}^T\mathbf{Fg}(t) \tag{20}$$

Therefore, a typical modal equation, n, can be written as

$$\Psi_n\ddot{Y}_n(t) + [1 - \rho\Psi_n]Y_n(t) = \phi_n^T\mathbf{Fg}(t) \tag{21}$$

The number of static shape functions is equal to the number of zero diagonal terms in the matrix $\mathbf{\Psi}$. For the static modes Ψ_n is equal to zero and the solution is written as

$$Y_n(t) = \phi_n^T\mathbf{Fg}(t) \tag{22}$$

For the dynamic elastic modes the generalized mass for each mode is Ψ_n and the classical free-vibration frequencies (radians per second) and the periods of vibrations (seconds) can be calculated from

$$\omega_n = \sqrt{\frac{1}{\Psi_n} - \rho} \quad \text{and} \quad T_n = \frac{2\pi}{\omega_n} \tag{23}$$

Note that the eigenvalue Ψ_n always has a finite numerical value; however, the frequency ω_n and period T_n can have infinite numerical values and cannot be numerically calculated directly for all modes. For example, Table 1 summarizes the eigenvalues, frequencies and periods for the simple beam shown in Figure 1.

Table 1. Eigenvalues for Simple Beam for $\rho = 0.01$

Mode Number	Eigenvalue Ψ_n	Frequency $\omega_n = \sqrt{\frac{1}{\Psi_n} - \rho}$	Period $T_n = \frac{2\pi}{\omega_n}$
1	100	0	∞
2	100	0	∞
3	0.826	0.995	6.31
4	0	∞	0
5	0	∞	0
6	0	∞	0

The generalized stiffness and mass for the normalized vectors are as follows:

$$\phi_n^T \mathbf{K} \phi_n = 1 - \rho\, \Psi_n = \begin{cases} 0 & \text{for rigid - body modes} \\ \omega_n^2 \Psi_n & \text{for dynamic modes} \\ 1 & \text{for static modes} \end{cases} \tag{25}$$

$$\phi_n^T \mathbf{M} \phi_n = \Psi_n = \begin{cases} 1/\rho & \text{for rigid - body modes} \\ \Psi_n & \text{for dynamic modes} \\ 0 & \text{for static modes} \end{cases} \tag{26}$$

Therefore, it is necessary to save both the generalized stiffness and generalized mass, for each mode, in order to determine the static, dynamic or rigid-body response analysis of the mode.

The solution for the dynamic modes can be obtained using the piece-wise exact algorithm [1]. For all rigid-body modes Ψ_n will equal $1/\rho$. Therefore, their response can be calculated by direct, numerical or exact, integration from

$$\ddot{Y}_n(t) = \rho\,\phi_n^T\,\mathbf{R}(t),\ \dot{Y}_n(t) = \int \ddot{Y}_n(t)\,dt\,,\ \text{and}\ Y_n(t) = \int \dot{Y}_n(t)dt \tag{24}$$

The sum of the static, dynamic and rigid-body responses produces a unified method for the static and dynamic analysis of all types of structural systems.

3.6. MATHEMATICAL CONSIDERATIONS

Except for reference to the Jacobi and piece-wise integration methods[2] Wilson, 2003, the numerical method for the generation of stiffness and mass orthogonal vectors presented is based on the fundamentals of mechanics and requires no additional references to completely understand. However, it is very interesting to note that the method is nothing more than the application of several well-known numerical techniques:

First, the change of variables introduced by Equation (3) is an application of the standard method of solving differential equation and is also known as the ***separation of variables*** in which the solution the solution is expressed in terms of the product of space functions and time functions. A special application of this approach in classical structural dynamics is called the ***mode superposition method*** in which the static mode response is neglected.

Second, the addition of the term $\rho\mathbf{M}u(t)$ to the stiffness matrix is called an eigenvalue shift in mathematics. However, it is worth noting the zero eigenvalues associated with the static modes are not shifted.

Third, the recurrence relationship, Equation (10), is identical to the inverse iteration algorithm for a single vector[2]. Therefore, the approach is a ***power method*** that will always converge to the lowest eigenvalues of the system.

Fourth, the series of vectors generated by the inverse iteration method is known as the ***Krylov Subspace***. A. N. Krylov, 1863-1945, was a well-known Russian engineer and mathematician who first studied the dynamic response of ship structures. However, Krylov did not include static modes in his work.

Fifth, orthogonality is maintained, Equations (14), by the application of the ***modified Gram-Schmidt algorithm***. Theoretically, after the initial block of orthogonal vectors are calculated, it is only necessary to make each new displacement vector orthogonal with respect to the previous two Krylov vectors. However, after many years of experience with the dynamic analysis of very large structural systems, we have found that it is necessary to apply the Gram-Schmidt method to all previously calculated vectors in order that the same vectors are not regenerated.

Sixth, the performance of the algorithm is improved if the load vectors $\mathbf{F}^{(i)}$, for each block, are made orthogonal with respect to the previously calculated displacement vectors, Equation (16), prior to the solution of the equilibrium equations. This additional step has made the algorithm ***unique and very robust***.

3.7. LOAD PARTICIPATION RATIOS AND ERROR ESTIMATION

In the analysis of structures subjected to three base accelerations there is a requirement that one must include enough modes to account for 90 percent of the mass in the three global directions. However, for other types of loading, such as base displacement loads, there are no guidelines as to how many modes are to be used in the analysis. The purpose of this section is to define two new load participation ratios, which can be calculated during the generation of the LDR vectors, to assure that an adequate number of vectors are used in a subsequent static or dynamic analysis.

From Equation (24), a typical modal equation n for load pattern j, can be written as

$$\Psi_n \ddot{Y}(t)_n + \omega_n^2 \Psi_n Y(t)_n = \phi_n^T \mathbf{F}_j g(t)_j \qquad n = 1 \text{ to } N \tag{27}$$

The error indicators are based on the two different types of load functions $g(t)_j$. In one case the loads vs. time excite the low frequencies; and, in the other case the high frequencies are excited.

3.7.1. *Static Loads*

The first error estimator is a measure of the ability of a truncated set of mode shapes to capture the static response of the structural system. For this case the load function $g(t)_j$ is applied linearly from a value of zero at time zero to a value of 1.0 at the end of a very large time interval. Therefore, the inertia terms can be neglected and Equation (27), evaluated at the end of the large time interval, is

$$\omega_n^2 \Psi_n Y_{ns} = \phi_n^T \mathbf{F}_j \qquad n = 1 \text{ to } N \tag{28}$$

Therefore the ***static mode participation*** can be written as

$$Y_{nj} = \frac{\phi_n^T \mathbf{F}_j}{\omega_n^2 \Psi_n} \qquad n = 1 \text{ to } N \tag{29}$$

From Equation (3) the approximate static displacement response of the structure due to N modes is

$$\overline{\mathbf{u}}_j = \sum_{n=1}^{N} \phi_n Y_{nj} \tag{30}$$

The approximate strain energy associated with the displacement defined by Equation (30) is

$$\begin{aligned}\overline{E}_{sj} &= \frac{1}{2}\overline{\mathbf{u}}_j^T \mathbf{K}\, \overline{\mathbf{u}}_j = \frac{1}{2}\sum_{n=1}^{N} Y_{nj}\phi_n^T \mathbf{K}\phi_n Y_{nj} = \frac{1}{2}\sum_{n=1}^{N} \omega_n^2 \Psi_n \phi_n Y_{nj}^2 \\ &= \frac{1}{2}\sum_{n=1}^{N} \frac{(\phi_n^T \mathbf{F}_j)^2}{\omega_n^2 \Psi_n}\end{aligned} \tag{31}$$

The exact static displacement due to the load pattern can be calculated from the solution of the following static equilibrium equation:

$$\mathbf{K}\,\mathbf{u}_j = \mathbf{F}_j \tag{32}$$

The exact strain energy stored in the structure for the load pattern is calculated from

$$E_{sj} = \frac{1}{2}\mathbf{u}_j^T \mathbf{K}\,\mathbf{u}_j = \frac{1}{2}\mathbf{u}_j^T \mathbf{F}_j \tag{33}$$

The ***static load participation ratio*** is defined as the ratio of the strain energy captured by the truncated set of vectors, $\overline{E}_j$, to the total strain energy, E_j. For the typical case where $\rho = 0$ the ratio is

$$r_{sj} = \frac{\sum_{n=1}^{N} (\phi_n^T \mathbf{F}_j)^2}{\mathbf{u}_j^T \mathbf{F}_j} \tag{34}$$

It must be pointed out that for LDR vectors, this ratio is always equal to 1.0. Whereas, the use of the exact dynamic eigenvectors may require a large number of vectors in order to capture the static load response. Also, if the static mode shapes are excited it is not possible for the exact dynamic eigenvectors to converge to the exact static solution.

3.7.2. *Dynamic Response*

The ***dynamic load participation ratio*** is based on the use of the application of the static loads as a delta function at time zero that produces an initial condition for a free vibration response analysis of the total structural system. It is well known that any type of time function can be represented by the sum of these impulse functions applied at different points in time. This type of loading will produce an initial velocity at the mass points of $\dot{\mathbf{u}}_j = \mathbf{M}^{-1}\mathbf{F}_j$. Therefore, the total kinetic input to the system, for a typical load vector j, is given by

$$E_{kj} = \frac{1}{2}\dot{\mathbf{u}}_j^T\mathbf{M}\dot{\mathbf{u}}_j = \frac{1}{2}\mathbf{f}_j^T\mathbf{M}^{-1}\mathbf{F}_j \tag{35}$$

From Equation (3) the relationship between initial node velocities and the initial modal velocities is

$$\dot{\bar{\mathbf{u}}}_j = \sum_{\mathrm{n=1}}^{\mathrm{N}} \phi_{\mathrm{n}}\dot{Y}_{nj} \tag{36}$$

Therefore, the kinetic energy associated with the truncated set of vectors is

$$\overline{E}_{kj} = \frac{1}{2}\dot{\bar{\mathbf{u}}}_j^T\mathbf{M}\dot{\bar{\mathbf{u}}}_j = \frac{1}{2}\sum_{n=1}^{N}\Psi_n\dot{Y}_{nj}^2 \tag{37}$$

The initial modal velocity $\dot{Y}_{nj}$ is obtained from the solution of Equation (27) as

$$\dot{Y}_{nj} = \frac{\varphi_n^T\mathbf{F}_j}{\Psi_n} \tag{38}$$

Substitution of Equation (38) into Equation (37) yields

$$\overline{E}_{jk} = \frac{1}{2}\sum_{n=1}^{N}\frac{(\varphi_n^T\mathbf{F}_j)^2}{\Psi_n} \tag{39}$$

The ***dynamic load participation ratio*** is defined as the ratio of the kinetic energy captured by the truncated set of vectors, $\overline{E}_{kj}$, to the total kinetic energy, E_{kj}. For the typical case where $\rho = 0$ the ratio is

$$r_{dj} = \frac{\sum_{n=1}^{N}(\omega_n\phi_n^T\mathbf{F}_j)^2}{\mathbf{F}_j^T\mathbf{M}^{-1}\mathbf{F}_j} \tag{40}$$

A dynamic load participation ratio equal to 1.0 assures that all the energy input is captured for the dynamic load condition j. In the case of base acceleration loading where the three load vectors are the directional masses, the dynamic load participation ratios are identical to the mass participation ratios.

3.7.3. *Automatic Termination of LDR Vectors*

Since the LDR vector algorithm starts with a full set of static vectors the static load participation factor will always equal 1.0. Equation (40), the dynamic load participation factor can be evaluated after each block of vectors is generated. Therefore, this factor can be computed as the vectors are calculated and it can be used as an indicator to automatically terminate the generation of LDR vectors. Based on experience, a dynamic load participation ratio of at least 0.95, for all load patterns, will assure accurate results for most types of loading. This is a very important user option since the number of vectors requested need not be specified prior to the dynamic analysis.

3.8. USE OF THE LDR ALGORITHM TO CALCULATE EIGENVECTORS

The LDR vector algorithm, as presented in this paper, generates the complete Krylov subspace for a specified set of load vectors and errors in the resulting dynamic response analysis are minimized. If one examines the frequencies associated with the LDR vectors it is found that all the lower frequencies are identical to the frequencies obtained from an exact eigenvalue analysis. Since the approach is related to the power method this is to be expected. The higher modes produced by the LDR vector algorithm are linear combinations of the exact eigenvectors and components of the static response vectors. The complete set of LDR vectors is the optimum set of vectors to solve the dynamic response problem associated with the specified static load patterns. Therefore, the number of LDR vectors required will always be less than if the exact eigenvectors were used.

If, for some reason, one wishes to calculate the exact eigenvalues and vectors the same numerical method can be used. The initial displacement vectors need only be set to random vectors. If, during the generation, vectors are generated which are identical to previously calculated vectors they can be replaced with new random displacement vectors. The procedure can be terminated at any time; however, the higher frequencies will not be exact. The introduction of iteration for each block can be used to calculate the exact eigenvalues and vectors. Note that if the system contains M masses, the method will generate M exact eigenvectors; nevertheless, if random load vectors are used directly, instead of $\overline{\mathbf{F}}^{(i)} = \mathbf{MV}^{(i-1)}$, the algorithm can continue and will produce N_d-M static response vectors which have infinite frequencies and zero periods.

3.9. SUMMARY OF THE COMPLETE LDR VECTOR ALGORITHM

The use of exact eigenvectors to reduce the number of degrees of freedom required to conduct a dynamic response analysis has significant limitations. The effects of the application of static loads to massless DOF cannot be taken into account. In addition, for certain types of loading a large number of vectors are required. On the other hand, a large number of exact eigenvectors may be calculated that are not excited by the loading on the structure.

The use of static and dynamic LDR vectors, presented in this paper, eliminates the problems associated with the use of the exact eigenvectors. In addition, the LDR vector algorithm produces a unified approach to the static and dynamic analysis of many different types of structural systems. In addition, it is possible to check if an adequate number of vectors are generated prior to the integration of the equations of motion.

4. The Fast Nonlinear Analysis Method for Seismic Analysis

The FNA method can be applied to both the static and dynamic analysis of linear or nonlinear structural systems. A limited number of predefined nonlinear elements are assumed to exist. Stiffness and mass orthogonal LDR vectors of the elastic structural system are used to reduce the size of the nonlinear system to be solved. The forces in the nonlinear elements are calculated by iteration at the end of each time or load step. The uncoupled modal equations are solved exactly for each time increment.

The computational speed of the new FNA method is compared with the traditional "brute force" method of nonlinear analysis in which the complete equilibrium equations are formed and solved at each increment of load. For many problems the new method is several magnitudes faster.

The response of real structures when subjected to a large dynamic input often involves significant nonlinear behavior. In general, nonlinear behavior includes the effects of large displacements and/or nonlinear material properties.

The use of geometric stiffness and P-Delta analyses includes the effects of first-order large displacements. If the axial forces in the members remain relatively constant during the application of lateral dynamic displacements, many structures can be solved directly without iteration.

The more common type of nonlinear behavior is when the material stress-strain, or force-deformation, relationship is nonlinear. This is because of the modern design philosophy that "a well-designed structure should have a limited number of members which require ductility and that the failure mechanism be clearly defined". Such an approach minimizes the cost of repair after a major earthquake.

4.1. FUNDAMENTAL EQUILIBRIUM EQUATIONS

The FNA method is a simple approach in which the fundamental equations of mechanics (equilibrium, force-deformation and compatibility) are satisfied. The ***exact*** force equilibrium of the computer model of a structural, at time *t*, is expressed by the following matrix equations:

$$\mathbf{M}\ddot{\mathbf{u}}(t)+\mathbf{C}\dot{\mathbf{u}}(t)+\mathbf{K}\mathbf{u}(t)+\mathbf{R}(t)_{\mathrm{NL}}=\mathbf{R}(t) \tag{41}$$

where $\mathbf{M}$, $\mathbf{C}$ and $\mathbf{K}$ are the mass, proportional damping and stiffness matrices, respectively. The size of these three square matrices is equal to the total number of unknown node point displacements N_d. The elastic stiffness matrix $\mathbf{K}$ neglects the stiffness of the nonlinear elements. The time-dependent vectors $\ddot{\mathbf{u}}(t)$, $\dot{\mathbf{u}}(t)$, $\mathbf{u}(t)$ and $\mathbf{R}(t)$ are the node point acceleration, velocity, displacement and external applied load, respectively. And $\mathbf{R}(t)_{\mathrm{NL}}$ is the global node force vector due to the sum of the forces in the nonlinear elements and is computed by iteration at each point in time. This approach was first applied to non-proportional damping systems in reference 10.

If the computer model is unstable without the nonlinear elements one can add "effective elastic elements" (at the location of the nonlinear elements) of arbitrary stiffness. If these effective forces, $\mathbf{K}_{\mathbf{e}}\mathbf{u}(t)$, are added to both sides of Equation (1) the exact equilibrium equations can be written as

$$\mathbf{M}\ddot{\mathbf{u}}(t)+\mathbf{C}\dot{\mathbf{u}}(t)+(\mathbf{K}+\mathbf{K}_{\mathbf{e}})\mathbf{u}(t)=\mathbf{R}(t)-\mathbf{R}(t)_{\mathrm{NL}}+\mathbf{K}_{\mathbf{e}}\mathbf{u}(t) \tag{42}$$

where $\mathbf{K}_{\mathbf{e}}$ is the effective stiffness of arbitrary value. Therefore, the ***exact*** dynamic equilibrium equations for the nonlinear computer model can be written as

$$\mathbf{M}\ddot{\mathbf{u}}(t)+\mathbf{C}\dot{\mathbf{u}}(t)+\overline{\mathbf{K}}\mathbf{u}(t)=\overline{\mathbf{R}}(t) \tag{43}$$

The elastic stiffness matrix $\overline{\mathbf{K}}$ is equal to $\mathbf{K}+\mathbf{K}_{\mathbf{e}}$ and is known. The effective external load $\overline{\mathbf{R}}(t)$ is equal to $\mathbf{R}(t)-\mathbf{R}(t)_{\mathrm{NL}}+\mathbf{K}_{\mathbf{e}}\mathbf{u}(t)$ which must be evaluated by iteration. If a good estimate of the effective elastic stiffness can be made the rate of convergence may be accelerated since the unknown load term $-\mathbf{R}(t)_{\mathrm{NL}}+\mathbf{K}_{\mathbf{e}}\mathbf{u}(t)$ will be small.

Using the complete set of LDR orthogonal stiffness and mass orthogonal vectors, a set of uncoupled modal equations can be written as for the dynamic modes as

$$\ddot{Y}_n+2\xi_n\omega_n\dot{Y}_n+\omega_n^2Y_n=f_n^i \quad n=1......N \tag{44}$$

The complete solution algorithm is given in chapter 18 in reference 2. Note that all mode equations are solved simultaneously and that model equations are coupled at each time step by the iterative term f_n^i.

5. Solution for Wind and Wave Loadings

The recurrence solution algorithm, summarized by Equation 13.16 in reference 2, is a very efficient computational method for arbitrary, transient, dynamic loads with initial conditions. The algorithm is unconditionally stable and exact for linear variation of load within a time increment. Also it is possible to use this same simple solution method for arbitrary periodic loading as shown in Figure 3. Note that the total duration of the loading is from $-\infty$ to $+\infty$ and the loading function has the same amplitude and shape for each typical period T_p. Wind, sea wave and acoustic forces can produce this type of periodic loading.

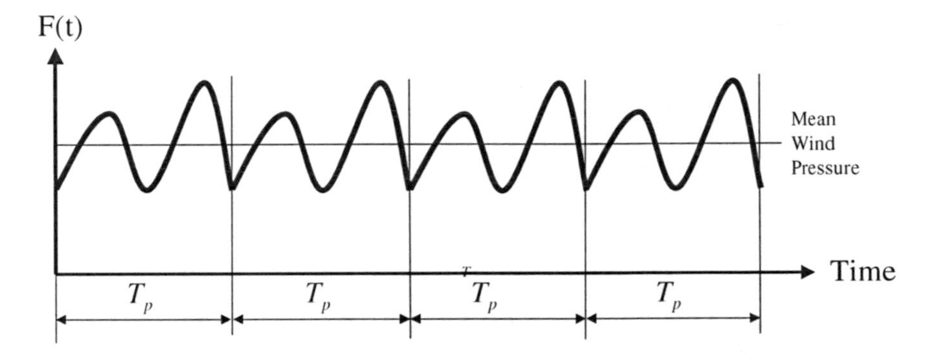

Figure 3. Example of Periodic Loading.

For a typical duration T_p of loading, a numerical solution, for each mode, can be evaluated by the application of Equation (45) without initial conditions. This solution is incorrect since it does not have the correct initial conditions. Therefore, it is necessary for this solution $y(t)$ to be corrected in order that the exact solution $z(t)$ has the same displacement and velocity at the beginning and end of each loading period. In order to satisfy the basic dynamic equilibrium equation the corrective solution $x(t)$ must have the following form:

$$x(t) = x_0 A_1(t) + \dot{x}_0 A_2(t) \tag{45}$$

where the dynamic functions are defined in Table 2.

Table 2 Summary of Notation used in Dynamic Response Equations.

CONSTANTS	
$\omega_D = \omega\sqrt{1-\xi^2}$ $\overline{\omega} = \omega\xi$	$\overline{\xi} = \dfrac{\xi}{\sqrt{1-\xi^2}}$
FUNCTIONS	
$S(t) = e^{-\xi\omega t}\sin(\omega_D t)$	$C(t) = e^{-\xi\omega t}\cos(\omega_D t)$
$\dot{S}(t) = -\overline{\omega}S(t) + \omega_D C(t)$	$\dot{C}(t) = -\overline{\omega}C(t) - \omega_D S(t)$
$A_1(t) = C(t) + \overline{\xi}S(t)$	$A_2(t) = \dfrac{1}{\omega_D}S(t)$
$\dot{A}_1(t) = \dot{C}(t) + \overline{\xi}\dot{S}(t)$	$\dot{A}_2(t) = \dfrac{1}{\omega_D}\dot{S}(t)$

The total exact solution for displacement and velocity for each mode can now be written as

$$z(t) = y(t) + x(t) \tag{46a}$$

$$\dot{z}(t) = \dot{y}(t) + \dot{x}(t) \tag{46b}$$

In order that the exact solution is periodic the following conditions must be satisfied:

$$z(T_p) = z(0) \tag{47a}$$

$$\dot{z}(T_p) = \dot{z}(0) \tag{47b}$$

The numerical evaluation of Equations (47) produces the following matrix equation which must be solved for the unknown initial conditions:

$$\begin{bmatrix} 1-A_1(T_p) & -A_2(T_p) \\ -\dot{A}_1(T_p) & 1-\dot{A}_2(T_p) \end{bmatrix} \begin{bmatrix} x_0 \\ \dot{x}_0 \end{bmatrix} = \begin{bmatrix} -y(T_p) \\ -\dot{y}(T_p) \end{bmatrix} \tag{48}$$

The exact periodic solution for modal displacements and velocities can now bc calculated from Equations (46a and 46b). Hence, it is not necessary to use a frequency domain solution approach for periodic loading as suggested in most textbooks on structural dynamics.

6. Personal Remarks

The majority of the population in the United States has an obsession with the danger of experiencing an earthquake. It is very common to meet someone from the East, South or Midwest, where hurricanes and tornadoes are very common, who states that he or she would not consider living in California because of the fear of earthquakes. However, if one looks at the *facts,* the fear of earthquakes is not justified.

During the past 500 years fewer than 2,000 people have been killed by earthquakes in the United States. This figure is extremely small compared to many other typical causes of accidental deaths. The largest causes are automobile crashes (over 30,000 each year), fire, wind and many other types of accidents. *In fact, during the past 500 years, several times more people have died in the United States from insect bites than from earthquakes. Also, lightning has killed more people than earthquakes.*

Tornadoes and minor flooding are natural disasters that kill hundreds of individuals each year. However, most of these isolated incidences are not reported by the local or national news media; whereas, a small earthquake in the San Andreas Fault, which measured four on the Richter Magnitude Scale and caused no property damage or personal injuries, is often reported in the national and international press.

Hurricanes Katrina, Rita and Wilma in 2005 devastated New Orleans, Southern Florida, and the Gulf Coastline killed over 1000 people and significant property damage. In recent years, loss of life from hurricanes has been minimized because the magnitude and time of the hurricane can be predicted. The largest number of fatalities from one hurricane in the US is estimated at 6,000 in Galveston Inland, TX in 1900. In the US wind and floods have caused approximately 100 times more damage and loss of life than earthquakes. However, several times more money is spent on earthquake research than research on the design of wind resistant structures.

One cannot disregard, however, that one of the largest recorded natural disasters in recent years was the Tangshan earthquake (Richter Magnitude 7.9) in eastern China in 1976 in which over 600,000 people were killed. During the past 50 years a large number of earthquakes, which have occurred outside the US, have killed over 10,000 people per earthquake. The most recent October 8,

2005 (magnitude 7.6) affected the Kashmir regions of Pakistan and India. Initial estimates are that it killed over 50,000 people. Nearly all these earthquakes have occurred in areas of the world with grossly inadequate design and construction standards. However, the failure of a dam with a large reservoir during an earthquake could cause a large number of fatalities. Or, a large tsunami, generated from an earthquake off the west coast of the US, could kill thousands of people

Most major structures, which are damaged during earthquakes, are designed by Civil Engineers, who have special training in Structural Engineering. All ground-supported structures are designed in the vertical direction to support their own weight, which is commonly referred to as 1.0g in the vertical direction. The present earthquake design specifications for most structures in the San Francisco Bay Area are less than 50 percent of the weight of the structure or 0.5g applied in the horizontal direction. In aerospace engineering it is common to design structures to carry loads over 10g. Therefore, the common statement, it is not possible to design structures to resist earthquakes, is not true. We have the technology to design earthquake resistant structures and it is an economic decision whether or not to obtain this goal. In addition, earthquake resistant design can place limitations on the architectural form of the structure.

Finally, it is apparent that there is a need to increase funding for applied research and development for wind and coastal engineering. The fundamental research in computational aero and fluid dynamics has been completed. We must create computer programs that are usable by the profession in order to improve the design of structures in wind and coastal areas.

References

1. R. W. Clough and J. Penzien, *Dynamics of Structures*, Computers and Structures, Inc., 1995 University Avenue, Berkeley, CA, 94704, ISBN 0-923907-50-5, (2003)
2. E. L. Wilson, *Three Dimensional Static and Dynamic Analysis of Structures,* Computers and Structures, Inc., 1995 University Avenue, Berkeley, CA, 94704, ISBN 0-923907-03-3, (2003).
3. J. Penzien, EERI Oral History Series, 2004
4. E. L. Wilson, SAP A General Structural Analysis Program, Report UCSESM, (September 1970)
5. K. J. Bathe and E. L. Wilson, "Large Eigenvalue Problems in Dynamic Analysis", Proceedings, American Society of Civil Engineers, Journal of the Engineering Mechanics Division, EM6, (December 1972) pp. 1471-1485.
6. K. J. Bathe, K. J., E. L. Wilson and F. E. Peterson, "SAP IV – A Structural Analysis Program ', Report EERC 73-11, (June 1973)

7. K. J. Bathe, E. L. Wilson and R. H. Iding, "NONSAP--A Structural Analysis Program for Static and Dynamic Response of Nonlinear Systems" UCB/SESM Report No. 74/3, Berkeley, (February 1974).
8. E. L. Wilson, M. Yuan and J. Dickens, "Dynamic Analysis by Direct Superposition of Ritz Vectors," Earthquake Engineering and Structural Dynamics, Vol. 10, (1982)
9. P. Léger, E. L. Wilson, Clough, R. W. "The Use of Load Dependent Vectors for Dynamic and Earthquake Analysis". Report UCB/EERC-86/04, Berkeley, (March 1986).
10. K. J. Joo, E. L. Wilson and P. Léger, "Ritz Vectors and Generation Criteria for Mode Superposition Analysis", Earthquake Engineering and Structural Dynamics, Vol.18, pp.149-167.(1989)
11. A. Ibrahimbegovic, H. Chen, E. Wilson and R. Taylor, "Ritz Method for Dynamic Analysis of Large Discrete Linear Systems with Non-Proportional Damping", Earthquake Engineering and Structural Dynamics, Vol. 19, 877-889, 1990.

Appendix: STATISTICS ON EXTREME LOADING CONDITIONS FOR CROATIA

SECURITY ISSUES IN CROTAIAN CONSTRUCTION INDUSTRY: BASES OF STATISTICAL DATA FOR QUANTIFYING EXTREME LOADING CONDITIONS

P. Marovic (marovic@gradst.hr),
HERAK-MAROVIC
University of Split, Faculty of Civil Engineering and Architecture, Matice hrvatske 15, HR-21000 Split, CROATIA

Abstract. This paper presents some general considerations about the security issues in the Croatian construction industry, in view of diversity of extreme loading conditions that characterize this country. The bases of the statistical data for quantifying the extreme loading conditions are given. Namely, the influence of the strong stormy winds, destructive earthquake motions, great temperature changes, fire and large snowfalls are shortly described considering statistical bases for quantifying the values at the national level within National Application Documents of the appropriate Eurocodes.

Key words extreme loading statistics, winds, earthquakes, fires

1. Introduction

Security is everyone's responsibility. Individuals, business and industry, government agencies, and private organizations share the obligation to protect people from needless injury and death as well as their properties from destruction [1]. So, this paper presents some general considerations about the security in the Croatian construction industry.

The new concept of codes, known as Eurocodes, is going to be implemented in the Croatian construction legislative. As each country has some own speciality, so Croatia has its too. These specialities are introduced through the values at the national level within National Application Documents of the appropriate Eurocodes [2-4]. These specialities are especially present in the field of the strong stormy winds (cold north wind called Bora) and destructive earthquake motions influencing the dynamic analyses of the engineering structures and in the field of great temperature changes (daily, seasonal, yearly as well as fire effects) and large snowfalls not influencing the dynamic analyses.

Part of these investigations has been carrying out for several years at the Faculty of Civil Engineering and Architecture, the University of Split, especially in the field of bearing structures for the case of extreme actions of the Bora wind load [5-10] and accidental fire load [10-13].

A. Ibrahimbegovic and I. Kozar (eds.),Extreme Man-Made and Natural Hazards in Dynamics of Structures, 383–395.

As the new approach to the security of engineering structures is based on the mathematical statistics and the theory of probability it is necessary to collect a large number of data through field investigation and to teach engineers to use principles of those theories.

In the following section the statistical bases for quantifying the values at the national level within National Application Documents of the appropriate Eurocodes are shortly described together with the drawn conclusions.

2. Bases of the Statistical Data for Quantifying the Extreme Loading Conditions

In this section the statistical bases for quantifying the values of the extreme loading conditions at the national level within National Application Documents of the appropriate Eurocodes will be shortly described together with the drawn conclusions.

2.1 STRONG STORMY WINDS

Croatia is a small country but its countryside varies from plains in the north, the Dinaride Mountains in the central part, to the Adriatic coast and the Adriatic Sea as its southern border. Vicinity of the Alps in the north and the dynamic changes in its relief make the airflow very complex. The winds Bora (local, cold, catabatic, changeable, north) and Jugo (warm, constant, south) are the most characteristic winds along the Adriatic coastal region.

In the last two decades the intensive and long-lasting field investigations have been performed in order to obtain meteorological bases for national values of the wind loads [14-18]. After first investigations the reference wind velocity was defined as a maximum 10-minutes average velocity to be expected in a 50 years period.

Also, two main geographic areas with different characteristics of extreme wind distributions were distinguished: (i) continental part of Croatia where the maximum 10-minutes wind velocity varies from 9,7 m/s to 22,0 m/s predominantly in W-NW-N direction; and (ii) coastal part of Croatia, predominantly from NE-NNE (Bora) or ESE-SE (Jugo), where the maximum 10-minutes wind can reach 43,5 m/s (maximum gust 69,0 m/s) at the location of the Maslenica Bridge (December 21, 1998) [10].

These investigations have shown that the 10-minutes averaging period is not sufficient to give correct presentation of the turbulent characteristics of Bora wind in the Croatian coastal region and that the shortest averaging period is necessary with the knowledge of the vertical profile changes. This initiated the new investigations [17].

As the example of the obtained results, the profile changes of the wind velocity in the 1-second interval for two characteristic cases at the location of the Dubrovnik Bridge is shown in Figure 1. The measurements were taken at the altitudes of 10, 52 and 140 meters. More detailed description of the performed investigations and obtained results can be found in Refs. [10].

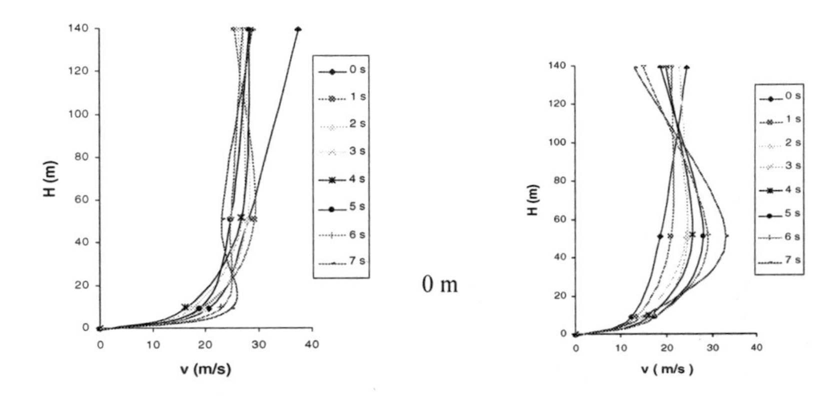

Figure 1. Profile changes of the wind velocity in the 1-second interval for two characteristic cases [10].

These investigations also confirmed that for the turbulent wind component it was necessary to have measurements of wind velocity with the shortest possible averaging periods, i.e. 1-second interval. This is illustrated in Figure 2 showing and comparing 1-second, 10-seconds, 60-seconds=1-minute and 600-seconds=10-minutes averaging intervals at the measurement point on the top of the Dubrovnik Bridge tower 140 meters high [10, 19].

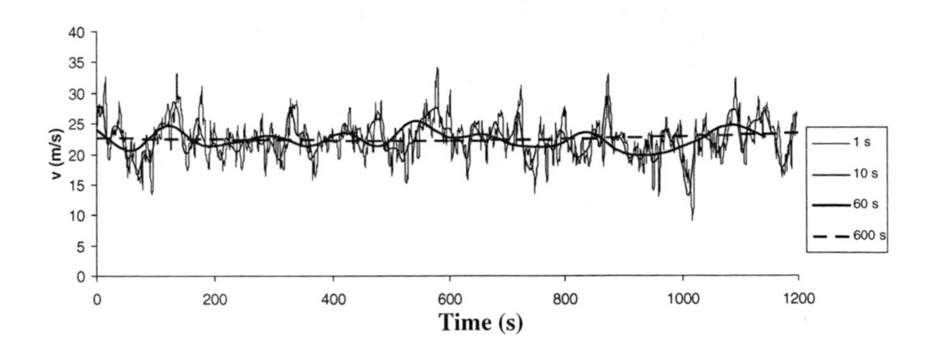

Figure 2. Comparison between the wind velocities at different averaging intervals at the measurement point on the top of the Dubrovnik Bridge tower 140 meters high [10].

All these investigations resulted in the proposal of the new wind velocity map of Croatia implemented in National Application Document of

Eurocode 1991-2-4 [20] as shown in Figure 3 but, unfortunately, for the present situation, with 10-minutes averaging interval. Wind velocity has to be measured at the altitude of 10 meters from flat ground and the return period of 50 years has to be taken.

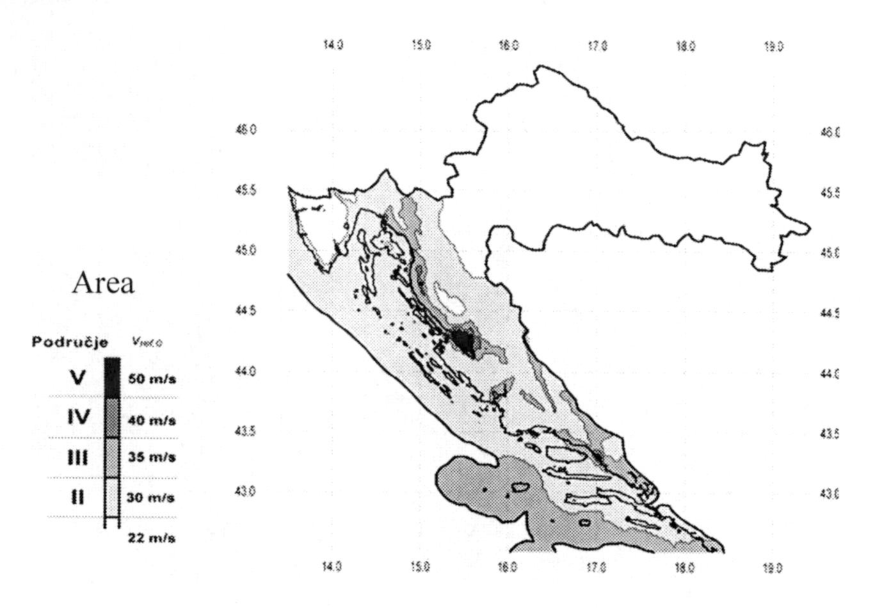

Figure 3 Wind velocity map of Croatia [20].

Table 1. Wind loaded areas related to basic comparative ($v_{ref,o}$) wind velocity and actual ($v_{ref,x}$) wind velocity [20].

Area	$v_{ref,0}$ (m/s)	$v_{ref,x}$ (m/s)
I	22	35
II	30	45
III	35	55
IV	40	65
V	50	75

Considering the influence of the wind load on structures it can be concluded that the 10-minutes averaging interval as recommended by Eurocodes as a reference velocity for structures loaded by wind is not sufficient to determine the correct values of the turbulent characteristics of the Bora wind and that they should be determined by the 1-second averaging interval obtained by field measurements. This is especially important for the high slender steel structures [6, 10].

On the other hand, wind has an impact on traffic conditions on roadways especially on bridges causing traffic closure [8, 21]. Namely, wind has an impact on the stability of vehicles on bridges and roadways if its lateral component of velocity exceeds 30 m/s. Since there is no standard defining wind velocity that is a criterion for traffic closure on a road or a bridge, these velocities are usually defined upon experience [21].

2.2 DESTRUCTIVE EARTHQUAKE MOTIONS

For the purpose of the Eurocode 8 [4], the national territory of Croatia is divided by the National Authorities into seismic zones, depending on the local hazard [22]. By definition, the hazard within each zone is assumed to be constant.

For most of the applications of Eurocode 8 [4], the hazard is described in terms of a single parameter, i.e. the value a_g of the effective peak ground acceleration in rock or firm soil, henceforth called 'design ground acceleration'.

So, Croatian National Authorities divided national territory into seismic zones, Figure 4, with appropriate design ground acceleration a_g belonging to each zone given in Table 2. Also, the design ground acceleration for each seismic zone corresponds to a reference return period of 500 years. To this reference return period an importance factor equal to 1,0 is assigned.

Table 2. Design ground acceleration a_g for seismic zones according to Figure 4 [22].

Earthquake intensity according to MKS-64 scale	**Designed ground acceleration a_g as part of gravity g**	**Designed ground acceleration a_g expressed in m/s^2**
6	0,05	0,5
7	0,10	1,0
8	0,20	2,0
9	0,30	3,0

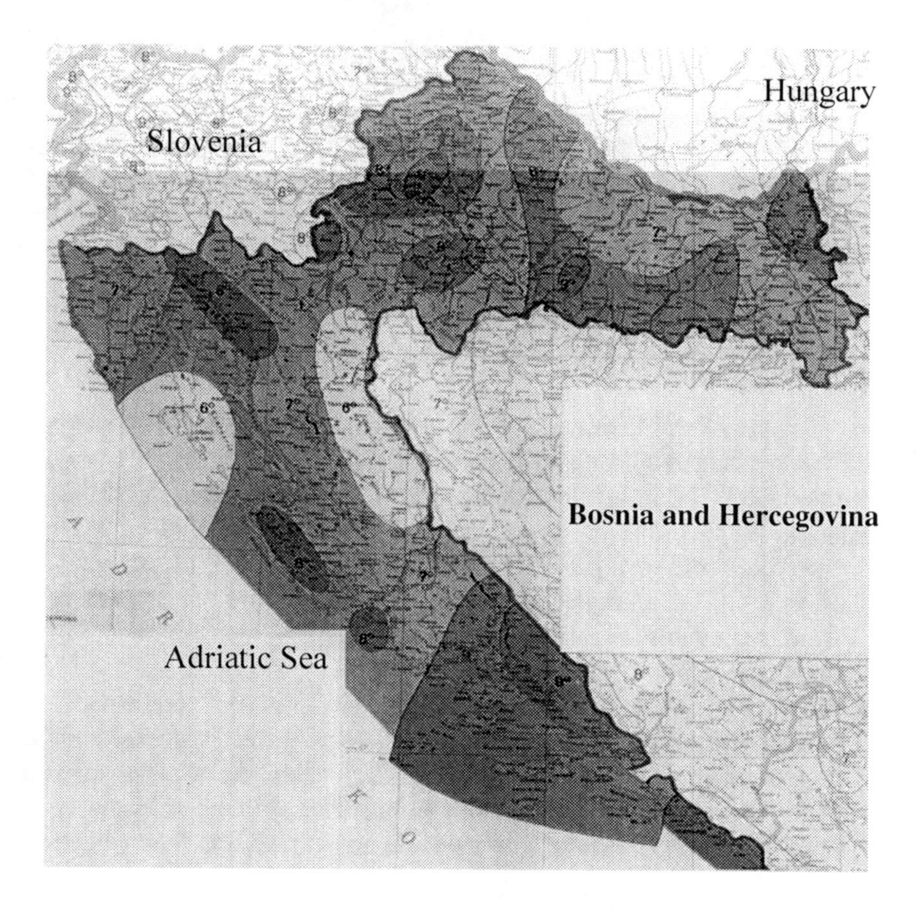

Figure 4 Seismic zone map of Croatia [22].

Additional parameters required for specific types of structures are given in the relevant Parts of Eurocode 8 [4].

2.3 TEMPERATURE CHANGES

The elements of the load bearing structure have to be checked to ensure that the thermal movement will not exceed limit states of the structure, either by the provision of expansion joints or by including the thermal effects in the design [23]. The analyzed thermal actions arise only from climatic effects due to the variation of shade air temperature and solar radiation. The Eurocode 1991-2-5 [23] provides rules as a guide to the matters which should be considered in the design. In this section only the values at the national level [23] are given that are obtained after intensive and long-lasting (the 1961-1990 period) field investigations [15, 24]. The temperature-based zoning of Croatia was conducted according to vertical gradients of parameters T_{max50} and T_{min50}, and based on their dependence on altitude [24]. Therefore, in Figure 5 the isotherms of maximum shade air temperature in accordance with the altitude are given while in Table 3 the appropriate numerical values are presented. The same is given in Figure 6 and Table 4 for minimum shade air temperature.

Figure 5. Isotherms of maximum shade air temperature in °C [23].

Table 3. Maximum shade air temperature in °C [23].

Altitudes up to (m)	**Area I**	**Area II**	**Area IIII**	**Area IV**
100	39	38	42	39
400	36	36	39	39
800	33	34	36	39
1200	30	32	34	--
1600	28	30	31	--

Table 4. Minimum shade air temperature in °C [23]

Altitudes up to (m)	Area I	Area II	Area III	Area IV	Area V
100	-26	-26	-17	-10	-16
400	-23	-26	-19	-13	-18
800	-20	-26	-21	-17	-19
1200	-17	-26	-23	-20	-21
1600	---	-26	-24	-24	-23
>1600	---	-26	---	-26	-24

Figure 6 Isotherms of minimum shade air temperature in °C [23].

2.4 FIRE

Structural fire design involves applied actions for temperature analysis and actions for structural analysis [25]. Depending on the representation of the thermal actions in design, the following two procedures are distinguished: (i) nominal temperature-time curves which are applied for a specified period of time (usually 30, 60 and 90 minutes), and for which the structures are designed by observing prescriptive rules, including tabulated data, or by using calculation models; (ii) parametric temperature-time curves, which are calculated on the basis of physical parameters and for which the structures are designed by using calculation models.

Parametric temperature-time or fire exposure curves have been only initiated and it has been left to each country to solve this within their National Application Document. Large investigations have been initiated at the Faculty of Civil Engineering and Architecture, the University of Split [10-13], and so far the obtained results show that the use of model of zones, based on approximate formulations, for different fire loads, obtained by varying the surface of ventilation openings, it was possible to obtain the temperature curves as a function of time which were used as the extreme temperature influence upon the load-bearing structures especially the steel ones. Furthermore, the obtained results show that the application of the standard temperature curve is not acceptable for large steel compartments since it is evident that the temperature action differs in space and depend upon the mentioned parameters so that it is necessary to perform computations for each actual fire load [12].

2.5 SNOWFALLS

Snow can be deposited on a structure in many different patterns depending on the upper shape of the structure, its thermal properties, the roughness of its surface, the amount of heat generated under the upper surface, the proximity of nearby buildings, the surrounding terrain and the local meteorological climate; in particular its windiness, temperature variations, and likehood of precipitation (either as rain or as snow). Furthermore, the deposits of snow can be due to an accumulation of snow from different directions or to one or more falls of snow within an individual weather system [26].

In this section only the values at the national level are given which are obtained after intensive and long-lasting field investigations [15]. Therefore, in Figure 7 the snow map of Croatia is given considering the altitude variations while in Table 5 the appropriate numerical values are presented.

Figure 7. Snow map of Croatia [26].

Table 5. Snow loads for altitudes over 100 m [26].

Altitudes up to (m)	Zone A	Zone B	Zone C	Zone D
100	1,10	1,10	0,45	0,35
200	1,30	1,40	0,80	0,50
300	1,55	1,75	1,20	0,70
400	1,80	2,20	1,65	0,90
500	2,05	2,65	2,15	1,15
600	2,35	3,15	2,70	2,70
700	2,65	3,70	3,30	3,30
800	2,95	4,25	3,95	3,95

900	3,25	4,90	4,65	4,65
1000	3,60	5,55	5,40	5,40
1100	3,95	6,25	6,20	6,20
1200	4,30	7,00	7,05	7,05
1300	--	7,80	7,95	7,95
1400	--	8,65	8,90	8,90
1500	--	9,50	9,90	9,90
1600	--	10,40	10,95	10,95
1700	--	11,40	12,05	12,05
1800	--	--	13,20	13,20

3. Conclusions

This paper presents some general considerations about the security in the Croatian construction industry. The bases of the statistical data for quantifying the extreme loading conditions are given. Namely, the influence of the strong stormy winds, destructive earthquake motions, great temperature changes, fire and large snowfalls are shortly described considering statistical bases for quantifying the values at the national level within National Application Documents of the appropriate Eurocodes.

After intensive and long-lasting field investigations [14-17, 20] a large number of data was obtained upon which relevant suggestions for national values within appropriate national codes were proposed and accepted to be tested in everyday engineering practice.

Acknowledgements

The partial financial support, provided by the Ministry of Science, Education and Sports of the Republic of Croatia under the project *Numerical and Experimental Models of Engineering Structures*, Grant No. 0083061, is gratefully acknowledged.

References

[1] Ch.C. Vance, Safety, The World Book Encyclopaedia, World Book Inc., Chicago, Vol. 17, pp. 9-16, 1996.

[2] ENV 1990: Eurocode 0: Basis of structural design, European Committee for Standardization, Brussels, 2002.

[3] ENV 1991: Eurocode 1: Basis of design and actions on structures, Croatian Standards Institute, Zagreb, 2005. (in Croatian)

[4] ENV 1998: Eurocode 8: Design provisions for earthquake resistance of structures, Croatian Standards Institute, Zagreb, 2005. (in Croatian)

[5] B. Peroš, Modelling of the Bora effects upon the lower layer, *International Journal for Engineering Modelling*, Vol. 7, No. 3-4, pp. 81-95, 1994.
[6] B. Peroš, Reliability of Structures with a Dominant Wind Load, Ph.D. Thesis, University of Zagreb, Faculty of Civil Engineering, Zagreb, 1995. (in Croatian)
[7] B. Peroš, Construction steel design for structures with a dominant wind Bora load, *Journal of Constructional Steel Research*, Vol. 46, No. 1-3, Paper No. 146, 1998.
[8] A. Bajić and D. Glasnović, Impact of severe Adriatic Bora on traffic, Proc. 4th European Conf. on Applied Meteorology, Norrkoping, 1999.
[9] B. Peroš, T. Šimunović and I. Boko, Modelling of the dynamic action of the Bora wind upon high slender structures, Proc. of the XL 2003 Conf. on Response of Structures to Extreme Loading, Toronto, August 2003., Eds. A. Ghobarah and Ph. Gould, Elsevier, Paper O17, 2003.
[10] B. Peroš, I. Boko and T. Šimunović, Safety of structures under the influence of extreme loading, Proc. of the NATO-ARW PST.ARW980268 Workshop on Multi-physics and Multi-scale Computer Models in Non-linear Analysis and Optimal Design of Engineering Structures under Extreme Conditions, Bled, June 2004., Eds. A. Ibrahimbegović and B. Brank, Faculty of Civil Engineering and Geodetic of University of Ljubljana, Ljubljana, pp. 585-612, 2004.
[11] I. Boko, Safety of Steel Structures under the Influence of Fire Loads, M.Sc. Thesis, University of Split, Faculty of Civil Engineering, Split, 2001. (in Croatian)
[12] I. Boko, Determination of the Safety Degree of Steel Structures under the Influence of Fire Loads, Ph.D. Thesis, University of Split, Faculty of Civil Engineering and Architecture, Split, 2005. (in Croatian)
[13] I. Boko and B. Peroš, Safety of steel structures under the influence of fire action, Proc. of the XL 2003 Conf. on Response of Structures to Extreme Loading, Toronto, August 2003., Eds. A. Ghobarah and Ph. Gould, Elsevier, Paper O97, 2003.
[14] A. Bajić, Severe Bora on the Northern Adriatic, Part 1: Statistical analysis, *Papers*, Vol. 24, 1989.
[15] A. Bajić, M. Gajić-Čapka, L. Srnec, V. Vučetić, K. Zaninović and Z. Žibrat, Meteorological bases for Croatian standards – snow loads, extreme air temperatures and wind loading, Report No. 2100-99/00, Croatian Hydrometeorological Institute, Zagreb, 2000. (in Croatian)
[16] A. Bajić, B. Peroš, V. Vučetić and Z. Žibrat, Wind load – a meteorological basis for Croatian standards, *Građevinar*, Vol. 53, No. 8, pp. 495-505, 2001. (in Croatian)
[17] A. Bajić, V. Vučetić and Z. Žibrat, Maximum recorded and expected wind velocities at the Croatian national territory, Report No. 360/03, Croatian Hydrometeorological Institute, Zagreb, 2004. (in Croatian)
[18] A. Bajić and B. Peroš, Meteorological basis for wind loads calculations in Croatia, *Wind and Structures*, Vol. 8, No. 6, pp. 389-406, 2005.
[19] A. Bajić and B. Peroš, Reference wind velocity – influence of averaging period, *Građevinar*, Vol. 53, No. 9, pp. 555-562, 2001.

[20] J. Radić, P. Sesar and A. Krečak, Wind on Croatian Adriatic bridges, Proc. Int. Conf. on Durability and Maintenance of Concrete Structures, Dubrovnik, October 2004., Ed. J. Radić, SECON HDGK, Zagreb, pp. 289-296, 2004.

Author Index

Printed in the United States
69183LVS00001B/73-90

9 781402 056550